HOW TO SUCCEED IN ORGANIC CHEMISTRY

HOW TO SUCCEED IN ORGANIC CHEMISTRY

MARK C. ELLIOTT
Cardiff University

Great Clarendon Street, Oxford, OX2 6DP,
United Kingdom

Oxford University Press is a department of the University of Oxford.
It furthers the University's objective of excellence in research, scholarship,
and education by publishing worldwide. Oxford is a registered trade mark of
Oxford University Press in the UK and in certain other countries

Impression: 1

Published in the United States of America by Oxford University Press
198 Madison Avenue, New York, NY 10016, United States of America

British Library Cataloguing in Publication Data
Data available

Library of Congress Control Number: 2019952902

ISBN 978–0–19–885129–5

Printed in Great Britain by
Bell & Bain Ltd., Glasgow

For my wife, Donna, and
my daughter, Kirsten

CONTENTS

FOREWORD

This is a textbook aimed primarily at undergraduate chemistry students in their first year of study. The goal of this textbook is very simple—to ensure that the reader exits with a solid understanding of the **principles** of organic reaction mechanisms.

Let me explain the motivation behind my writing of this book.

Students of organic chemistry face some very particular challenges. Every year I see exam scripts containing carbon atoms with five bonds, or curly arrows that start at H^+. It is really important to get rid of these errors very early in your journey. Success in organic chemistry requires students to become comfortable with drawing structures, stereoisomers, and reaction mechanisms with total confidence.

Chances are, if you make one mistake while drawing a mechanism, you will then have to make another mistake in order to dig yourself out of a hole!

My students are not stupid! Far from it! The students who make these mistakes do not lack understanding. They just haven't developed the correct **habits**. Let me give you a few analogies. Bear with me!

A young Geraint Thomas[1] decides he wants to be a professional cyclist. He gets on a bicycle at the age of three, learns to pedal, gets the stabilizers off, and rides round his local park a couple of times. Several years later, a few weeks before the big race, he gets on a bike again, and 'revises' his cycling. I think you know how this story ends, and it isn't with him winning anything!

*He still **knows** how to ride a bike, but he didn't do the right things at the right time to become good at it!*

Perhaps you have already learned to drive a car. Recall your first driving lesson. You wouldn't know how far to turn the steering wheel. You would struggle to change gear, and the car would bunny-hop down the road. We all did it! And yet, if you persevered, you would find that very quickly, you barely even noticed changing gear—it just 'happened automatically' when the engine sounded 'right'! I bet you can't even get a car to bunny-hop now!

The point is, you simply practised the same skills, over and over, until you got good at it. The even bigger point is that you accepted that this was the best (indeed only) way to learn. And you don't need to know how the engine actually works.

It's the same with learning to play a musical instrument. You first learn the mechanics of the instrument, and then you practise scales and simple tunes until you get more

[1] Tour de France winner, 2018.

fluent or more competent. You would recognize that this isn't the end goal,[2] but you would also recognize that it is a necessary step along the way. Even experienced musicians warm up by playing a few scales.

Everyone who has played the guitar will recall the first time they played a B-flat chord at the first fret. Trying to avoid the top E string with your ring finger was painful. But you probably persevered and can now do this without any conscious thought. What was once difficult has become natural!

I bet you don't even know what you are doing differently that allows you to succeed—and that is the beauty of it!

In these examples, you would be doing something that you recognize that you would get better at with practice! You might never win the Tour de France, but you can still become a good cyclist.

Training Versus Learning

Organic chemistry is a difficult, technical subject, right? Of course it is! I'm not saying you don't have some difficult concepts to understand.

But many of the things you need to do as an organic chemist can become ***habits****—things that you do with little or no conscious thought—but that takes time and effort.*

However, it doesn't take any more time and effort than revising shortly before an exam, but it produces better results. The key to success is to practise the right things at the right time, in order that they become **internalized**.[3]

You need to be able to draw structures of molecules confidently—to ensure that each carbon atom has exactly four bonds.

As a key part of this journey, you will find that when you are presented with a structure with a mistake, you know there is something wrong before you work out *exactly* what the problem is—it will just 'look wrong'.

You need to be able to draw stereochemical representations, and to be able to visualize what the representations mean.

We use various types of diagram to indicate the arrangement of atoms in space. You can learn what each one means in a matter of minutes, but what you really need to do is become so used to them that you see a molecule on a page, and you visualize what it looks like in three dimensions.

[2] You can only play 'Twinkle Twinkle Little Star' so many times!

[3] I will be using this word a lot!

When you read the text on this page, the letters and words only have 'meaning' because you have completely internalized that meaning. The sound of the words pops into your head as you are reading, and you can do it really quickly and accurately.

When you learned to read, the process of practising 'rewired' your brain to become good at this. Training does that! You need to rewire your brain, through a process of training—over a period of time—so that the key skills associated with organic chemistry become natural.

I could give other examples related to organic chemistry, but I think I have made the point. Instead of relying on revision and short-term memory, you need to structure your learning so that the necessary skills become so ingrained that they are part of your process—something you cannot forget and do not need to revise.

INSPIRATION

For me, the classic organic chemistry text, when it comes to learning the fundamentals of reaction mechanisms, is '*A Guidebook to Mechanism in Organic Chemistry*' by Peter Sykes. The 6th and final edition of this book was published in 1986! It contains a lot of information, presented in a lucid style. I suspect the reason so many of us were drawn to this book is that, at 400 pages, it isn't too intimidating.

I found books that required a level of interaction to be particularly useful. The 'workbook' style texts by Stuart Warren ('*Chemistry of the Carbonyl Group: A Programmed Approach to Organic Reaction Mechanisms*' and '*Designing Organic Syntheses: A Programmed Introduction to the Synthon Approach*') were ground-breaking for their interactive style. A new edition of the carbonyl book was published in 2018.

There are many textbooks simply called 'Organic Chemistry'. They are weighty tomes, generally over 1000 pages, and they contain a huge amount of information and many examples of reactions. If you are a serious student of organic chemistry, you should own one of these books. My preferred choice is the one written by Clayden, Greeves, and Warren (yes, same Warren!).

As a student, I dipped into the larger textbooks, but I read the smaller books cover-to-cover. I am not sure whether I considered the reasons for this at the time. I think it is more than just the size of the book. A larger book does appear more daunting, but a smaller book 'fits into the hand' nicely. There is a reason why the paperback book format is successful! I want this to be a book that will spend a lot of time in your hands, but I also want it to be a book that you will actively engage with, rather than passively reading.

The next inspiration is probably going to seem strange. The thriller writer James Patterson writes books (loads of them!) with very short chapters. Patterson knows the formula for writing a 'page-turner'. It is very easy to continue reading well into the night, because there is a low 'activation barrier' to reading another chapter—you know you will finish it!

This book also contains many more, but shorter, chapters than most organic chemistry textbooks that cover the same material. I hope that this will encourage you to re-read chapters until you have fully understood everything.

Of course, I have taken all of this inspiration and turned it into something that I hope is my own. In a sense, this book is autobiographical. In remembering what I found difficult to learn, I can suggest how you should approach the same material to achieve a better outcome. While this is a traditional organic chemistry textbook, I do hope it reads as a personal document, and that you feel it is connecting with you, the reader.

ACKNOWLEDGEMENTS

It is not easy to single out individuals who have guided my chemistry career over the last 30 years, but I would particularly like to thank Professor Chris Moody, Professor Harry Heaney, Professor David Knight, and Dr Michael Hewlins for their support and encouragement at various stages. In recent years I have enjoyed a very productive research collaboration with Professor Keith Smith, which has helped me think about chemical reactivity in a much more rigorous way.

As I have moved from a focus on research to a focus on education in recent years, the late Professor Chris Morley contributed far more than he ever knew. Although Chris did not see any of the book in its advanced form, his support and influence are ever present.

I have learned a great deal about education from the many martial arts instructors I have been privileged to train with over the years. I have learned that a student with no natural talent (me!) can continue to improve, as long as they persevere and keep making the small individual adjustments that eventually become larger improvements. All of these instructors have been inspirational in their own way, and it would be inappropriate to single out any individuals.

I am grateful to Dr Andrew Roberts for his indulgence and philosophical discussions. Chemists and architects are not so different after all! The publisher and I also appreciate the feedback from the anonymous reviewers whose constructive comments on the draft manuscript doubtless improved it. We also thank Professor Mark Bagley, University of Sussex, for reading through the final manuscript and making further helpful comments and suggestions.

I think it is fair to say that I have bored my colleagues rigid with the progress of this book over the last 12 months. I thank them for not showing their boredom, and for many helpful discussions. In particular I am grateful to Dr Niek Buurma for his incisive insight into organic reaction mechanisms and to Dr Ben Ward for his assistance and discussions of some of the computational chemistry.

Perhaps though, I have learned the most from the many students I have taught over the last 22 years. The discussions I have with students are the most rewarding part of my job.

It is almost inevitable that, even with the best of intentions, some errors will remain. These are entirely the responsibility of the author. I hope that you will forgive me for any errors, that they don't confuse you, and that you tell me, so I can correct them!

ADDITIONAL RESOURCES

In order to help you engage with the book, there are several online resources that you can access.

First of all, the book has its own Facebook page, **https://www.facebook.com/shuhariorganic**, which I encourage you to 'like'. I post updates related to the book, as well as links related to educational philosophy. You will be able to post on the page and ask questions. Perhaps we can build a community that will answer the questions, or perhaps they will inspire me to provide additional material.

Speaking of which ...

I've produced a series of **videos** to be watched alongside reading this book. Some are answers to questions posed in the book. Others are based on different problems, but they are analysed in the same style.

Look out for the icon as you read through the book.

To access the videos, simply use the access code provided inside the front cover, and follow the instructions provided there.

There is another distinct reason for providing answers and additional material as videos. I have tried to write the book in a more conversational style than is the norm. When my students read my lecture notes, they tell me that they can (rather scarily) hear my voice! This has always been deliberate, and I can only connect with you if I provide you with resources that will allow you to 'get to know me'.

So, there isn't, and won't be, an answer book. For the more complicated stuff, answers (or at least guidance) are already incorporated. For the basic stuff, you don't need the answers. You don't need me to tell you if you have drawn a five-valent carbon atom. You just need to be looking for it, until you spot your mistakes without conscious effort. As you progress, you will find that the things that were difficult at first will become second nature.

STYLE AND SUBSTANCE

The organization of this book

No matter how you learn organic chemistry, you need to be able to access the basic ideas and theories. All this information is present. Information is rather 'compartmentalized', into short chapters, and with lots of cross-references to highlight where one area of organic chemistry impacts on another.

This is the problem with organic chemistry. Everything is interlinked.

There are seven sections to the book. The first section **'Laying the Foundations'** is exactly what it says it is. We deal with the absolute basics—drawing, functional groups, curly arrows. We cover the basics of bonding in there, perhaps a little later than in other textbooks. In the second section, we are **'Building on the Foundations'**. We continue to draw structures, but we do more with them, and we start to consider energetics and stability. We inevitably start to consider the shape of organic molecules. We then formalize some of this in the third section, **'A Focus on Shape'**, where we try to build good habits for drawing the shape of molecules, but we also start to consider the implications of shape in reactivity.

Section 4, **'Types of Selectivity'** is a very short section. The real point of this section is that selectivity is selectivity. There may be different outcomes, but they are all governed by the same ideas. The fifth section focuses on the fact that **'Bonds Can Rotate'**, and that this has implications for stability and reactivity. Of course, the rotation of bonds is considered in the earlier sections, but we formalize the discussion here.

In Section 6 we are **'Eliminating the Learning'**. Of course, this is a rather cheap play on words. Most of the discussion (but not all) is about elimination reactions, but the point is that if you are applying principles, there isn't actually that much to learn.

This is the beauty of organic chemistry. Everything is interlinked.

The earlier sections stand alone, but it all leads to the final section, **'Building Skills'**. This is where we consolidate the learning. You will have practised drawing cyclohexanes. Now it's time to understand how the various conformations influence reactivity. We will see why one stereoisomer reacts more slowly than another. You will have practised drawing Newman projections for chiral compounds, and you now have to use them to work out why you form one regioisomer/stereoisomer of an alkene more readily than another. You will have considered the energy differences between conformers/carbocations/carbanions and you will have to draw plausible reaction energy profiles for (reasonably) complex processes in order to understand what is happening to the energy as we proceed from starting materials to products.

All of these examples will contain variable amounts of guidance and signposting to earlier relevant material so that you will see how it all fits together.

This is all about building your skills, judgement, and instinct so that you know how to approach rather difficult organic chemistry problems without too much effort.

Underlying all of this is a 'nag' factor, 'gently encouraging' you to practise more, and telling you what you need to do with the information, and how it links to things you have already seen. The 'secret' to success in organic chemistry is making connections. Initially, it seems as though there is a vast amount of information, but with experience all of the reactions can be seen as applications of a few principles. I am trying to guide you on the journey.

Types of chapter

I have tried to classify the chapters.

Basics These chapters deliver information, as well as explaining what you need to do with it.

Habits These are still delivering basic information, but there is more emphasis on skills that need to become natural to you. When you first learn these skills, they may seem difficult, but with practice you will be able to do them without really thinking about them. And you need to. I want to be clear about that, so you spend the time developing the skills.

Applications These chapters take the basic information and apply it to various reactions. You may feel that there is new information, and perhaps there is. I would argue that it is the application of the information that is new. You need to follow the 'method'.

Common Error There are many common errors that students make, and these are discussed throughout. However, it's nice to give you discrete chapters that focus only on this one idea.

Practice This one is obvious. It's where you get to develop your skills. Working through one example after a lecture isn't practice. You will get it right because it is fresh in your mind. And then it won't be! You need to keep coming back to these chapters and repeating the exercises until you are bored rigid! And then you will know that you've got it!

Perspective These chapters go a little bit beyond what most textbooks at this level cover. They are intended to add depth to the discussion, but they do contain some difficult ideas. If you find them intimidating, skip them, and come back later.

Fundamental Reaction Type Just the absolute basics about the relevant reaction. I want you to have something to refer back to if you get confused. It's always good to be able to return to the most basic level. You cannot afford to be confused about which is substitution and which is elimination!

Reaction Detail Here, we fill in the details, whether it is about the reactions themselves, or their stereochemistry, or whatever else. A lot of this could be considered 'Applications' as well, because if you understand the basics, you will be able to predict how changing a parameter will change the outcome!

Fundamental Knowledge Recap There are some things that you do just need to learn. Even these things will become easier to remember as you use them. But in the first instance, you need to know that you need to know them!

Worked Problem *and* **Solution to Problem** And if you cannot predict how changing a parameter will change the outcome, these chapters will help you identify that and fix the problem. There are questions, guidance, and answers, given at various levels of depth.

Formatting

I have tried to keep it simple. Sometimes, when I need to refer to compounds in the text, the structure in the diagram has a compound number. These numbers are only continuous within the one chapter.

There are various styles of text and highlight throughout the book.

You have already seen the blue text. This is used to pull out important points, and to nag/encourage you. Perhaps in some cases it just breaks up the monotony of a page of text!

Orange text is used to ask you questions or to indicate activities, and is accompanied by an icon:

Where I pose a question for you to consider you'll see this icon.

Where I suggest an activity for you to work through you'll see this icon.

In **Section 7**, *when we really are making connections with earlier material, I introduce the green bar alongside text to highlight the links. Of course, all the sections have cross-links to other chapters, but in* **Section 7**, *while you are working through the problems, I really want you to go back and re-read the relevant previous chapters. That's how your knowledge and understanding will transfer to your long-term memory.*

How should you study?

I would strongly encourage working with a group of your peers to check and challenge one another's understanding and answers. Chances are, if you all get the same answer to a problem, you will all be right. If you disagree, this is the time to step back and look at how you got there.

I know this seems as though I am placing the onus on you to 'teach yourself' organic chemistry. The reality is, that's what you have to do. Perhaps it's a little more complicated than that. I can give you an explanation. You will understand it when you read it. But then, you will have forgotten it two weeks later, or you won't understand how to apply it to a new scenario.

By talking to your peers, you will start 'speaking the language' of organic chemistry. One of you will say something like 'what if we add an extra methyl group – will it change anything?' and someone else will say 'are you looking for a steric or an electronic effect?'

*Hopefully everyone will shout '**both!**' And then you've got it!*

Wait a minute, where are all the reactions?

Organic chemistry has loads of reactions. They are hard to learn. All we are covering here are nucleophilic substitution reactions and elimination reactions. Are you being short-changed? I don't think so. This was a conscious decision. Your first goal is to 'start thinking like an organic chemist'. You need to understand the types of problems addressed in organic chemistry and to develop your ability to apply basic principles to these problems. You don't need loads of reactions in order to learn to do this. There is a very real risk that if you see loads of reactions first, you won't know which bits of information to apply. Once you've got good at applying the basics, then you can learn more reactions and solve problems that use them.

SECTION 1

LAYING THE FOUNDATIONS

INTRODUCTION

There are certain key skills that you need to have in order to succeed at organic chemistry. You need to be able to draw good structural representations, and to see exactly what those structures mean. You need to know what a structure is called, and what structure to draw from a chemical name.

The eventual goal is that you see the name of a compound written, and you visualize the complete molecular structure.

In considering the various different structures with the same molecular formula, we will establish key concepts such as isomerism, and double bond equivalents.

Perhaps most importantly, we will use the convention of curly arrows to draw the 'mechanisms' of organic reactions.

Initially, we won't worry too much about what is happening at a molecular orbital level, but we will establish the rules that good curly arrow drawing must follow.

In this first section, we will establish many of the basic ideas that we will be building on in later sections. There will be quite a few 'Practice' chapters here, as this is how you get comfortable with the basics.

Don't just read this book. Draw the structures. Make notes from it. Have a go at the problems, and discuss your answers with your friends. If you are getting it wrong, someone will tell you. If you are getting it right, you'll be able to persuade someone else that you have the right answer.

Learning organic chemistry is not easy. There will be pain! That feeling of frustration as you draw yet another compound with a five-valent carbon, yet another curly arrow that doesn't make sense.

All of that is a necessary part of the journey. You didn't learn to ride a bicycle without falling off!

Gradually you will find that those mistakes disappear. Don't expect to not make the mistakes at all. Just feel confident that drawing a structure or a reaction mechanism again and again will internalize the knowledge. Drawing one or two extra examples will help it become instinctive.

If you don't master the basics, it will be impossible to draw the mechanism of a complex reaction. Even then though, you can use that mechanism as practice in order to 'fix' your knowledge of the basics.

Trust me. You have already learned to do many complicated things. Understand **how** you did it, and then stop worrying. Let it happen. There is a wonderful Japanese proverb. 'If you have one eye on your destination, you only have one eye left to find the way.' The exam is not the goal!

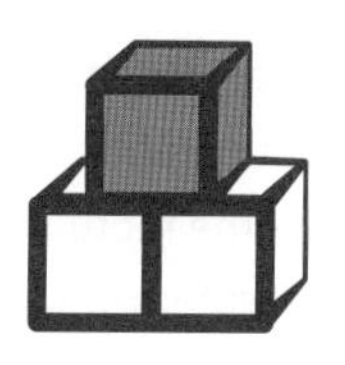

BASICS 1

Structures of Organic Compounds

Let's start with the absolute basics—drawing structures. We need to consider the different conventions that are in common usage, and what the benefits and disadvantages are of each.

Simply reading through this chapter will not make you good at drawing structures. Every reader will probably *understand* everything in this chapter. But some people will still make mistakes when they are under pressure. We are going to need to establish some habits and 'ground rules' for how you should approach your studies and this book.

NUMBER AND TYPE OF BONDS

The vast majority of organic compounds contain only the elements carbon, hydrogen, oxygen, and nitrogen. **In a stable, neutral molecule, carbon forms four bonds, hydrogen forms one bond, oxygen two bonds, and nitrogen three bonds.** Most organic chemistry textbooks begin with some of the theory of bonding in organic compounds. I want to take a slightly different approach. For now, let's simply state that we can have single bonds, double bonds, and triple bonds. We represent a single bond by a single line, a double bond by a double line, and a triple bond by a triple line.

H–C≡C–H

ethane ethene ethyne

Clearly, since hydrogen can only form one bond, it cannot form a double bond! Carbon can form four bonds—a triple bond 'counts' as three bonds, so carbon can form single, double, or triple bonds. Nitrogen can form three bonds, so it can also form single, double, and triple bonds. Oxygen can only form two bonds, so it can only form single and double bonds. Here are examples with all permutations.

NH_2–CH_3 H_3C–N=C(CH_3)$_2$ H_3C–C≡N

H_3C–OH H_3C–O–CH_3 H_3C–C(=O)–CH_3

1

Most of the time, organic chemists will draw structures by habit. They do understand the bonding, but they don't actually need this understanding to draw the correct number of bonds. So, let's focus on building the habits. We will do this by looking at some factual information, and a little guidance. We will build this up over a couple of chapters, and then provide some examples to work through.

The more structures you draw, the better you will get, but I want to ensure that you undertake some reflection on what you have drawn.

If you are drawing structures with the wrong number of bonds to some atoms, you can already check this yourself. The act of checking will make it less likely for you to make the same mistake next time. You might want me to give you the answers to all of the problems. My preference is to provide you with the tools to allow you to check your own structures.

Hydrogen might seem like the simplest of all, and there is little to explain really. A hydrogen atom is the thing you add on to satisfy the valence of other atoms. However, don't get complacent about this simplicity. Most of the mistakes you make will be because you forgot where the hydrogen atoms were. Many important reactions feature either the addition or removal of a hydrogen atom (normally in the form of a proton). This is *really important!*

Everyone can learn these basic rules. You ***will*** *make mistakes, and you will draw the occasional carbon atom with five bonds. The important thing is to keep looking for them, and eventually you will find you don't draw them anymore.*

First of all, there are some conventions that we need to be aware of, and we will expand on these here.

REPRESENTATIONS OF ORGANIC STRUCTURES

We use several different representations for organic structures. It is important that you recognize all of them. For example, the following three structures are all representations of the same compound, *n*-butane (more about the names later).

$H_3C{-}CH_2{-}CH_2{-}CH_3$

1 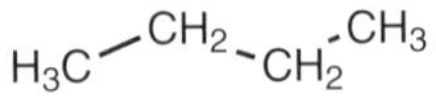 2 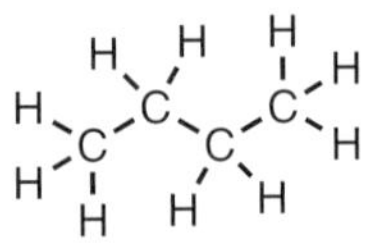 3

Which of these is clearest? Most year 1 undergraduate students will feel that structure **3** is the clearest. All hydrogen atoms are explicitly drawn, along with all bonds. It is easy to work out the formula. Importantly, you'll never forget where a hydrogen atom is, or draw an incorrect mechanism as a result of this.

However, this structure can look quite cluttered. If you have even a moderately complex structure, this type of representation will look very messy, and you'll find that you don't get an overall feeling for the key features of the structure.

Structure **2** is almost as clear. I have explicitly drawn H_3C on the left, so it is clear that the bond to the next carbon is from the carbon. However, the next carbon is a CH_2 group, so something has to be on the left and something on the right. We would never think that because the H is on the right, the bond to the next carbon is from hydrogen. We will look at this convention at various points over the next few chapters.

For me, structure **1** is actually the clearest. I can see at a glance that this is a saturated (no double or triple bonds) compound with four carbon atoms.

I can only see this because of the conventions used.

- Each line represents a chemical bond.
- Where the lines join or terminate represents a carbon atom.
- If no label is used for the atom, then a carbon atom can be assumed. We are organic chemists after all!

1.1

For most organic chemistry textbooks, this is the most common sort of representation. This is the one you need to get used to. So, you need to very quickly dispense with representation **3** and practise drawing all structures in representation **1**.

*If you find yourself drawing structures in representation **3**, **STOP** and redraw in representation **1**. Perhaps draw representation **3** and representation **1** alongside one another for a while, or draw (part of) a structure in representation **3** if you cannot see what is going on. But every time you do this, redraw the same structure and mechanism in representation **1**.*

You probably read the above point and thought 'but I like representation **3**' or 'I'm comfortable with representation **3**'. Of course you are! It's easier to learn at first. But I can assure you that representation **1** can become just as comfortable. I'll go a little further than this. If you don't make the effort to become comfortable with representation **1**, you're going to be wasting time with the rest. You'll know when you are getting good at this, in the same way that you know you are getting good at playing a musical instrument. It will just feel right,[1] and you will instinctively know that you've drawn a nice clear correct structure. Don't take short cuts!

[1] Or sound right, in the case of a musical instrument!

EMPHASIZING VALENCE AND FURTHER CONVENTIONS

Carbon is always tetravalent.[2] It forms four bonds. Once we take this into account, structure **1** needs a total of 10 hydrogen atoms.

You should be careful to use consistent representations. You would never draw *n*-butane as in structure **4** or structure **5**.

4 **5**

In both cases the second carbon atom from the left is explicitly drawn but without hydrogen atoms. This is just confusing. If you are going to explicitly write 'C' then you need to indicate the attached hydrogen atoms.

Conversely, if you see a structure with the atoms explicitly drawn, make sure you understand the bonding. Don't add other atoms to satisfy what you think the bonding is. Have a look at **Common Error 1** for an example of this.

It may be that you use the less cluttered representation, but you need (perhaps when drawing a reaction mechanism) to show a particular hydrogen atom. This is fine. Just draw something like structure **6** below.

Just because you explicitly draw the H, you don't need to draw the C.

6

Again, there is little to understand. Every student who studies organic chemistry at university can 'get' all of this.

Let's add a bit more basic information.

CHARGES ON ATOMS

We aren't going to worry about positive and negative charges on carbon just yet. That will come up soon enough.

[2] There are exceptions. They are rare, and we don't need to concern ourselves with them at this point. If you develop the right habits, you'll be prepared when you encounter the exceptions, and they won't confuse you. For now, four bonds to carbon is where we are at.

We have established that oxygen forms two bonds. This is only the case if it is neutral.[3] An oxygen atom with one bond will normally have a negative charge and an oxygen atom with three bonds will normally have a positive charge. So, if I draw structure **7** below, then you would know that there was something wrong with the oxygen atom.

$H_3C{-}O$

7

You either need to add a hydrogen atom, as in **8**, or a negative charge, as in **9**. Which you do will depend on what the correct structure is, but it certainly isn't **7**!

$H_3C{-}OH$ $H_3C{-}O^{\ominus}$

8 **9**

There is something else we haven't added to the oxygen atom—lone pairs. The oxygen in **8** has two lone pairs. We would show them as in **10**.

$H_3C{-}\ddot{\underset{\cdot\cdot}{O}}H$

10

At a basic level, you know why this is. Oxygen is in group 6 of the periodic table, and therefore starts with six outer-shell electrons. It gains one extra electron from its bonding with carbon and one from hydrogen, giving an octet (eight electrons). Four of these are used in bonding, leaving four (two pairs) that are not bonding. A more rigorous discussion of bonding is coming in **Basics 6**.

But, to reiterate a point I made before, organic chemists will draw lone pairs on oxygen by habit rather than by thinking of the bonding. In my experience this happens whether you learn about the bonding first or not. So, let's make 'building habits' the major focus.

Remember that the main focus of this discussion is the conventions used to draw organic structures. I would be perfectly happy drawing structure **8** for methanol. But, if I needed to do something with it that required me to 'use' the lone pairs on oxygen, then I would draw structure **10**. The key point is that either representation is perfectly correct. There are always two lone pairs on the oxygen. We just don't always draw them. It is important to state this, because the sooner you start to get used to this, the faster you will learn. At no point can you say 'I forgot that there was a lone pair on that oxygen!'

A neutral nitrogen atom forms three bonds. It will have one lone pair. Again, we very often don't draw it unless we are going to use it in a reaction.

[3] If you are ever getting confused, think of a simple compound. For me, water is my 'go to' molecule with oxygen. It's important to have a 'go to' strategy.

MULTIPLE BONDS

We can have double bonds between carbon and carbon, or carbon and other atoms (most commonly oxygen). We can even have carbon–carbon triple bonds. We will look at the strengths of these bonds in due course, along with their bonding. Let's just look at these briefly for now in order to make two more points.

$H_2C{=}CH_2$ — ethene

$(H_3C)_2C{=}O$ — propanone (trivial name acetone)

H−C≡C−H — ethyne (trivial name acetylene)

I have used two different representations above. For ethyne, I used the simplest skeletal representation, but I explicitly drew the hydrogen atoms in. I didn't need to do this. The representation below is perfectly correct.

ethyne oct-4-yne

It is quite common in the early stages to get confused with alkynes and to count the number of carbon atoms incorrectly. The structure above right is oct-4-yne and has a total of eight carbon atoms. Count them! There is one at each end of the triple bond. This is easier to see than to explain. Look at it carefully for now, and we will include some examples in the exercises.

*The reason behind this problem is a tendency to draw the two bonds coming **from** the alkyne much shorter, and then to be confused about whether the two carbon atoms are **attached** to the alkyne, or are **actually** the alkyne carbons. This then leads to the assumption that this compound has only six carbon atoms. We will emphasize the importance of correct drawing in **Habit I**.*

If you do make this mistake, you will never be able to draw a correct structure for any reaction product derived from the alkyne.

You don't need to be clever to get the number of carbon atoms right. You just need to have invested the effort at the right time to get used to it.

1

AROMATIC COMPOUNDS

When we put three double bonds in a six-membered ring, it gives it a special type of stability known as aromaticity. We will explain this later, but for now it is enough to mention the pattern. The simplest aromatic compound is benzene (**11**).

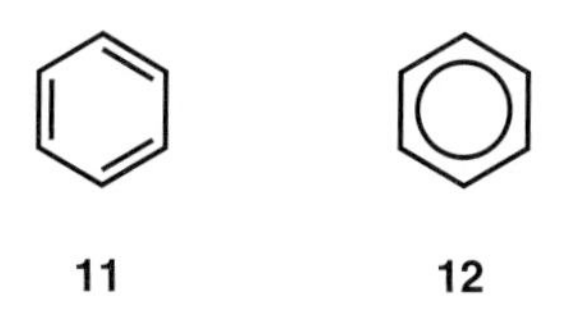

*Benzene does not have alternating single and double bonds. All of the carbon–carbon bonds are the same length. You have probably been told that structure **12** is a better representation, as it shows that all of the bonds are the same length. When we draw reaction mechanisms, we will find that representation **11** is better because it allows us to keep track of electrons, and it allows us to draw curly arrows (more about these later—much more!). When you get used to it, you will always recognize that benzene has six identical C–C bonds in structure **11**. Again, it simply needs to become a habit.*

Remember that not all rings are aromatic! Cyclohexane (**13**) is a perfectly good organic molecule. We use the prefix '*cyclo*' in the name to indicate that there is a ring.

13

Here we have two hydrogen atoms attached to each carbon. One of the most common mistakes I see is students drawing the hexagon, and then automatically drawing the circle that would turn it into benzene. Remember that cyclohexane is not a single lady. Even if you like it, don't put a ring on it![4]

CONCLUSION

I'm hoping by now that you're starting to realize why this book is different. My focus is on telling you how you should study, what mistakes you will make, and how to correct them.

[4] There will be quite a few bad jokes throughout this book. My book, my rules! If you don't get the joke, just Google 'Beyonce'. If you don't know what Google is, then the book has surpassed my wildest hopes with its longevity!

Don't beat yourself up about the mistakes. The only reason I know to highlight these mistakes is because I have seen them all, many times. Just be mindful of them, and you'll find it takes less time to spot the mistakes, right up to the point where you stop yourself before you actually draw an incorrect structure.

As we start to encounter various problems in organic chemistry, I will try to give you good problem-solving strategies.

You will also notice that this book is written in the first person. I am writing this for you, the reader. I want to use a writing style that makes it more personal—to achieve a direct connection.

HABIT 1

Always Draw Structures with Realistic Geometry

In the previous chapter, I mentioned the importance of forming good habits. I want to formalize this by including short chapters focused only on this one theme. Here is the first one!

REALISTIC GEOMETRY

Now we have looked at the different types of bond, we can consider the typical bond lengths and angles.

Here are the structures of ethane (**1**), ethene (**2**) and ethyne (**3**, nobody ever calls it ethyne—it is always acetylene). I've added the bond lengths in angstrom units (Å). There are two absolute key points to recognize at this level.

- Pretty much all chemical bonds in organic compounds are between 1 and 2 Å long.
- For a given type of bond, triple bonds are shorter than double bonds, which in turn are shorter than single bonds.

If you look closely at the numbers, you will notice some trends. Don't worry about them for now. We will add the details later.

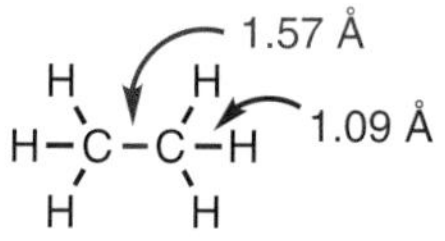

1.57 Å
1.09 Å
1
1.35 Å 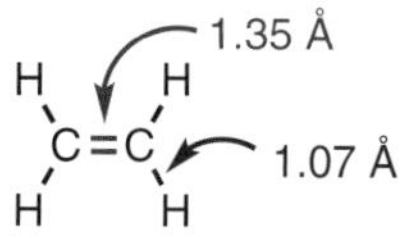
1.07 Å
2 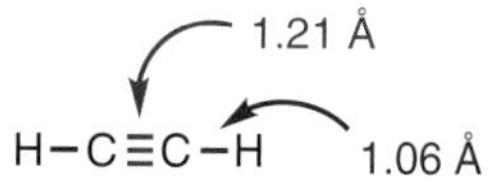
1.21 Å
1.06 Å
3

And here are the bond angles. In ethane, all bond angles are essentially tetrahedral, 109.5°. For ethene, the atoms are all in a plane, with approximate angles of 120°. For acetylene (ethyne) the bonds are linear, 180°.

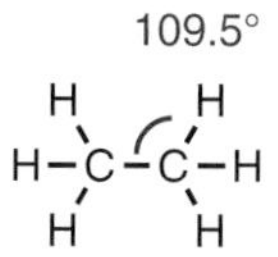

109.5°
1 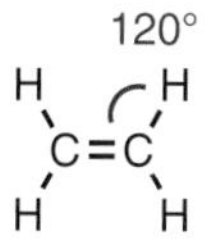
120°
2 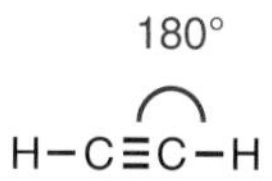
180°
3

1.2

It is important to develop the habit of always trying to draw structures with realistic bond lengths and angles where possible. As you can see with ethane, this is not very easy! After a while, you will get used to the situations in which it is not possible to draw all of the bond lengths and angles realistically, and you will correctly interpret a given structure.

WEDGES AND DASHES—REPRESENTING SHAPE

We simply cannot draw methane on a flat page and have all bond angles be 109.5°.

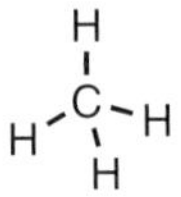

If we have one of the H–C–H angles close to 109.5°, all the others will look smaller. The key point in this representation is to 'internalize' the drawing of the structure, so that you instinctively recognize what the actual shape is, no matter how it is drawn.

When I see the structure above, I 'see' a tetrahedral structure. That's because, deep down, I know that it cannot be anything other than tetrahedral. Right now, you are reading a paragraph of text. You never stop to ask 'why is this the letter "a"?' You just 'know' it, and as a result, the words leap out of the page and into your head. You've done it so many times that it has become second nature. And you started doing that when you were three years old. So why on earth would you question your ability to do the same with molecular structures? All you need to do is practise enough.

We also have certain conventions that we can apply. We use a wedged bond to show that an atom is coming out of the page, and a dashed bond to show that an atom is going into the page.

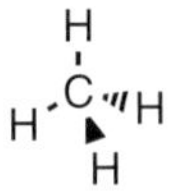

There isn't much to 'understand' here. It is what it is! You need to get used to these representations, so that when you see it, you can visualize the molecule in three dimensions.

Use molecular models to help with this, particularly as the structures get more complex.

We aren't going to directly do much with stereochemistry until Section 3. However, in the various 'Practice' chapters, I will be drawing structures with wedged and dashed bonds. I will do that because the molecules are three-dimensional.

When you get to these chapters, don't ignore the stereochemistry. Start to 'absorb' it. Just don't stress about it. Once you've read Section 3, come back to these and have another go. You'll 'see' more in the structures.

BASICS 2
Functional Groups and 'R' Groups

INTRODUCTION

Functional groups are the parts of an organic compound that lead to its reactivity. There are many different functional groups. They all have names. The important thing is to know what each functional group is, what it is called, and not to confuse different functional groups.

In this section, I will list several important functional groups, and show how they are commonly drawn (and how they should **not** be drawn). Please read this section very carefully, so that you don't accidentally learn to draw them the wrong way.

Before we look at functional groups, we will look at the carbon chains and rings that they are attached to. Before that, though, a cautionary note.

If you associate a functional group with the name as a word, there is a risk that you will confuse it with another (sometimes very different) functional group with a similar name. Always try to associate the name with the structure. The best analogy I can give for this is learning a new language. At first, you will think of an object in your native language and ask 'what is the word for X?' Eventually you will see the object and the word will automatically come to you in your new second language.

ALKYL GROUPS—*PRIMARY, SECONDARY,* AND *TERTIARY*

We might draw a structure such as **1** to mean any aliphatic alcohol.

R−OH

1

The 'R' refers to a carbon chain in a very general sense. We need to look at these carbon chains. We refer to the 'R' group as an alkyl group. There are a great many of these, and we will look at how we name them, and their parent hydrocarbons, in **Basics 3**. For now,

though, we will look at a few of them and classify them according to the nature of the attached carbon atom. The ones we need are summarized in the table below:

Abbreviation	Name	Structure
Me	Methyl	CH_3–
Et	Ethyl	CH_3CH_2–
n-Pr	Propyl	$CH_3CH_2CH_2$–
i-Pr	Isopropyl	$(CH_3)_2CH$–
n-Bu	Butyl	$CH_3CH_2CH_2CH_2$–
t-Bu	*tert*-Butyl	$(CH_3)_3C$–

Note: in isopropyl and *tert*-butyl, we need to use the brackets to make it clear that the CH_3 groups are attached to the same C.

Here is the key distinction.

Ethyl, n-propyl, and n-butyl are primary alkyl groups. The point of attachment is a CH_2. Isopropyl is a secondary alkyl group. The point of attachment is a CH. tert-Butyl is a tertiary alkyl group. The point of attachment has no attached hydrogen atoms.[5]

Can you work out what a *sec*-butyl group looks like? If you can, you will probably see why I didn't include it in the table.

AROMATIC RINGS

Benzene, **2**, is an aromatic compound. We will encounter the definition of aromaticity in **Basics 10**. For now, let's just focus on recognizing patterns.

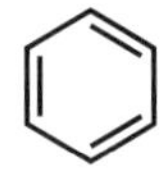

2

If we have 'some sort of' aromatic ring attached to (for example) an amino group, we can refer to the aromatic ring as 'Ar', and we can abbreviate the structure as in **3** below.[6]

$$Ar-NH_2$$

3

[5] Is methyl a *primary* alkyl group? Methyl is methyl. If you had to classify it, you would say it was *primary*. Methyl is small and unhindered. It is better to understand the effect of different groups with different branching rather than to lump methyl in with all the other possible *primary* alkyl groups.

[6] You probably don't want to do this if you work on organoselenium chemistry!

Aromatic compounds have very different reactivity to aliphatic compounds, so it is important that within these 'generic' representations, we can make this important distinction.

We will look at the naming of aromatic compounds in **Basics 3**.

ALCOHOLS AND PHENOLS

Structure **4** is methanol, an alcohol. It has an OH group attached to carbon. Compound **5** is phenol. In this case, phenol is the name of the specific compound, but also the name of the functional group. All compounds with an OH group attached to a benzene ring are known as phenols. Perhaps the most commonly encountered phenol is 2,4,6-trichlorophenol (**6**), a fungicide/antiseptic.[7]

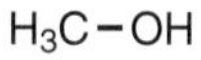

H_3C-OH OH Cl OH Cl Cl

4 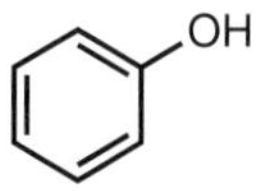 **5** 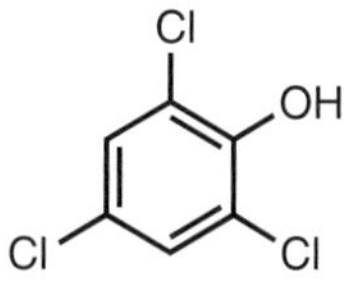 **6**

AMINES

Compounds **7**, **8** and **9** are all amines. They don't have to have a hydrogen atom attached to the nitrogen, but they can have.

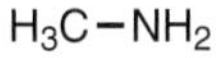

H_3C-NH_2 $H_3C-NH(CH_3)$ $H_3C-N(CH_3)_2$

7, methylamine
a *primary* amine

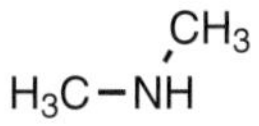

8, dimethylamine
a *secondary* amine

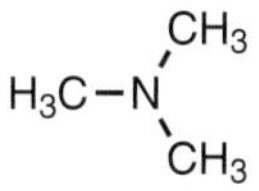

9, trimethylamine
a *tertiary* amine

PRIMARY, *SECONDARY*, AND *TERTIARY*—A NOMENCLATURE MINEFIELD

Before we get too much further, I want to illustrate the problems you face with chemical nomenclature—and the solutions.

[7] There seems to be somewhat of an urban myth that the antiseptic solution 'TCP' is so-called because it contains 2,4,6-trichlorophenol. In fact, TCP is an abbreviation for trichlorophenylmethyliodosalicyl, the active ingredient prior to the 1950s. This name makes no chemical sense, and as far as I know, nobody knows what the structure of the active ingredient actually was!

A *tertiary* alcohol has **one** *tertiary* alkyl group attached to the oxygen. A *tertiary* amine has three alkyl groups attached to nitrogen. These alkyl groups could be *primary, secondary*, or *tertiary*.

Of course, since an alcohol has an OH group, and oxygen can only form two bonds, there can only ever be one alkyl group attached. If we refer to a tertiary alcohol, we ***must*** *be talking about the alkyl group!*

What is the solution to this problem? At a simple level, just keep reading, looking at how people describe structures, and absorb the terminology.[8]

KEEP YOUR DATA CONNECTED!

Every alcohol has a hydrogen attached to oxygen. Not every amine has a hydrogen attached to nitrogen.

While I don't want to say too much about infrared spectroscopy, an N–H bond gives quite a characteristic peak. It is quite common for students to interpret an infrared spectrum and state that the compound concerned is an amine because they see an N–H peak. They will then draw a structure such as **9** which is indeed an amine but doesn't have the N–H bond that led to the conclusion in the first place. It is really important to keep track of those hydrogen atoms and what they mean.

ALKENES AND ALKYNES

A double bond between two carbon atoms is known as an alkene. A triple bond between two carbon atoms is known as an alkyne.

$H_2C{=}CH_2$ H–C≡C–H

ethene (an alkene)

ethyne (an alkyne)

CARBONYL COMPOUNDS

A double bond between carbon and oxygen has the general name 'carbonyl', but there are so many different types of carbonyl compound depending on what else is attached.

[8] Here's a confession. I don't think I had really thought about this contradiction until I started writing this book. But I still used the terminology correctly. I had internalized the learning to the point that I didn't think there was anything to worry about. This is the best sort of learning!

They all have their own names. Structure **10** is an aldehyde. Structure **11** is a ketone. Structure **12** is an ester. Structure **13** is an amide. Again, it doesn't need to be NH_2 to be an amide. There can be other alkyl groups attached to the nitrogen atom.

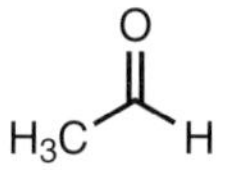

10 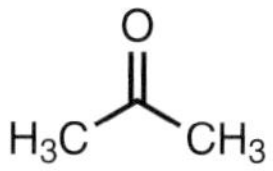 **11** 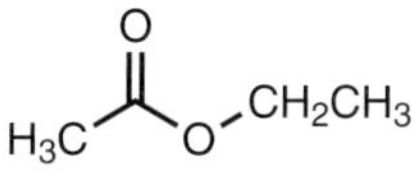 **12** 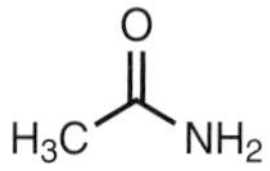 **13**

We can abbreviate the aldehyde **10** as CH_3CHO. We would not normally abbreviate it as CH_3COH. This can be a bit confusing. The oxygen and hydrogen atoms are both bonded to carbon, so it isn't clear which should be first. However, in the second representation (RCOH) it looks as though the H is bonded to O. Basically, CH_3CHO is right and CH_3COH is wrong. Sometimes it is best to accept it and try to get used to it.

In some representations, we might want the aldehyde on the left, as in $OHCCH_3$. Again, you would not normally write $HOCCH_3$.

With an ester, the point of attachment becomes more important. If we were to abbreviate the above example (**12**), we would write $CH_3CO_2CH_2CH_3$. We would not write $CH_3OCOCH_2CH_3$. This would be a different compound, as it implies that the CH_3 group is directly bonded to oxygen.

This is a lot to take on board. There are many places you can make mistakes. However, if you practise, this will all become second nature. If you don't practise, you will always be trying to remember how to draw something or what to call it. Succeeding in organic chemistry is all about doing the right things at the right time.

Here is one final point, with one more functional group.

ETHERS

Compound **14** is an ether. It has two alkyl groups (or they could be aryl groups) attached to one oxygen atom. This one is diethyl ether, a common solvent in the lab.

$$H_3CH_2C-O-CH_2CH_3$$

14

ETHERS AND ESTERS—A POINT OF CONFUSION

So, here is a very important point. It is not uncommon for students to confuse an ester and an ether. After all, the names are very similar. Most of the time this doesn't cause any problems—they just use the wrong word, but they still know which functional group they are talking about.

But in other cases, a student may correctly interpret an infrared spectrum (don't worry if you don't know about infrared spectroscopy) as having an ester functional group due to the characteristic peak around 1740 cm^{-1}. They will then draw an ether as a possible structure for the compound.

This is because they are associating the data with the word rather than with the molecular feature.

It is important that you start, very early on, to rely on structures as your primary tool for all organic chemistry.

All I can do is caution you against these mistakes. The whole point is that you need to study correctly and internalize certain information and patterns. There is no way I can do this for you, but I can (and will) give you exercises for practice.

HISTORICAL PERSPECTIVE

There is another important point.

In the lab, you will use diethyl ether and petroleum ether as solvents. Diethyl ether is an ether. Petroleum ether is not an ether. It simply refers to the petroleum fraction distilling in a particular range (*e.g.* petroleum ether 40 – 60 is the fraction boiling in the range 40 to 60 °C. It is a hydrocarbon (alkane)).

The term 'ether' is a very old one, and it doesn't always refer to the specific functional group.

What a nightmare! And yet if you just relax and focus on understanding molecular structure, the rest of it will take care of itself in time. Don't get overwhelmed by any of this. Just come back to this chapter in a couple of weeks and have another read!

Now we have got a selection of functional groups covered, we can start to think about systematic nomenclature.

BASICS 3

Naming Organic Compounds

NOMENCLATURE OF ORGANIC COMPOUNDS

We have to have clear and understandable names for chemicals. We have already used some of these names. You will find that in most instances you learn the names of compounds as you progress through the book. We will reinforce this with practice.

Whenever you encounter a compound name, you should take a few minutes to ensure that you understand why this name is correct for a given molecular structure. It doesn't make any sense to try to learn how to name all possible compounds.[9] Understanding the names is important, but it's more important to understand what the compounds do and why they do it. Just 'absorb' nomenclature as you encounter it.

For now, we will simply start with some basic rules, and expand these as we look at various functional groups.

ALKANES

For simple saturated hydrocarbons, the name is made up of a component that refers to the number of carbon atoms, followed by 'ane'. 'Saturated' simply means that there are as many hydrogen atoms as the number of carbon atoms can take. We will come back to this point later.

Here is a table.

Number of Carbon Atoms	Formula	Structural Formula	Name
1	CH_4	CH_4	Methane
2	C_2H_6	CH_3CH_3	Ethane
3	C_3H_8	$CH_3CH_2CH_3$	Propane
4	C_4H_{10}	$CH_3CH_2CH_2CH_3$	Butane

[9] I've got a book that is just about naming of organic compounds! Does that make me sad?

3

Number of Carbon Atoms	Formula	Structural Formula	Name
5	C_5H_{12}	$CH_3(CH_2)_3CH_3$	Pentane
6	C_6H_{14}	$CH_3(CH_2)_4CH_3$	Hexane
7	C_7H_{16}	$CH_3(CH_2)_5CH_3$	Heptane
8	C_8H_{18}	$CH_3(CH_2)_6CH_3$	Octane
9	C_9H_{20}	$CH_3(CH_2)_7CH_3$	Nonane
10	$C_{10}H_{22}$	$CH_3(CH_2)_8CH_3$	Decane

You will notice two new aspects of structure representation in the above table. First of all, while we normally draw a single bond as a line between two atoms, we can 'write' the structure with no need for the lines.

We can actually represent alkenes and alkynes using this convention as well. Writing '$CH_3CHCHCH_3$' would give us but-2-ene (more about naming alkenes later in this chapter). There must be a double bond between the second and third carbon atoms, as we are explicitly writing the hydrogen atoms and there aren't enough for a saturated structure. Personally, I would still make it clearer by writing '$CH_3CH{=}CHCH_3$'.

You may notice that in the structural formula column, after butane I started to use brackets to indicate the total number of CH_2 groups rather than drawing them out. This is another convention that you should get used to.

If we have branched alkanes, we identify the longest carbon chain. This becomes the 'parent' name. Substituents along the chain are then identified by the type of substituent and their position along the chain. If there is more than one of the same type of substituent, we use prefixes such as 'di', 'tri', *etc*. This is best seen with examples.

4,5-dimethyldecane

3-ethyl-4-methylheptane

Notice in the second example, there are actually two different 7-carbon chains that we could have considered. Actually, in this case it doesn't matter. The substituents are listed in alphabetical order, and numbering is done to give the lowest possible numbers (*i.e.* 3 and 4 rather than 4 and 5 if you counted from the other end).

PREFIXES FOR ALKYL CHAINS

We saw some simple alkyl substituents above. These are named from the parent alkane. Here they are again. I'm willing to bet that you can add more lines to this table.

Abbreviation	Name	Structure
Me	Methyl	CH_3–
Et	Ethyl	CH_3CH_2–
n-Pr	Propyl	$CH_3CH_2CH_2$–
i-Pr	Isopropyl	$(CH_3)_2CH$–
n-Bu	Butyl	$CH_3CH_2CH_2CH_2$–
t-Bu	*tert*-Butyl	$(CH_3)_3C$–

Note: in isopropyl and *tert*-butyl, we need to use the brackets to make it clear that the CH_3 groups are attached to the same C.

We defined the terms *primary, secondary,* and *tertiary* in **Basics 2**. We now need to consider how we would name compounds with various alkyl groups. Here is an example.

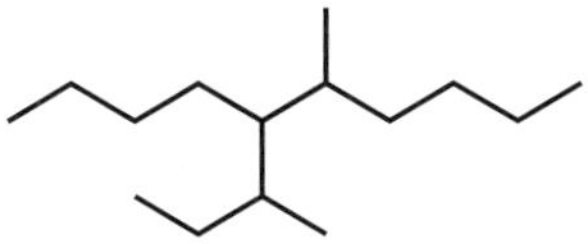

The longest alkyl chain has 10 carbon atoms. It is a derivative of decane. There is a methyl group on C5 or C6 depending on which end you start the numbering at. Then there is a *sec*-butyl group[10] on the other of these two carbon atoms. We could simply call it a *sec*-butyl group, or we could call it 'but-2-yl', where the '2-yl' bit makes it clear that the butyl group is attached via **its** 2-position.

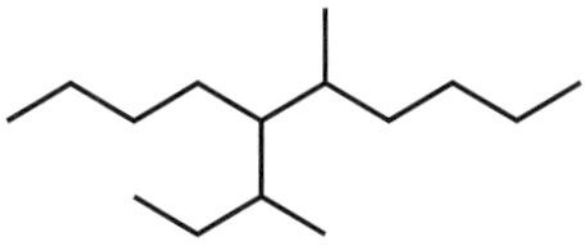

5-(but-2-yl)-6-methyldecane

There isn't much more to say about this. I could write out a dozen more examples, but you wouldn't learn anything new from them. I would much prefer to set them as problems, so you will be aware that you are getting better at naming compounds.

HALOALKANES

Most of the time, we simply consider the halogen to be a substituent. As above, we put substituents into alphabetical order. If there is only one place to put a substituent, there is no need for the number.

[10] Did you work it out correctly in **Basics 2**? Make sure you refer back. This constant reinforcement is how you will learn.

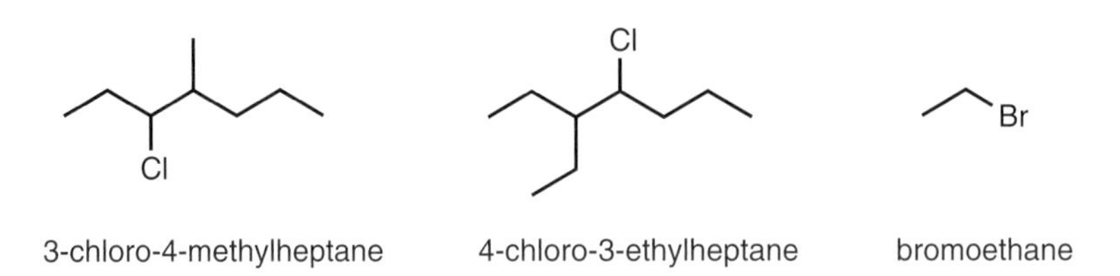

3-chloro-4-methylheptane 4-chloro-3-ethylheptane bromoethane

Trivial names are commonly used. Most people would name bromoethane as ethyl bromide.

The main point about nomenclature is that if you write a compound name, someone reading it could only draw one possible structure. If you didn't put the substituent names in alphabetical order, this wouldn't be a massive problem. Sure, it wouldn't be strictly correct, but it would still do the job.

ALKENES

Here, the name of the compound ends with 'ene'. Since a double bond is always between two adjacent carbon atoms, you don't need to specify both carbon atoms.

In simple compounds, it is okay to put the number at the beginning. As names get more complicated, it is sometimes necessary to put the number directly before 'ene' to make sure it is clear that the number refers to the double bond position.

1-octene (or oct-1-ene) (*E*)-2-octene or (*E*)-oct-2-ene

2-Octene could exist as two double bond isomers, depending on whether the methyl and pentyl groups are on the same side of the double bond or on opposite sides. We use (*E*) and (*Z*) to describe these. (*E*) comes from the German word '*entgegen*', meaning 'against'. (*Z*) comes from the German word '*zusammen*' meaning 'together'. When you only have two substituents on a double bond it is clear which one to use. When you have three or four substituents it is less straightforward. We will come back to this in Habit 6.

ALKYNES

The naming of alkynes is similar to that of alkenes. Here are a couple of examples.

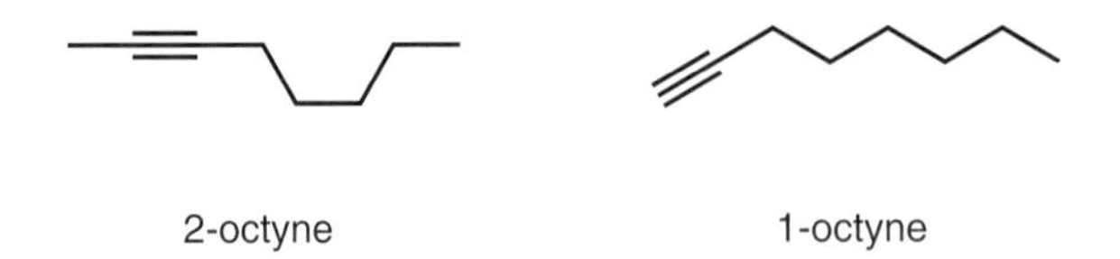

2-octyne 1-octyne

*As I mentioned in **Basics 1**, the alkyne functional group confuses a lot of students. Take a moment to look at these structures and count the atoms. Make sure you understand why this representation shows eight carbon atoms in each case.*

AROMATIC COMPOUNDS

First of all, there is just one more abbreviation we need to add, Ph. Ph is short for phenyl, formula C_6H_5. A phenyl group is a benzene ring with one hydrogen atom removed to allow something else to be attached. For example, structures **1** and **2** represent the same structure.

Ph—NH2

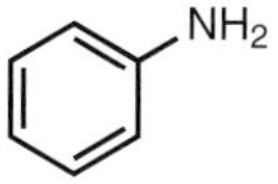

NH2

1 **2**

*There is one common point of confusion with phenyl. It sounds a bit like 'phenol'. I find that many students talk about 'the phenol group' when they mean 'the phenyl group'. Phenol is Ph-OH. I've tried to think what I can write to make this point any clearer. There isn't anything. Phenyl is phenyl, **and you just need to know this**. I will remind you of this lots, to make sure it becomes habit.*

Now we have got that out of the way, what happens if we have more than one substituent on a benzene ring? At that point, we can have **constitutional isomers** (**Basics 4**). Here are three phenol examples.[11]

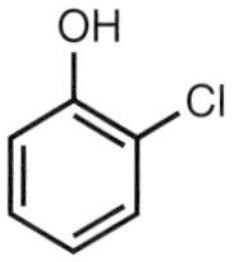

OH Cl

OH Cl

OH Cl

2-chlorophenol
o-chlorophenol

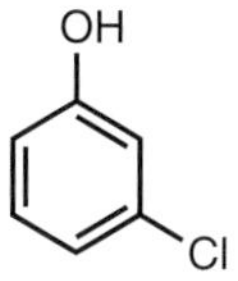

3-chlorophenol
m-chlorophenol

4-chlorophenol
p-chlorophenol

There are two conventions that we use. The first is numbering. In the compound on the left, the chlorine is in the 2-position relative to the OH group being on the 1-position. We don't need to specify both numbers. We can also use the terms *ortho, meta*, and *para* (usually abbreviated as *o*-, *m*-, and *p*- in the compound name) to indicate the relative placement of one substituent relative to another.

[11] These ones really are phenols!

This book doesn't cover reactions of aromatic compounds, but you still need to be familiar with the nomenclature.

ALCOHOLS

Depending on the compound, you can specify an alcohol with the suffix 'ol' or the substituent prefix 'hydroxy'. You would use the prefix if there is another substituent that is 'more important'. It takes a while to get used to which substituent will be more important, and you tend to learn by example rather than formally understanding. In the second example below, we have added an extra functional group.

OH

heptan-1-ol (or 1-heptanol)

OH

(*E*)-hept-4-en-1-ol
(or (*E*)-1-hydroxy-4-heptene)

In the second compound above, we only need to use one number to define the alkene position. If the alkene bond was between C3 and C4, then it would be a hept-3-ene, not a hept-4-ene.

ALDEHYDES

These are named with the suffix 'al'. Note that in most cases we don't need to specify a number for the aldehyde group. If it wasn't at the end, it wouldn't be an aldehyde. Everything else is numbered relative to the aldehyde being position '1'.

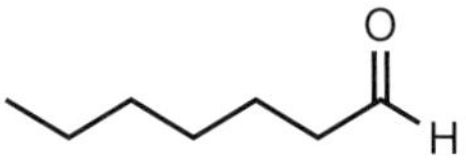

O
H

heptanal

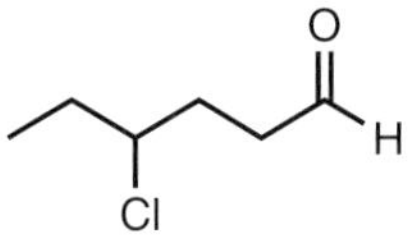

O
H
Cl

4-chlorohexanal

KETONES

Here you definitely do need to specify the position of the ketone group, unless it is symmetrical. Even then, it is better to specify the number (see heptan-4-one below) so it is clear that you didn't just forget!

heptan-2-one

heptan-4-one

SUMMARY

This discussion is really just to get you started. There are many more functional groups to consider, but they follow broadly similar rules. As with the structural representations, once you understand the principles and apply them with regular practice, you'll start to come up with plausible compound names without too much effort.

Fundamentally, a compound name needs to describe one compound and one compound only. Any name that is ambiguous **must** be wrong.

There's an easy way to check if you have named a compound sensibly. Take the name and draw a structure from it (or ask someone else to do this). If it is only possible to draw one structure, the name is good. We will use this approach in ***Practice 2****.*

PRACTICE 1

Drawing Structures from Chemical Names

THE PURPOSE OF THE EXERCISE

The point of this section is not really about drawing the correct structure from a chemical name. This is only one aspect of it. The main focus is to practise drawing good skeletal structures of organic compounds.

If you need to draw a structure that shows every hydrogen atom at first, that is fine. However, then redraw it as the skeletal structure. Draw bond angles as close to reality as you can, while still considering the hydrogen atoms that are not shown.

If it's a good structure, you will be able to clearly see what the compound is. If you get confused, perhaps you need to draw the structure more clearly.

THE EXERCISE

Draw structures for the following compounds:

1. 3-(Hex-2-yl)heptane
2. 4-Chlorohexanal
3. 4-Bromo-7-methylnonan-2-one
4. 5-Chloro-2-propylphenol
5. 3-Ethylhexanoic acid
6. 7-Chlorohept-3-yn-2-ol
7. Methyl propanoate
8. Ethyl benzoate
9. 3-Hydroxyacetophenone (this is a non-systematic name—get used to it!)
10. 2-Aminoheptan-4-ol

11. 4-Chloro-1-methylcyclohexan-2-ol
12. *N*-Ethyl-*N*-propyl-(2-hydroxypentylamine) (this one is tricky—what does the '*N*' mean?)[12]

1

Note: You will probably need to look up some compound names or discuss them. This is fine. **Basics 3** doesn't cover all aspects of naming compounds.

Above all, don't get stressed by this. Remember, once again, the main point is that the structures you draw should be neat and tidy, and should be the more economical skeletal structures. You should be able to draw them quickly and confidently.

Get your friends to draw the structures, and compare your structures with theirs. Discuss which ones are best, and why.

For the compounds you have drawn, identify the functional groups present. Your answers may depend on how you interpret the term 'functional group'. I don't normally think of a methyl group as a functional group, but if you have a reaction that leads to its functionalization, perhaps it is! Discuss and debate your answers with colleagues. Don't try to learn in a vacuum!

For now, you need to be looking at organic structures and taking the time to try to see what is in them.

FOLLOW-UP EXERCISE

Give the structures you have drawn to a friend and ask them to propose systematic names for the compounds. Don't worry about number 9, and number 12 might be quite difficult. Once they have had a go, discuss the names with them.

Are the compound names in my list the best systematic names for all of the compounds?

Were there any that were particularly troublesome?

AND IN REVERSE!

If you can propose a structure from a chemical name, you should be able to propose a chemical name from a structure. Here are some structures of *relatively* simple hydrocarbons to practise with.

[12] This probably isn't a fair question, as I haven't explained this. But you can look at the rest of the name and make an educated guess. Also, I want you to get used to looking things up in a variety of sources.

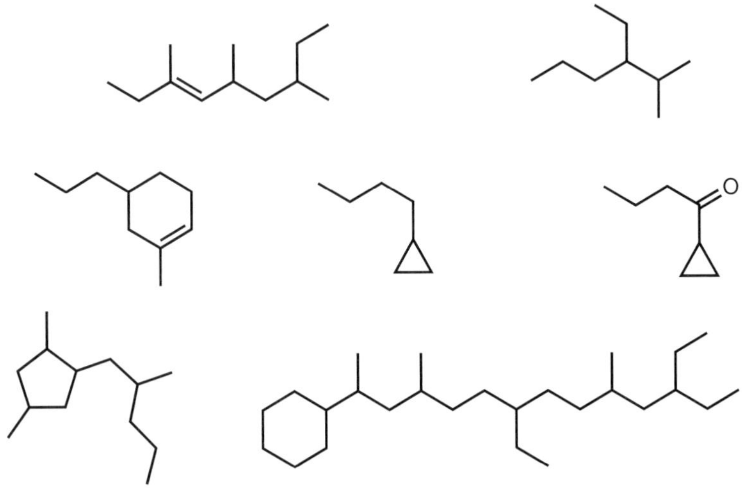

There are a few things to think about. In particular, compounds with rings can be challenging. Do you consider the ring to be the main feature, and everything else a substituent, or is it better to consider the ring itself to be a substituent? There isn't always a simple answer to this.[13]

1.4

Video 1.4 shows the worked answers to these problems. First of all, have a go and discuss your answers with friends.[14]

[13] Well, there are a set of definitive rules, but there is a lot to be said for giving the shortest possible **unambiguous** name for a compound. Remember why we are doing this!

[14] Not too much though! I'd like you to keep your friends!

BASICS 4

Isomerism in Organic Chemistry—Constitutional Isomers

WHAT IS ISOMERISM?

The concept of isomerism is a very important one in all areas of chemistry. There are quite a few different types of isomer, so first of all, let's state clearly what we mean by isomers.

Isomers are compounds that have the same molecular formula, but are not the same compound.

Within this very broad definition, there are quite a few ways in which isomers could differ. For now, we will focus on one of these. Once you've got the hang of this, we will add another one.

CONSTITUTIONAL ISOMERS

Constitutional isomers, also known as **structural isomers**, are compounds that have the same molecular formula—they have the same number of each type of atom within the molecule—but the atoms are connected in a different way.

EXAMPLES

Let's have a look at some alkane structures, along with their names.

CH_4

methane ethane propane

Note that I have used the most economical representation for each structure—lines to represent bonds. Where a line ends, or two lines join, there is a carbon atom. For methane, with only one carbon atom, we have to state 'CH_4'. We couldn't just draw a dot.

For these three compounds, we would find that if we were given a molecular formula, we could only draw these exact structures. There are no isomers. The molecular formula would be enough to fully define the structure.

If we have four carbon atoms, there are two possible structures. The first is where all four carbon atoms form a single chain. This is *n*-butane, or simply 'butane'. The second has a shorter carbon chain, with one carbon atom as a substituent. This is 2-methylpropane, because the longest carbon chain is three carbon atoms (propane). The methyl group is on the middle carbon, or the 2-position. A methyl group is 'CH_3'—never forget the attached hydrogen atoms. *n*-Butane and 2-methylpropane are constitutional isomers.

n-butane 2-methylpropane

The more carbon atoms we add, the more constitutional isomers there can be. For example, we have three isomers of pentane, although only one of them is actually called pentane.

n-pentane 2-methylbutane 2,2-dimethylpropane

We often call this isomer '*n*-pentane'. The '*n*-' refers to 'normal' which relates to the straight chain (rather than branched).

Cyclohexane and 1-hexene are isomers. Don't take my word for it. Check! This sort of thing is good practice. The key point here is that one of these has a carbon–carbon double bond and the other does not. The molecular formula alone (work it out!) doesn't tell you anything about the functional groups present.

cyclohexane 1-hexene

Here is another example. The two compounds below are isomers, and again they have different functional groups.

O OH

2-pentanone 2-penten-4-ol

This brings us to the next topic. Some molecular formulae make sense, while others do not. Although you can't tell the full structure of a compound from the formula, you can get some information from it. We will look at this next, after a short 'Practice' chapter.

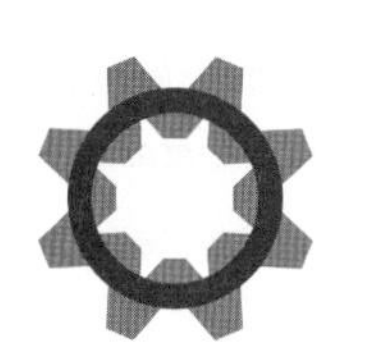

PRACTICE 2
Constitutional Isomers and Chemical Names

THE PURPOSE OF THE EXERCISE

It's really about getting more practice! Sometimes it's that simple. The added layer here is that it isn't always possible to draw a plausible structure for **any** given molecular formula. A good chemist will spot the impossible formulae quickly. There are several things to look out for. They are covered in the next chapter. But you already know enough to work them out for yourself.

QUESTION 1

How many different structures can you draw for the following molecular formulae?

1.5

1. $C_6H_{12}O$
2. $C_8H_{15}O$
3. $C_7H_{13}NO$
4. C_4H_7Cl
5. $C_6H_7NO_2$
6. $C_4H_{12}O$

 Can you actually draw a structure in every case?[15]

QUESTION 2

Propose a systematic chemical name for each of the structures you have drawn. Give the compound names to a friend and ask them to draw structures. They should draw the same structure you did, but it might not look exactly the same.

[15] You can bet the answer to this question is a resounding 'no'! The more important thing is learning to recognize the patterns that will allow you to spot an 'impossible' formula without even trying to draw a structure. Don't draw atoms with the wrong number of bonds just so that you have a molecule with the 'correct' formula.

Learning to spot two identical structures when they are drawn differently is a skill that cannot be overstated. No-one can teach you how to do this. You have to start with small structures and work up to the bigger structures.

You probably did 'spot the difference' puzzles as a child. After a while, you probably started to anticipate the parts of a picture where the artist would try to hide a difference. You would get better at solving the problem with experience. Funny, that!

QUESTION 3

For the structures you have drawn in **QUESTION 1**, identify all of the functional groups present. I cannot possibly give you all of the answers. You may need to look up some functional groups. It depends on how creative you have been!

I asked you to identify functional groups in ***Practice 1****. Are you tempted to ignore the question here because you have already done it? Don't be tempted! If it is easy, it won't take you long. If it takes you too long, then you definitely need to do it. No short cuts!*

HABIT 2

Identifying When a Formula is Possible

Let's start with something obvious. You cannot draw a structure for the formula 'CH_6'. This is because there are too many hydrogen atoms. Carbon can only form four bonds.

It's always good to reduce an idea to a simple and obvious example. Now let's develop the idea a bit further.

FORMULAE—WHAT IS POSSIBLE AND WHAT IS NOT?

If you have a compound with only carbon and hydrogen atoms, you must have an even number of hydrogen atoms. If the compound is completely saturated (as many hydrogen atoms as possible) and has no rings then the formula, for n carbon atoms, will be

$$C_nH_{2n+2}$$

You cannot add more hydrogen atoms than this.

Adding an oxygen atom doesn't change the maximum possible number of hydrogen atoms. Sometimes it is easiest to see this with structures.

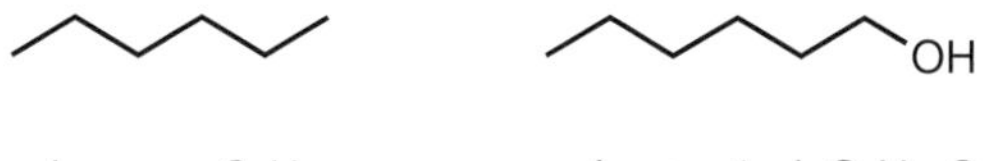

hexane, C_6H_{14} hexan-1-ol, $C_6H_{14}O$

As long as the compound is completely saturated, then adding an oxygen atom doesn't change the number of hydrogen atoms, and it doesn't change the fact that there need to be an even number of hydrogen atoms. We could insert an oxygen atom into any C–H bond or C–C bond and the rest of the formula would not change.

Adding a nitrogen atom does change things! If you have an odd number of nitrogen atoms, and only carbon, hydrogen, and oxygen otherwise, then you will have an odd number of hydrogen atoms.

Look at the following structure. It has a formula $C_6H_{15}N$. Make sure you know where all the hydrogen atoms are. In effect, you have taken hexane, removed one H and added an NH_2 group. The net effect is one additional hydrogen atom in the formula.

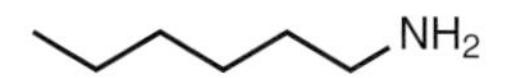

DOUBLE BOND EQUIVALENTS

When you make a compound, and you don't know the structure, the idea of 'double bond equivalents' is a very useful one. The only information you need is the formula. Take a look at the following structures.

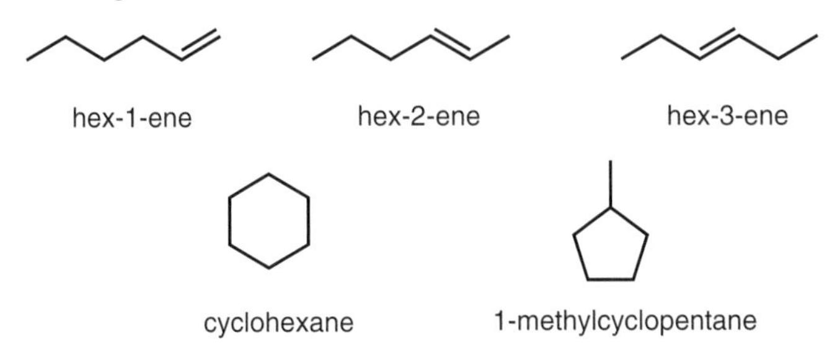

All of these structures have the formula C_6H_{12}. The saturated compound with six carbon atoms would be C_6H_{14}. You form a double bond by taking away two hydrogen atoms. You form a ring by taking away two hydrogen atoms and joining the ends. If we are given the information that a compound has the structure C_6H_{12}, we cannot tell whether it has a double bond or a ring. We just know that it must have one or the other.

There is a specific name for this lack of two hydrogen atoms in a formula. It is a ***double bond equivalent****.*

There is an equation that will allow you to calculate the number of double bond equivalents from the number of carbon, hydrogen, oxygen, and nitrogen atoms. I'm not going to show you this formula.

Don't look it up just yet!

I'd much rather you understand structure and be able to work it out. After all, the equation doesn't include any other elements, so you would need to be able to work out examples with things that are not in the equation.[16]

Let's see how this could work. As an example, we will use a formula $\mathbf{C_6H_5NO}$. The simplest question you can ask is 'what would be the formula if there were no double bond equivalents?' That is, 'how many hydrogen atoms would it need?'

The easy way to answer this question is to draw a structure with the carbon, nitrogen, and oxygen atoms, then add the right amount of hydrogen. In this case, I would draw

HO NH_2

[16] If you asked me to write down the equation right now, I'm not sure I could. I know what it looks like, but I will probably mix up the plus and the minus. I can take a bit of time, and I can work out the equation by using the principles (how many bonds each atom can form). But if I am going to do that, I might as well just skip the equation and apply the principles directly to a given molecular formula.

This has a formula $C_6H_{15}NO$. Therefore, C_6H_5NO is 'missing' 10 hydrogen atoms. Each double bond equivalent requires two hydrogen atoms, so that the formula has **5 double bond equivalents**. You can apply this method to any formula. Just make sure you draw a structure with no double bonds or rings, so your structure will have the maximum number of hydrogen atoms. You will find it doesn't matter where you put the oxygen or nitrogen!

TWO IMPORTANT POINTS

First up, if you are ever given a structure and asked to work out how many double bond equivalents it has, don't waste time determining the formula and then working it out.

Inspect the structure and identify the double bonds and/or rings. This shows that you actually understand what a double bond equivalent is.

Secondly, if you ever determine the number of double bond equivalents for a given structure or molecular formula, and your answer is half integral (something and a half!) then you have made a mistake. Either that, or the formula you have been given is nonsense. I do sometimes give my students nonsense formulae to check that they understand the principle.

PRACTICE 3
Double Bond Equivalents

THE PURPOSE OF THE EXERCISE

An organic chemist might have determined the molecular formula of something they have made, but they don't know the structure. Knowing how many double bonds or rings are present is useful information in determining the outcome of a reaction.

Beyond that, it is yet more practice and getting used to looking at increasingly complex molecules. The added layer of complexity is that the structures have wedged bonds and dashed bonds. We have looked at what these mean at the simplest level. We will sort out the detail in Section 3, but it's still worth getting used to seeing them now.

QUESTION 1

Here are the molecular formulae that we looked at in **Practice 2**. For each formula, determine the number of double bond equivalents present. Look back at your answers to **Practice 2**. Pick one or two of your structures and try to identify where the double bond equivalents are in the structure.

If the number of double bond equivalents looks 'wrong', what does this tell you about the structure? Were you able to draw any 'sensible' structures in those cases?

1. $C_6H_{12}O$
2. $C_8H_{15}O$
3. $C_7H_{13}NO$
4. C_4H_7Cl
5. $C_6H_7NO_2$
6. $C_4H_{12}O$

1.6

QUESTION 2

Now it's time to step it up a gear. Here are eight natural product structures. Identify the double bond equivalents.

Strychnine

Hemibrevetoxin B

Penicillin G

Artemisinin

Vigulariol

Pseudomonic acid C

Discodermolide

Doxorubicin

Do this in each of the following three ways.

1. Work out the molecular formula of the compound. Then work out how many hydrogen atoms would be present if the molecule had no double bond equivalents.
2. Draw the structure. Identify which features correspond to double bond equivalents by inspection.

3. Look up the equation that I referred to in the previous chapter[17] and use it to calculate the number of double bond equivalents.

Which method is easiest? It probably depends on whether you are given a formula or a structure? Will you remember the equation? Can you be sure you will get it right?[18]

This is not easy! Look at the structure of artemisinin. You have a seven-membered ring bridged by two oxygen atoms. How many rings is this? We could have the seven-membered ring, and then two different rings that contain the bridge with two oxygen atoms.

My 'fallback' position is to 'break' a ring and add a couple of hydrogen atoms. Then see how many rings are left. You can do this as many times as you like!

A COMMENT!

If you want to check your answers, the formula of each compound can be readily found online! Did you get them all correct? If not, the equation is of no use to you!

What do these compounds do? You need to appreciate why organic compounds are important! These compounds, and many thousands of other compounds, save and enhance lives. And all of these compounds have been synthesized in the laboratory. That's why you are studying organic chemistry!

[17] I know it would be easier if I gave it to you. But I really don't like it!
[18] I can't! But then again, I have viable alternatives.

COMMON ERROR 1

Formulae, Functional Groups, and Double Bond Equivalents

1.8

Here is the structure of lysergic acid diethylamide, LSD.

What is the molecular formula of this compound?

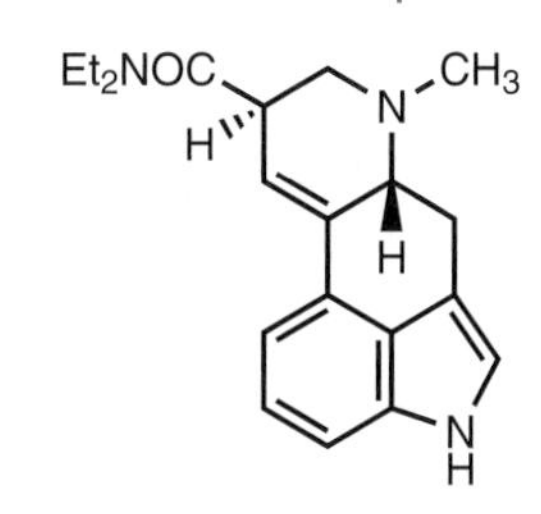

This is relatively straightforward, but there are a couple of bits that cause problems. The main one is the Et_2NOC substituent.

I have drawn it this way round so that it is clear that the carbon atom is joined to the ring.

There is a tendency to assume that the nitrogen atom is bonded to oxygen. After all, they are 'together' in the representation. This then leads to the following partial structure.

Et N O C Et

Note that I am using a wiggly line across the bond to indicate where it is attached to the rest of the molecule.

The problem here is that the carbon atom only has two bonds, not four.

How does this relate to the formula?

If you made this mistake, you might then add a couple of hydrogen atoms to the carbon to 'fix' the problem.

This doesn't fix anything.

This part of the molecule isn't a skeletal structure, so you cannot just add hydrogen atoms. You are explicitly told how many hydrogen atoms there are in this part of the structure.

Okay, so the nitrogen atom cannot be bonded to oxygen. It must be bonded to carbon. There is no other possibility. The oxygen must also be bonded to carbon, and to have two bonds to oxygen and four bonds to carbon, we need a double bond.

The next common error (at least in the very early stages) is very simple—not knowing that Et is C_2H_5 (or CH_3CH_2, which is the same thing). This error will disappear very quickly if you immerse yourself in molecular structures.

How many abbreviations do you need to know? You definitely need to know as far as butyl. We don't normally use two-letter abbreviations for alkyl chains longer than this.

So, combining these two errors, you have the possibility of determining a molecular formula that isn't $C_{20}H_{25}N_3O$ (the correct answer!).

Everyone can make a mistake. Hopefully you will spot all three nitrogen atoms. There are an odd number, so there must be an odd number of hydrogen atoms. This gives you an opportunity to spot any mistakes.

 What functional groups are present in the structure?

As you might imagine, if you misidentified Et_2NOC then you won't realize that this is an amide. There is also an amine, and we won't worry about the heterocyclic (has a ring with an atom that is not carbon) ring system at the bottom of the structure for now.

 How many double bond equivalents are present?

We established that there is an equation that allows you to calculate this based on the number of carbon, hydrogen, and nitrogen atoms. I also told you that you will probably remember the formula incorrectly.

Some readers will have seen the equation before—you will already feel comfortable with it. If you determined the molecular formula incorrectly, the equation will give you the wrong answer. If you learned the equation incorrectly, it will also give you the wrong answer. It's just an opportunity to make mistakes.

The structure has six double bonds (including the amide carbonyl that we discussed above). It has four rings. Each ring is a double bond equivalent. Six plus four equals ten. And that's it! Much easier than using an equation.

You're going to tell me that you wouldn't make these mistakes.

With enough practice, you probably won't! And then you will have three good ways to solve a problem of this type.

Until you have had enough practice, it is best to be aware of the pitfalls, and to know the more reliable methods for solving the problem.

HABIT 3

Ignore What Doesn't Change

This is an important idea, but one that is almost impossible to teach in a formal way.

In any reaction of an organic molecule, there will be a change to one functional group (or sometimes more), while the rest of the molecule is untouched. The key skill is to learn to recognize what has changed and what has not.

This will allow you to streamline your working, so that you spend your time looking at the right parts of the structure.

I'll show you an example. Here is a single step from the synthesis of an antibiotic compound, erythronolide B. The synthesis was reported in 1978.[19] The starting material is moderately complex, with 35 carbon atoms and 9 stereogenic centres. We will worry about stereogenic centres in Section 3.

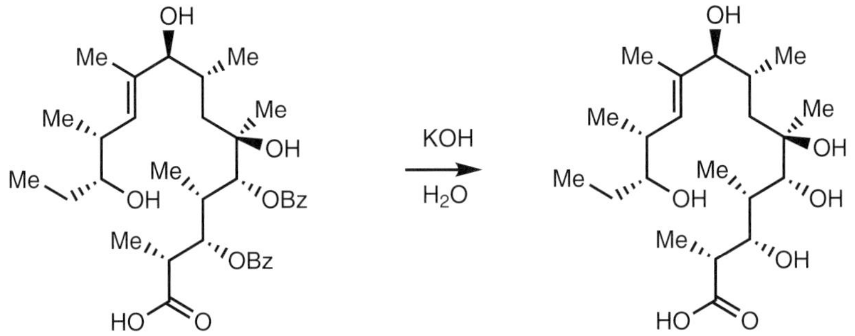

This is a reaction type that we aren't going to cover in this book. Don't worry about that! In fact, the only thing that is happening is that two benzoate esters (Bz is a standard abbreviation for benzoyl, PhCO—yes, I explained one abbreviation by using another abbreviation) are hydrolysed.

Here it is again with the relevant functional groups shown in purple.

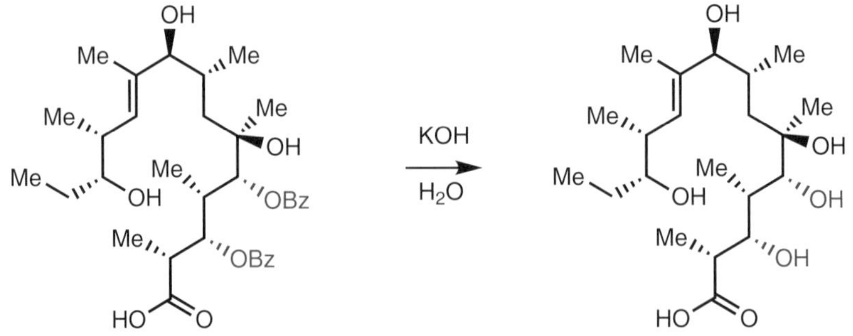

[19] You may think of this as ancient history. I remember it well!

When you look at more complex reactions, you need to recognize the need to be able to determine which parts of the molecule you need to worry about, and which parts you do not. For example, you might draw the reaction as follows.

R^1 Me OBz R^2 OBz

KOH
H_2O

R^1 Me OH R^2 OH

There are advantages to doing this. For a start, it is quicker to draw. But there are definitely disadvantages. You never quite know when one of the functional groups in one of the 'R groups' will be needed in a subsequent reaction. If you are planning a synthesis, you might find that you ignore the possibility that the reaction will not be as selective as you like. We will explore the concept of selectivity later.

For now, I simply want to highlight the nature of the problem. What can you do about it? The honest answer is 'not much'!

Initially, you need to be mindful of the problem, and you should take every opportunity to look at reactions of more complex molecules, rather than avoiding them in favour of simple examples.

And then gradually, you will get quicker at spotting the functional groups that have changed. It will just happen over time and with experience.

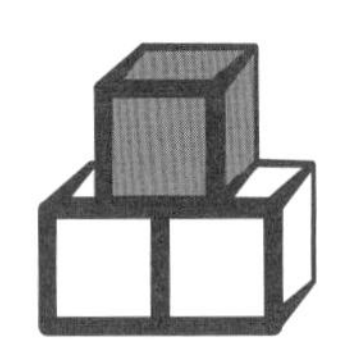

BASICS 5
Electronegativity, Bond Polarization, and Inductive Effects

When we look at a structure, we need to consider how the electrons are distributed within the structure, so that we can predict how the structure will react with different reagents under various different reaction conditions.

The simplest parameter that affects electron distribution is electronegativity.

ELECTRONEGATIVITY

Electronegativity is a fundamental concept in chemistry, and it has implications throughout. To give a very brief summary, electronegativity is a measure of how much a particular atom will pull electron density towards itself. Once we exclude the unreactive noble gases, the most electronegative elements are at the top right of the periodic table, and the most electropositive (least electronegative) elements are at the bottom left of the periodic table.

Let's have a look at a few values, and consider the implications of them.

Element	Electronegativity
C	2.55
H	2.20
N	3.04
O	3.44
Cl	3.16
F	3.98
Li	0.98
Mg	1.31

BOND POLARIZATION

First of all, carbon and hydrogen have very similar electronegativity values. Hydrocarbons have very little polarization of bonds.

It gets more interesting when we consider elements such as O, N, or Cl. If we consider the structure of CH_3–Cl, we can see that the Cl is considerably more electronegative than the C. This means that the Cl will withdraw electron density towards itself, so there will be a polarization in the bond. We can represent this as follows:

$$\overset{\delta+}{H_3C}-\overset{\delta-}{Cl}$$

The Greek letter delta is used to indicate a partial charge on the atom. The carbon atom has a partial positive charge, and is susceptible to attack by a **nucleophile**, something that has an excess of electron density. The term **nucleophile**, and the corresponding term **electrophile**, will be fully defined in **Basics 9**.

If we now consider a compound that you will probably not have encountered, methyllithium (CH_3–Li), the relative electronegativity of the elements tells us that the bond polarization will be as follows:

$$\overset{\delta-}{H_3C}-\overset{\delta+}{Li}$$

In this case the carbon has a partial negative charge, and it will tend to react with something that is deficient in electron density.

You need to be comfortable with trends in electronegativity, and to understand how they affect the electron distribution in organic compounds.

INDUCTIVE EFFECTS

This bond polarization is referred to as an **inductive** effect. This is a ground state polarization of the σ bonded framework of the molecule. It is a relatively short-range effect, and it diminishes within a couple of bonds.

We will encounter a corresponding electronic effect, the **mesomeric** (or resonance) effect in **Basics 10**.

It is important that you recognize and can distinguish these two types of electronic effect, and know how to represent them.

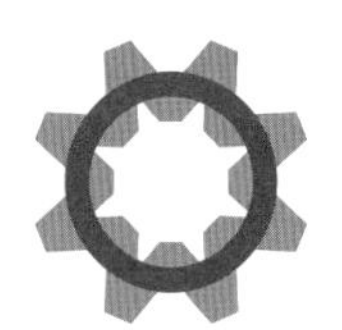

PRACTICE 4

Bond Polarization and Electronegativity

THE PURPOSE OF THE EXERCISE

You need to be thinking about the distribution of charge in organic structures. You need to be able to do this no matter how complex the structure. In order to do this, you will need to 'see' functional groups even when they are not explicitly drawn out. Strangely enough, it's all about practice.

THE EXERCISE

Here are the natural product structures we saw in **Practice 3**. Redraw these structures and indicate with a δ+ or δ– the partial charge on 'key' atoms.

You don't need to mess around 'balancing' the delta charges. It's perfectly okay to have one δ– and three δ+ if needed.

I want you to think about what I mean by 'key' atoms. Does it make sense to put a δ on every atom, or are some more important than others? We are going to come back to this in **Practice 6**.

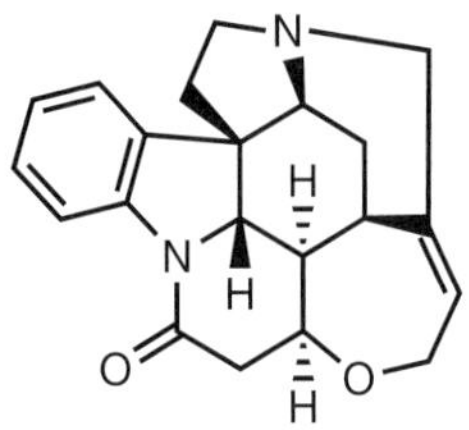

Strychnine

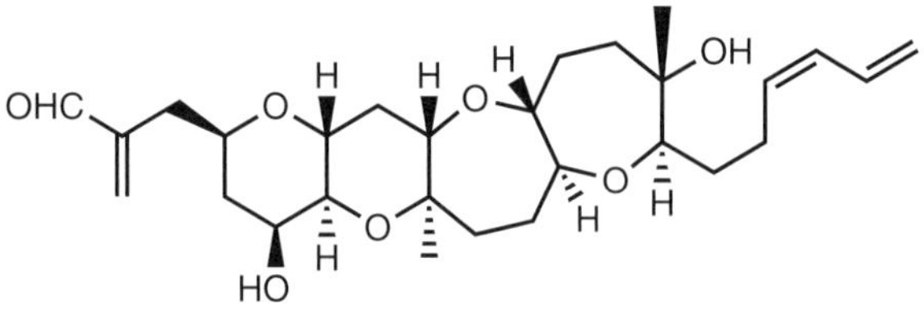

Hemibrevetoxin B

Penicillin G

Artemisinin

Vigulariol

Pseudomonic acid C

Discodermolide

Doxorubicin

You will see that some of the functional groups on some of the structures are drawn differently to the same compounds in **Practice 3**. This gives you a chance to check them!

1.9

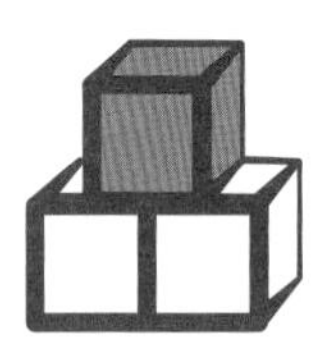

BASICS 6

Bonding in Organic Compounds

We need to cover the basics of bonding and molecular orbitals here, to give you the tools you need to describe bonds and their reactivity.

We are going to discuss the overlap of orbitals to form σ (sigma) and π (pi) bonds. Before we do that, we need to look at some trends in bond dissociation energy. The bond dissociation energy is the energy needed to break a bond. Actually, this is a bit of a simplification, but it will do for now. We will sort out the detail in **Basics 12**.

BOND STRENGTHS, LENGTHS, AND ANGLES

Here are the structures of ethane (**1**), ethene (**2**), and acetylene (**3**), with the bond dissociation energies in kJ mol^{-1} and the bond lengths in angstrom units (Å). So, most chemical bonds are between 1 and 2 Å long. Single bonds in organic compounds have a bond strength of about 400 kJ mol^{-1}. Make sure you are familiar with these numbers. It is so easy to be a factor of 10 or even 100 out, and that would make a massive difference in any calculations you do.

$H_3C{-}CH_3$ **1**: C–C 377 kJ mol^{-1} (1.57 Å); C–H 420 kJ mol^{-1} (1.10 Å)

$H_2C{=}CH_2$ **2**: C=C 728 kJ mol^{-1} (1.35 Å); C–H 458 kJ mol^{-1} (1.07 Å)

$H{-}C{\equiv}C{-}H$ **3**: C≡C 954 kJ mol^{-1} (1.21 Å); C–H 549 kJ mol^{-1} (1.06 Å)

Before we move on, I will state that these are homolytic bond dissociation energies. We aren't going to worry about this subtlety now, but we will come back to it in **Basics 12**. Furthermore, we will encounter 'average' bond dissociation energies in **Basics 13**, and find that the numbers are somewhat different to those above. We will explore the reasons for this in **Perspective 1**.

There are quite a few important points here. Let's try to summarize them, but also look a little deeper. Remember, we want to understand, not just to learn.

A C–C double bond is stronger/shorter than a C–C single bond. A C–C triple bond is even stronger/shorter.

This is all pretty obvious really. A C–C double bond is about 1.9 times as strong as a C–C single bond. A C–C triple bond is about 2.6 times as strong as a C–C single bond. So, on the face of it, each additional bond contributes less. This is true, but these numbers do not give the whole story.

A C–H bond in an alkene is stronger than a C–H bond in an alkane. A C–H bond in an alkyne is even stronger. Now this is a little surprising. Remember that we are only talking about the bonds to H directly from the alkene/alkyne carbon atoms.

And here are the bond angles, as we saw in **Habit 1**. In ethane, all bond angles are essentially tetrahedral, 109.5°. For ethene, the atoms are all in a plane, with approximate angles of 120°. For acetylene (ethyne) the bonds are linear, 180°.

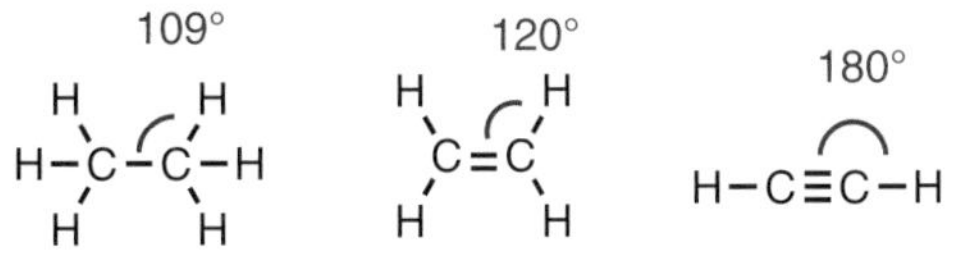

109° 120° 180°

H–C≡C–H

1 2 3

TYPES OF BOND IN ORGANIC COMPOUNDS

Before we go any further with theories of bonding, let's just make sure we know about the different types of bonds that we will encounter in organic compounds. There are plenty of textbooks that talk about atomic orbitals, so we are not going to explain them here (but we will use them).

It is easy to get bogged down with detail. Let's focus on one thing at a time. A C–H bond is a σ bond. A C–H bond in an alkene is stronger than a C–H bond in an alkane. A C–H bond in an alkyne is even stronger. We need an explanation for the bonding that leads naturally to this conclusion.

Let's look at the structures of ethane, ethene, and ethyne again, and focus on the conventions.

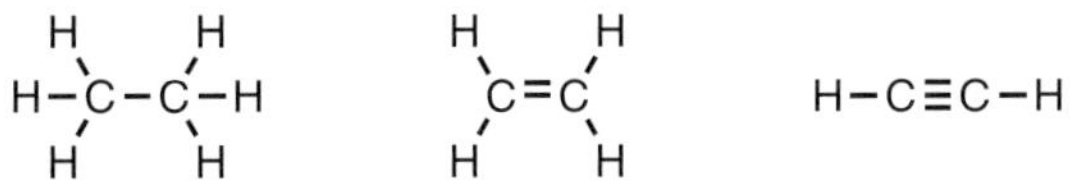

H–C≡C–H

1 2 3

If we have a single line between two carbon atoms, it is a σ bond. If we have two lines between two carbon atoms, it represents a σ bond and a π bond. If we have three lines between two carbon atoms, it represents a σ bond and two π bonds.

*Whatever else is happening, we **always** have a σ bond.*

Since hydrogen only ever forms one bond, it must be a σ bond.

ATOMIC ORBITALS

All of the bonds we will encounter in organic compounds are **covalent** bonds formed by the **sharing** of electrons. Electrons in atoms and molecules reside in **orbitals**. An orbital is quite complicated, but we can represent the different types of orbitals as a shape. The shape shows us where there is a high probability of finding an electron at any given time.

We have to talk about probability, since electrons in orbitals are governed by the laws of quantum mechanics, which tell us that we cannot tell exactly where an electron is at any given time. We can only tell where it is more likely to be.

As organic chemists, most of the time we only need to consider **s orbitals** and **p orbitals**.

An s orbital is spherically symmetrical. The probability of finding an electron is the same in any direction from the nucleus. We can show it as follows, where the nucleus (not shown) is at the centre of the sphere.

s orbital

A p orbital looks like a dumb-bell. The nucleus is situated between the two coloured **lobes**.

p orbital

A p orbital has a **node**.

A node is a plane (in this case, although it can be a different shape) in which the probability of finding an electron is zero.

There is a change of **symmetry** at the node, and we represent this by the two coloured lobes.

The symmetry is a rather abstract idea, but one that you need to get used to. Eventually, you will encounter reactions for which the orbital symmetry determines the outcome in various ways. In depicting p orbitals, I have coloured one lobe of the orbital in green and the other one in yellow. Other textbooks may use different colours, or even + and – signs for the same purpose.

*I don't like the use of + and – signs for this purpose. Let's be absolutely clear. Orbitals are filled with electrons, and electrons **always** have a negative charge. When we talk about a change of symmetry, we are **not** talking about a change of charge.*

In fact, atoms such as carbon (p block!) have three p orbitals, each of which is directed along a different Cartesian axis. The plane of the node is defined by the other two axes.

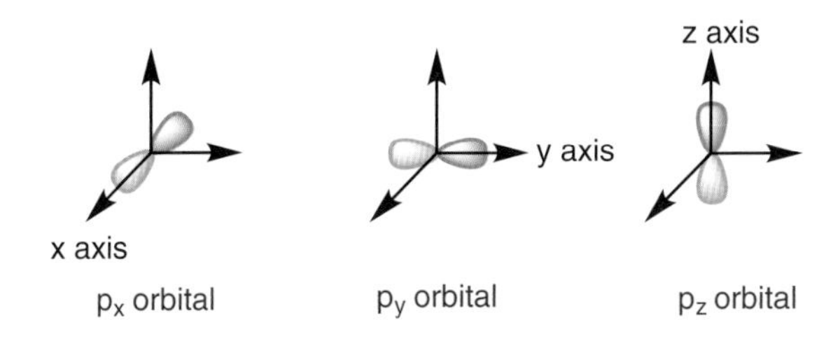

PI (π) BONDS

Let's start with π bonds. They are easier, and this will allow us to focus on the difference between the σ bonds in the various compounds. A π bond is formed from the overlap of p orbitals on two adjacent atoms. For now, we will assume that they are both carbon, although this is not always the case. We can show this in two different ways as follows:

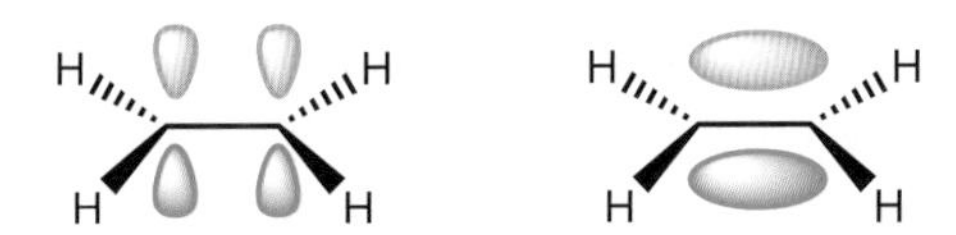

On the left, we are clear that this bonding is coming from the p orbitals on carbon. On the right we are showing that this overlap has occurred. Most organic chemists would use these representations interchangeably. We will look at π bonding in more detail later (**Basics 10**).

A π bond has a **node** in the plane of the atoms, and a change of **symmetry** which is a direct consequence of the symmetry of the p orbitals from which it is made.

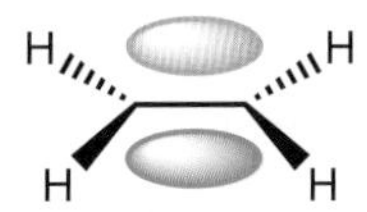

Now let's make a statement that is obvious but profound. If we have a π bond that is formed by overlap of p orbitals on adjacent carbon atoms, each carbon atom has a p orbital that is not involved in σ bonding.

We will see what difference this makes!

SIGMA (σ) BONDS

A σ bond is spherically symmetrical around the axis of the bond. We will see in a moment that the representation below is not quite complete, but it will do for now.

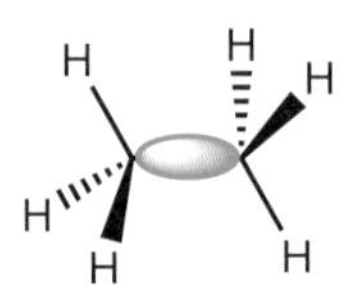

DESCRIPTIONS OF BONDING IN ORGANIC COMPOUNDS

There are many levels of theory that can be used to describe chemical bonds. Ultimately, everything derives from quantum mechanical equations that cannot be solved fully for any 'meaningful' molecule. Therefore, everything is a model or an approximation. Some are more approximate than others.

Most problems stem from the fact that we tend to look at the 'big world' and molecules and atoms are the 'small world'. Our own perspectives and language do not translate well to this world. It is very alien to us that we cannot know precisely where an electron is at any given time, and yet this is fundamental to quantum mechanics. We now have to think in terms of electron distributions and probability.

It helps to have a mental image of a chemical bond. **All** organic chemists use a theory called 'hybridization' to describe bonding. Even though this is only a model (and as such you can identify limitations with it, and eventually use other models of bonding where appropriate) it is very useful.

HYBRIDIZATION

Let's look at the fundamental problem. If we look at the structure of methane, we find that all C–H bonds are identical in length, and all H–C–H bond angles are identical (109.5°). However, we have four atomic orbitals on carbon, one spherically symmetrical s orbital and three p orbitals, which are at 90° to one another.[20] So how can we use these orbitals to make bonds at 109.5° to one another?

[20] We are ignoring the '1s' electrons, and focusing on the '2' shell which is what we will be using to form bonds.

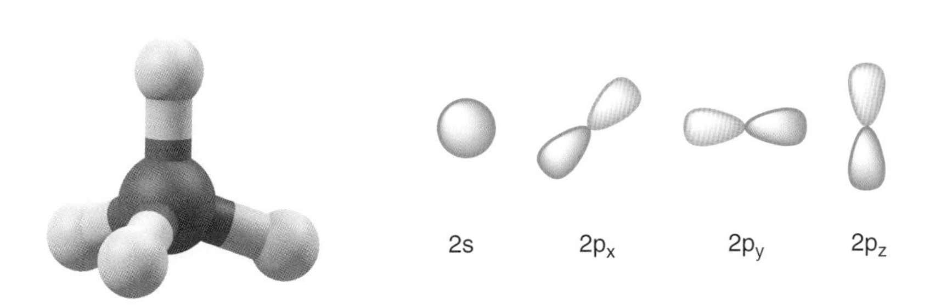

In order to get around this problem, we use a **model** known as hybridization.

What we do, conceptually, is to make 'new' orbitals from the original atomic orbitals.

This may seem like quite a random thing to do, but it turns out that the hybridized orbitals and the original orbitals are all valid solutions to the quantum mechanical equations, so there is no reason not to do this. Remember, this is ***only a model*** *for the bonding.*

Hybridization is easier to explain using an example. We will continue to use the simplest possible hydrocarbon, methane, as our compound in the first instance, before looking at more complex molecules.

HYBRIDIZATION IN ALKANES—sp^3 HYBRIDIZATION

Carbon is in group IV of the second row of the periodic table. Therefore, a carbon atom has the electronic configuration

$$1s^2 2s^2 2p^2$$

We will ignore the 1s shell, as it does not get involved in bonding (at least in any way we need to worry about).

So, there are 4 'outer' electrons. As there are three 2p orbitals available, and they can each have 2 electrons, the '2' shell requires 4 more electrons to completely fill it with 8 electrons—an octet. It can accomplish this by gaining one electron from each of four hydrogen atoms (or indeed from other atoms). Carbon can therefore form four bonds.

Imagine that we promote an electron from the 2s orbital to a 2p orbital. We now have one electron in the 2s orbital and three electrons in the 2p orbitals—we will have one electron in each of p_x, p_y and p_z.

So, we now think of carbon as having the electronic configuration

$$1s^2 2s^1 2p_x^{\,1} 2p_y^{\,1} 2p_z^{\,1}$$

Now the funky bit! We **hybridize** the orbitals. Conceptually, we can add up the 2s and the three 2p orbitals, divide by 4 and get 4 new orbitals. Since each of these is derived

6

from one s and three p orbitals, we will refer to them as sp^3 hybrid orbitals. We can view these orbitals as 25 per cent s and 75 per cent p as far as the carbon atom is concerned (we only have the 1s orbital on H). In effect, we now have an electronic configuration of

$$1s^2 2(sp^3)^4$$

Here is a picture of methane with one of the sp^3 hybrid orbitals drawn in. It does look a lot like a p orbital. However, there is more electron density between the C and the H. Rather than being symmetrical, like a p orbital, the lobes of the orbital are distorted, with only a small lobe behind the carbon.

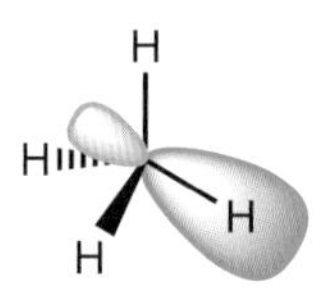

As a result of hybridization, we have four new bonding orbitals that are disposed tetrahedrally, whereas the p orbitals that we started with were at 90° to one another and the s orbital was spherically symmetrical.

This is the bit that I didn't like! If we have three orbitals that are at 90° to one another, how do we end up with four orbitals at 109.5° to one another?

Remember that the hybridized molecular orbitals are a perfectly valid solution to the quantum mechanical equations for methane. They are as good as 'any other' orbitals. However, if you really don't like this, there is an alternative model presented in **Perspective 3**, in which the tetrahedral shape arises more naturally from the atomic orbitals.

*I hope that when you read **Perspective 3**, you will relate it to hybridization, and you will become more comfortable with hybridization as a result.*

HYBRIDIZATION IN ALKENES—sp^2 HYBRIDIZATION

What about the bonding in alkenes? The double bond in an alkene is made up of a σ bond and a π bond. The π bond is formed from the overlap of p orbitals on the two carbon atoms. We have already seen what these orbitals look like.

Now we should think about the σ bonded framework. Let's have another look at our electronic configuration after we promoted an electron. We have

$$1s^2 2s^1 2p_x^1 2p_y^1 2p_z^1$$

Let's 'reserve' the $2p_z$ orbitals on the carbon atoms to form the π bond. So, we have the 2s, $2p_x$, and $2p_y$ orbitals left. Let's now add them up and divide by three. This will give us three new orbitals that we would describe as sp^2 hybrid orbitals (33 per cent s and 67 per cent p).

$$1s^2 2(sp^2)^3 2p_z^1$$

We can now do what we did before, and form bonds using the 1s orbitals of hydrogen and now the sp^2 orbitals on carbon. We also need to form a C–C σ bond using one sp^2 orbital on each carbon atom. We might draw the sp^2 hybrid orbitals in much the same way as we drew the sp^3 hybrid orbitals.

There isn't an easy way to show the difference between 25 per cent s and 33 per cent s.

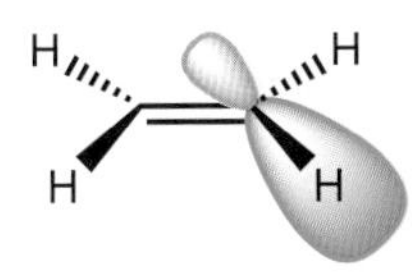

We would do the same thing for a C=O carbonyl double bond. We will come back to this.

HYBRIDIZATION IN ALKYNES—sp HYBRIDIZATION

The triple bond in an alkyne is made up of one σ bond and two π bonds. We will treat the π bonds as above, but this time we reserve two of the p orbitals, so that all we have left are the s orbital and one of the p orbitals. The π bonding in an alkyne can be shown schematically as follows.

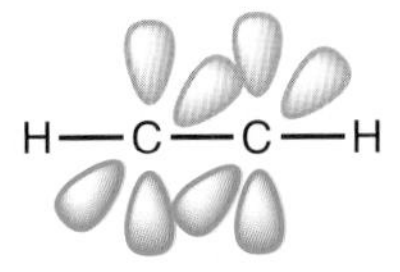

The σ bond is then made up by 'hybridizing' the s orbital and the remaining p orbital. We get two sp hybrid orbitals (50 per cent s, 50 per cent p) that are at an angle of 180° to one another. The alkyne is linear.

THE ADVANTAGE OF HYBRIDIZATION

Hybridization allows us to describe a single entity that we can call 'the bond'. We can think about two electrons that are associated with 'this' bond and with no other, and we can think about the shape of the bonding (and as we will see, the antibonding) orbital.

As we will see in due course, there are other ways to describe the bonding in organic compounds. **For now, accept hybridization as a fact.**

6

HYBRIDIZATION AS PART OF YOUR LANGUAGE

When you listen to organic chemists discussing a compound, or pointing out one specific carbon atom, you will hear expressions such as 'the sp^3 hybridized carbon there'. In this situation, they are probably not trying to tell you anything about the bonding *per se*. They could equally say 'the tetrahedral carbon', or 'the saturated carbon'. This is really what they are meaning.

There is a consequence to this. As you become more confident talking about the hybridization state of a carbon atom, you will use the terminology fluently, but often without thinking about the actual bonding.

Personally, I think this is a good thing.

If you want to talk about specific bonding and antibonding hybridized orbitals, it doesn't really matter much whether they are sp, sp^2, or sp^3 hybridized. The general shape and symmetry (which is what really matters) are the same.

HOW DOES HYBRIDIZATION AFFECT SIGMA BOND STRENGTH?

Now we can return to the bond lengths/strengths that we saw at the start of the chapter. The C–H bond in acetylene is considerably stronger than the C–H bond in methane. Methane is sp^3 hybridized, so as far as the carbon atom is concerned, the C–H bond is 25 per cent s and 75 per cent p. Acetylene is sp hybridized, so as far as the carbon atom is concerned, the C–H bond is 50 per cent s and 50 per cent p. The s orbitals are lower in energy and more tightly held by the nucleus. Therefore, the more s character a bond has, the stronger the bond. This leads us to what we saw above. The C–H bond in an alkyne is stronger than that in an alkane.

$H_3C–CH_3$: C–C 377 kJ mol^{-1} (1.57 Å); C–H 420 kJ mol^{-1} (1.10 Å)

1

$H_2C=CH_2$: C=C 728 kJ mol^{-1} (1.35 Å); C–H 458 kJ mol^{-1} (1.07 Å)

2

H–C≡C–H: C≡C 954 kJ mol^{-1} (1.21 Å); C–H 549 kJ mol^{-1} (1.06 Å)

3

A VERY SUBTLE POINT

We can draw a further conclusion from this. The C–C bond in ethane has a bond dissociation energy of 377 kJ mol^{-1}. That of the C=C bond in ethene is 728 kJ mol^{-1}. However, this doesn't mean that the strength of the π bond is 351 kJ mol^{-1} (the difference between these numbers) because the σ bond component of the double bond in ethene is stronger than the C–C σ bond in ethane. It might feel a bit odd at first thinking about the two separate components of the bond. However, you'll get used to it soon. The strength of the π bond in ethene is considerably less than 351 kJ mol^{-1}.

HYBRIDIZATION OF NITROGEN AND OXYGEN

Hybridization isn't just something carbon does! Let's consider an amine, for example trimethylamine.

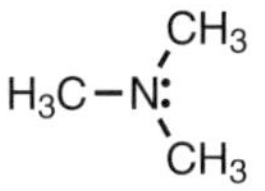

CH_3
$H_3C-\ddot{N}$
CH_3

Nitrogen has the electronic configuration

$$1s^22s^22p^3$$

Again, we can ignore the 1s shell, and focus on the outer electrons. It isn't as easy in this case to think about promoting one electron from 2s to 2p. Each of the different (x, y, and z) 2p orbitals are already singly occupied. I would take a direct step to sp^3 hybridization and then consider how many electrons will be in the orbitals. We get

$$1s^22(sp^3)^2(sp^3)^1(sp^3)^1(sp^3)^1$$

One of the sp^3 hybrid orbitals already has two electrons. It cannot participate in bonding. It is a lone pair! The other three sp^3 hybrid orbitals can form bonds to another element, carbon in the example we are using. This has clear implications for the shape of trimethylamine. It is tetrahedral.[21]

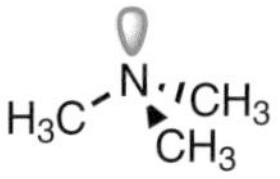

H_3C N CH_3
CH_3

We could do the same for oxygen in water or dimethyl ether, and we come to a similar conclusion.

[21] There is a little bit more complexity to this, but I don't want to get bogged down right now. I haven't told you any lies here. I just haven't told you the whole truth!

Have a go at working this out!

What about a double-bonded oxygen such as a carbonyl? The electronic configuration of oxygen is

$$1s^2 2s^2 2p^4$$

We could expand this as

$$1s^2 2s^2 2p_x^2 2p_y^1 \mathbf{2p_z^1}$$

If we want to form a π bond, we need to reserve a single electron in a p orbital. This is highlighted in purple above. We then have to hybridize the $2s^2 2p_x^2 2p_y^1$ part to give three new orbitals. The new electronic configuration would be

$$1s^2 2(sp^2)^2 2(sp^2)^2 2(sp^2)^1 2p_z^1$$

We have five electrons in the sp^2 hybrid orbitals because we have five electrons in the orbitals that we are constructing these from. Because two of these orbitals are full, we can only form one bond. The oxygen in a carbonyl group is sp^2 hybridized and is trigonal planar. The hybridized lone pairs are in the same plane as the methyl groups.

O

H_3C CH_3

The key point here is that there is no difference between hybridization of carbon and hybridization of other elements.

There is another way to think about bonding. We can use molecular orbital theory. We will see that this leads us to broadly the same conclusions, but with minor differences, in ***Perspective 3****.*

Whether you like hybridization or not, you should still find yourself using it as part of the language of organic chemistry. When you compare the molecular orbital theory explanation, you will find that it does not rely on some of the more 'random' parts of hybridization, but it loses some of the key advantages.

Can you remember what the advantages are? If you can't, re-read this chapter. They have been mentioned!

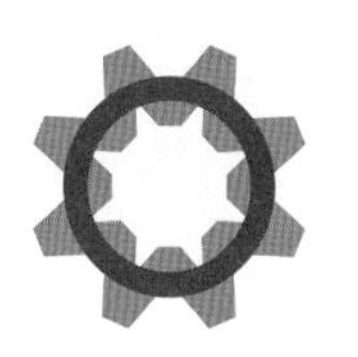

PRACTICE 5
Hybridization

THE PURPOSE OF THE EXERCISE

Most of the time, when an organic chemist states the hybridization of a carbon (or nitrogen or oxygen) atom in an organic compound, they are really describing the number of bonded atoms and the geometry. Stating the hybridization (it will always be 'sp^3', 'sp^2', or 'sp'!) needs to become automatic!

It's also a good point to take stock of where you are. By now, you should be drawing skeletal structures automatically. You will probably be drawing the structures faster, more confidently, and more accurately. You will still make mistakes (for example, the wrong length carbon chain) but you will notice and correct it without really thinking about it.

If you aren't 'there' yet, don't worry about it. Just keep going back to the earlier problems and keep doing them until it all clicks. It's important that you know how to measure your success.

THE EXERCISE

Here are some of the compound names we looked at in **Practice 1**. I have changed a couple of them. You will understand why once you look back at the original names, and you answer the questions below.

1. 4-Chlorohexanal
2. 4-Bromo-7-methylnonan-2-one
3. 5-Chloro-2-propylphenol
4. 4-Aminobenzonitrile
5. Methyl propanoate
6. Ethyl benzoate
7. 2-Hydroxyhept-5-yne
8. 2-Aminoheptan-4-ol
9. 4-Chloro-1-methylcyclohexan-2-ol
10. *N*-Ethyl-*N*-propyl-(2-hydroxypentylamine)

You might not need to draw them all out to answer the question. When you come back and repeat this (and repeat it!), you'll find that you instinctively go to the right answer.

Here are the questions.

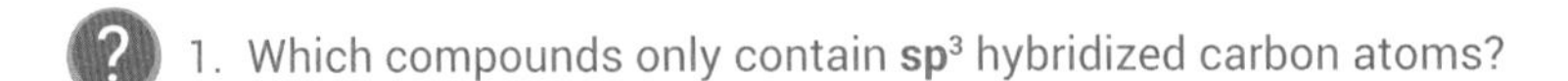

1. Which compounds only contain **sp**^{3} hybridized carbon atoms?

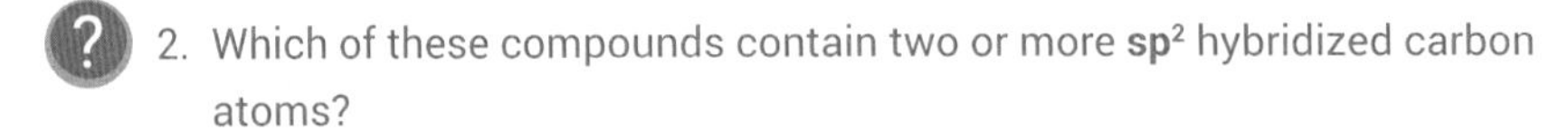

2. Which of these compounds contain two or more **sp**^{2} hybridized carbon atoms?

1.10

3. Which of these compounds contain one or more **sp** hybridized carbon atoms?

Next up, we will make some connections. Here are the natural product structures that we looked at in **Practice 3**. For each structure, determine the number of **sp**^{2} and **sp** hybridized carbon, nitrogen, and oxygen atoms by inspection.

Is it more difficult to spot them on the more complicated structures? It won't be if you practise.

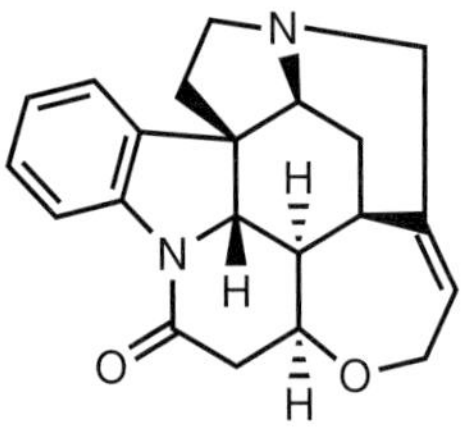

Strychnine

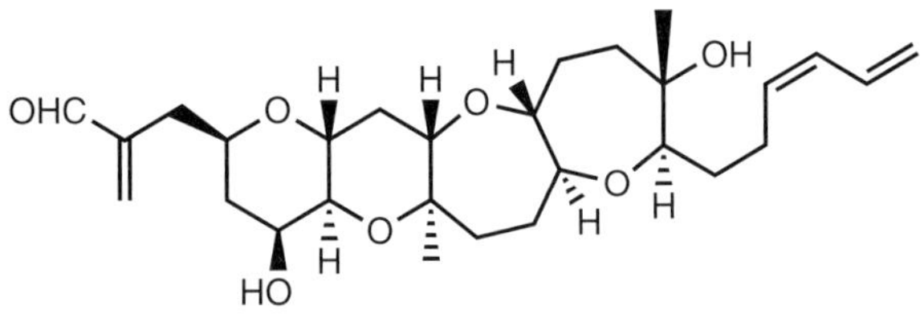

Hemibrevetoxin B

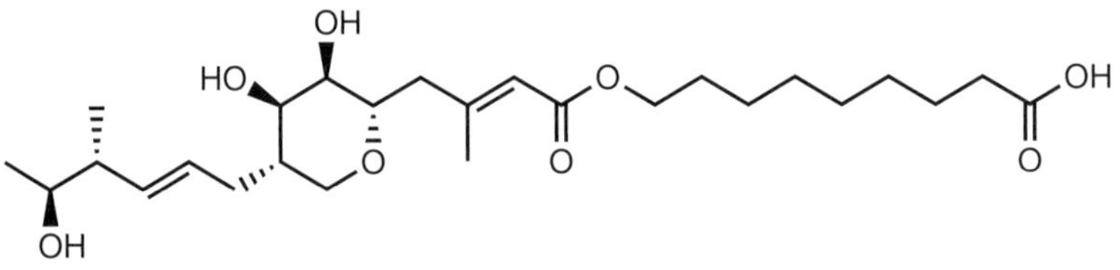

Pseudomonic acid C

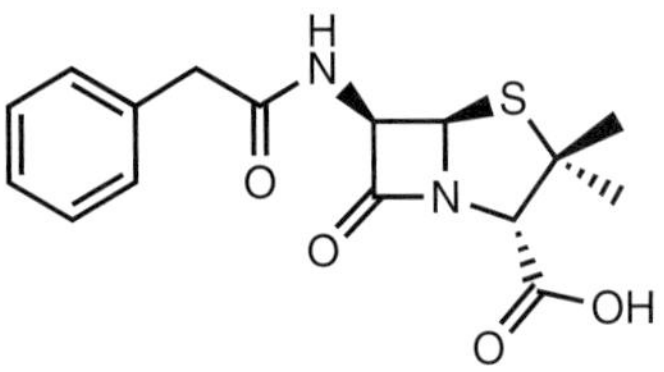

Penicillin G

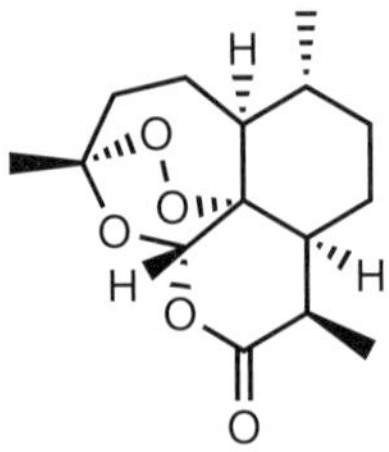

Artemisinin

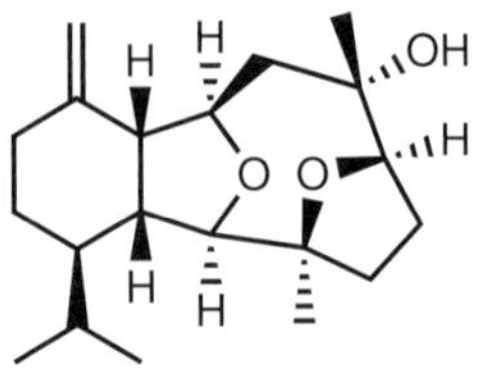

Vigulariol

Discodermolide

Doxorubicin

Now, use the number of sp^2 and sp hybridized atoms, along with the number of rings, to determine the number of double bond equivalents.[22]

1.11

This is quite an artificial problem. You don't really need to consider hybridization in order to work out how many double bonds you have. The real focus is getting you looking at more complex structures and doing 'something' with them.

[22] I haven't actually told you 'how' you should do this. That's okay. You can work it out! If you are struggling, talk to a friend or a lecturer.

7

BASICS 7

Bonding and Antibonding Orbitals

Electrons reside in orbitals. In stable molecules, these tend to be bonding orbitals—there is a high probability of finding electron density between the nuclei involved in the bond.

Whenever we form a bonding orbital by overlap of electrons from two atoms, we always form a corresponding antibonding orbital.

It turns out that the antibonding orbital is really important when we want to make the connection between curly arrow mechanisms (**Basics 8**) and molecular orbitals (**Basics 9**).

BONDING AND ANTIBONDING ORBITALS

Let's recap some simple molecular orbital theory. You know that hydrogen gas exists as a diatomic H_2 molecule, but helium is monatomic.

Here is a molecular orbital diagram showing the formation of H_2 from two hydrogen atoms, each with one electron.

We form a bonding (σ) orbital and an antibonding (σ*) orbital from the combination of two 1s atomic orbitals (two hydrogen radicals). Only the σ orbital is filled, so that energetically this situation is favourable compared to two separate hydrogen radicals.

You will understand the reference to hydrogen radicals when we talk about bond dissociation energy (**Basics 12**).

The bonding and antibonding orbitals of H_2 look like this:

1.12

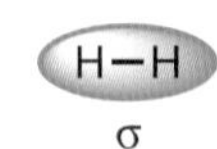

σ

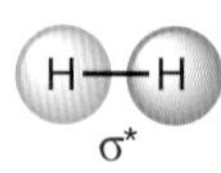

σ*

The antibonding orbital always has one more node, between the atoms, than the bonding orbital.

Let's compare this to the situation with a hypothetical diatomic helium (He_2) molecule.

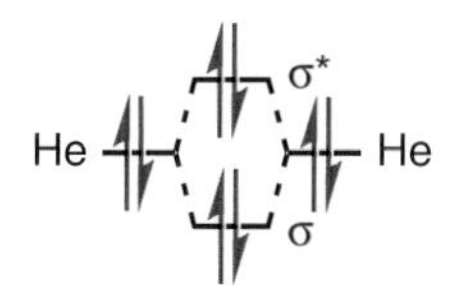

Each helium atom has two electrons, so now we must fill the bonding and the antibonding orbital. Whatever energy we 'gain' by putting electrons into the bonding orbital is offset by also putting electrons into the antibonding orbital.

Therefore, there is no net bonding.

BONDING AND ANTIBONDING ORBITALS IN ORGANIC COMPOUNDS—SIGMA BONDS

We saw sp^3 hybrid orbitals of methane in **Basics 6**, but we glossed over one important point. We didn't consider the formation of the bonding and antibonding orbitals from the sp^3 hybrid orbital on C and a 1s orbital on H.

When we explicitly consider the overlap, there are two possible outcomes, related by symmetry. These are shown below.

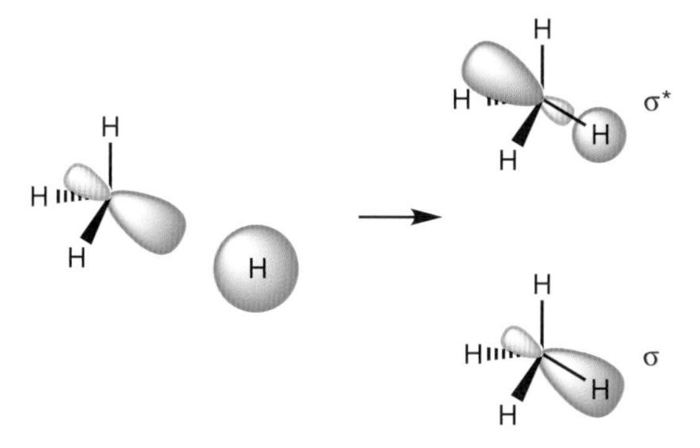

The antibonding orbital (σ*) has a significant lobe (orbital coefficient) behind the atom. There is also a node between the atoms.

We will see how this determines one key aspect of the outcome of S_N2 substitution reactions in due course.

MOLECULAR ORBITALS FOR ETHENE

We saw the π bonding molecular orbital for ethene in the previous chapter. Recall that the π bond is made by overlap of p orbitals on adjacent carbon atoms as shown below. If we consider the overlap of two p orbitals to make the π bond, there are two possibilities that differ in symmetry.

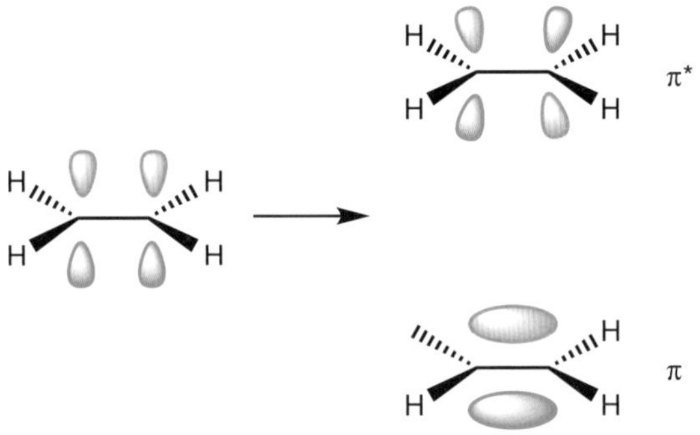

One (π) is bonding, and it has overlap of the p orbitals on the two carbon atoms. The other (π^*) is antibonding, and it has a node between the two atoms in addition to the node in the plane of the double bond. Only the bonding orbital is occupied normally.

We will look at the implications of antibonding orbitals in reaction mechanisms at various points. They really are important!

BASICS 8

Introduction to Curly Arrows

There are few ideas more important in organic chemistry than curly arrows. We use them to represent 'electron flow' in reactions.

A good organic chemist will draw a curly arrow mechanism with an understanding of the meaning, but a great deal of what they do will be borne of habit rather than direct application of understanding.

Let's take this from the absolute basics. A σ or π bond contains two electrons.[23] In principle, when we form a bond between two compounds, we could have both electrons coming from one compound, or one electron from each. For the vast majority of reactions, we can consider that two electrons come from one of the reacting partners.

We represent this with a curly arrow, showing the flow of electrons, and indicating where the bond is being formed.

Let's be absolutely clear about this. Most students who fail organic chemistry do so because they draw incorrect curly arrows. They will draw curly arrows going in the wrong direction, or draw curly arrows that would lead to impossible structures (e.g. those with five-valent carbon). It actually gets worse than that. Usually, if you draw one step in a mechanism incorrectly, you'll then need to draw another step incorrectly to get out of the hole you have dug yourself into. You absolutely need the correct drawing of curly arrows to become a habit.

For this reason, we are going to focus on the basic rules in this chapter, looking at only a small number of reaction types. Since there are only two fundamental reactions covered in this book, we won't have much opportunity to practise drawing curly arrows. As you progress in your studies, it will be important to keep thinking about what the curly arrows actually mean, rather than just memorizing them.

PROTONATION OF WATER

Let's start with a simple example, that isn't even an organic reaction (although the same process is found throughout organic chemistry). You know that if you add H^+ (acid) to water, you will protonate it. Here's an equation for this reaction.

[23] Let's assume hybridization (**Basics 6**), so we can think of a single bond being two electrons in one orbital that contributes fully to the bond.

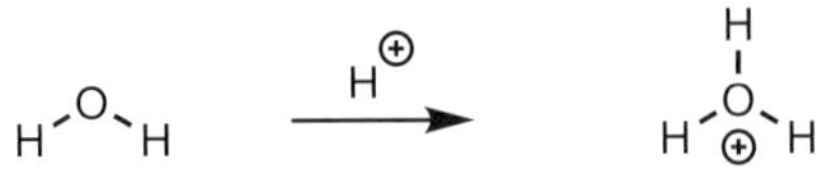

We are definitely making a new bond between O and H. We must be **sharing**[24] two electrons from the oxygen atom, because there is one thing that H^+ doesn't have, and that is electrons. We must be using lone pair electrons, because we are still keeping the two existing O–H bonds. Here is the equation again, with the lone pairs added. Note that the product does still have a lone pair.

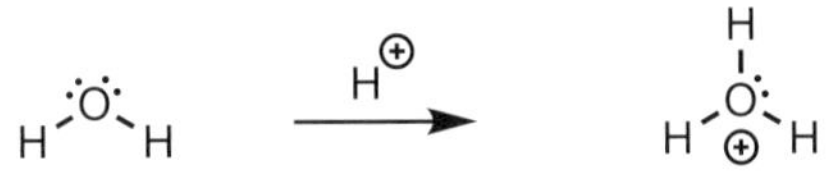

Now we have established where the electrons are coming from, we can add a curly arrow.

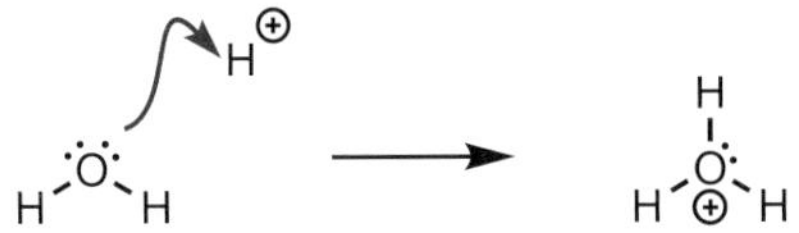

The curly arrow starts where the electrons are, and it finishes where the electrons are going. There isn't much more to it than this.

Curly arrows can start at non-bonded lone pairs of electrons.

DEPROTONATION OF THE HYDRONIUM ION

Now let's look at the reverse reaction.

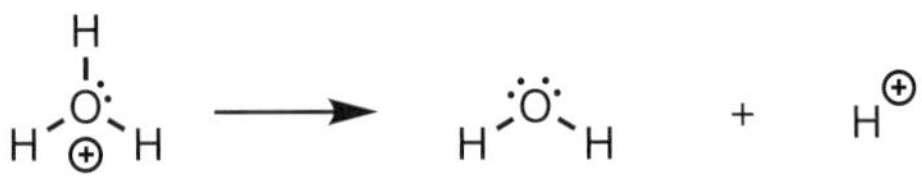

Here, we need to break an O–H bond, giving both electrons in this bond back to the oxygen atom. In saying this, we have established where the curly arrow needs to start (the bond) and where it needs to finish (the oxygen atom).

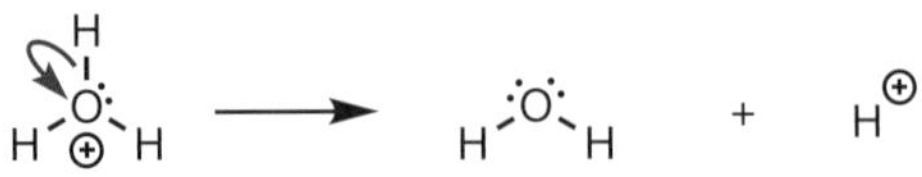

Curly arrows can start at bonds.

[24] I am being careful not to use words such as 'giving' or 'donating'. Since we are forming a bond between O and H, the two electrons are shared between the two atoms. The O retains a share!

We could add a little more complexity to this, by considering the possibility that the proton is removed by a base. We will use hydroxide as base. This will give us two water molecules as products.

$$H_3O^{\oplus} + {}^{\ominus}OH \longrightarrow H_2O + H_2O$$

Here we are forming a new bond from the hydroxide ion to the proton that is being 'lost'. The curly arrow to break the O–H bond is the same as before (purple). We just need to add one more curly arrow (blue) to form a new bond.

$$H_3O^{\oplus} + {}^{\ominus}OH \longrightarrow H_2O + H_2O$$

Curly arrows can start at non-bonded pairs of electrons in negative charges.

Whenever we have two curly arrows going to/from a single atom, we must count electrons. A hydrogen atom can have a share in two outer electrons (the 1s shell). The blue curly arrow above 'gives' the hydrogen atom a share in two more electrons. This cannot happen unless we take a share in two electrons away from it at the same time (the purple curly arrow).

We have now established the ground rules of curly arrow drawing, and we should consolidate these ideas with a couple more examples.

NUCLEOPHILIC SUBSTITUTION

This is one of the two reaction types that we are covering in this book, and you have almost certainly encountered it before. It takes place at saturated (sp^3 hybridized) carbon.

Here's a simple example of this reaction, with the hydroxide anion reacting with iodomethane.

$$HO^{\ominus} + H_3C{-}I \longrightarrow HO{-}CH_3 + I^{\ominus}$$

We are forming a new C–O bond and breaking a C–I bond. Looking at the charges, there is no doubt that we need a curly arrow going to the iodine. After all, it has a negative charge on the product side of the equation. Similarly, the oxygen on the left has a negative charge, but is neutral on the right. It needs to lose electrons.[25] Here is the reaction with curly arrows added.

[25] Here's an important point. If we consider the O to be 'giving away' two electrons, then it would go from minus to plus. Because it is sharing two electrons, it is (in effect) giving away one electron – hence the change from minus to neutral.

$HO^{\ominus}$ + H_3C-I ⟶ $HO-CH_3$ + $I^{\ominus}$

So far, what we have done is connect the given reaction outcome with the curly arrows necessary to achieve that outcome. This is important.

But there is something else to consider. The iodine atom in iodomethane is electronegative. It will have a δ– charge. The carbon will have a δ+ charge. It is far more natural to draw a reaction in which a nucleophile attacks the more δ+ centre, and 'gives' electrons to the more electronegative element.

*We will formally define 'nucleophile' in **Basics 9**, but you will already have the idea by then! That's the beauty of seeing nucleophiles in action and not worrying too much!*

It turns out that there is another way we can draw this reaction, which doesn't break and form bonds in the same step. We are going to deal with this one later (**Fundamental Reaction Type 1**), and we will build further on this once we have covered carbocation stability.

Now we are going to look at some curly arrows for reactions we won't be discussing in this book.

ADDITION REACTIONS OF ALKENES

If we consider the reaction of an alkene with HBr, the following outcome is typical.

R–CH=CH₂ + HBr ⟶ R–CH(Br)–CH₂–H

We should consider a number of factors. We are forming two bonds (C–Br and C–H) and breaking two bonds (H–Br and half of the C=C double bond). Do these events all happen at the same time, or in separate steps? If they happen in separate steps, which happens first, and which way do the electrons 'flow'?

Since we know that alkenes are electron-rich, it is natural to consider a curly arrow going from the C=C bond to somewhere. In fact, since we are breaking this bond, it is essential.

Similarly, we are breaking the H–Br bond. We know that Br is more electronegative than H, so that this bond is polarized as follows.

$$\overset{\delta+}{H}-\overset{\delta-}{Br}$$

If we are going to push a curly arrow from this bond, either to the H or the Br, it has to go to the Br, since this is the end of the bond that 'wants' the electrons (this is what electronegativity means!).

We are almost there now. We need to convert the C=C double bond into a single bond, so we need to start a curly arrow at this bond. We need to form a new C–H bond, so we will need to push this curly arrow to the H. We then need to break the H–Br bond. Putting all of this together looks like this:[26]

Finally, we need to form a C–Br bond. Because we got the first step right, there is only one way to do this.

Of course, if you got the first step wrong, and didn't realize it, you would probably get the second step wrong as well.

CARBONYL ADDITION REACTIONS

The carbonyl group is without doubt the most important functional group in organic chemistry. You can do so much with it. For now, we are just going to look at the curly arrows. Here is the reaction of the cyanide anion with an aldehyde.

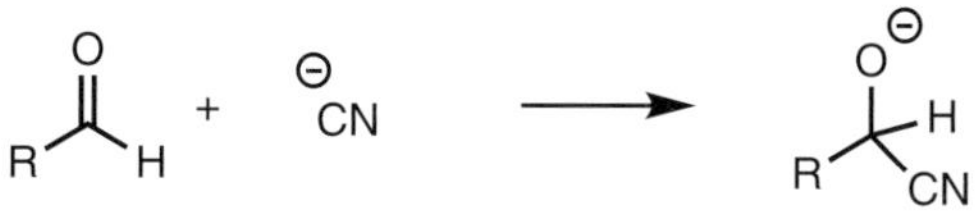

Oxygen is considerably more electronegative than carbon, so the carbonyl bond is polarized. The carbon has a partial positive charge and the oxygen has a partial negative charge. Therefore, the carbon atom is electron-deficient.

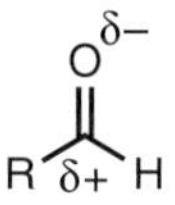

[26] We aren't going to worry about why the proton attaches to the carbon at the end, although perhaps you can propose a sensible reason for this once you have read **Basics 16**.

We have converted the C=O double bond into a single bond in this reaction, and the oxygen has gained a negative charge. There is little doubt that we need a curly arrow going from the carbonyl bond to the oxygen atom.

Of course we do! Oxygen is electronegative!

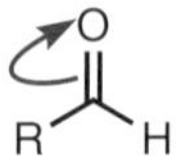

This would leave a carbon atom with only three bonds, and a positive charge. We know that we need to form a new bond from the cyanide carbon to this carbon atom, and this defines the next curly arrow. The complete curly arrow mechanism is shown below.

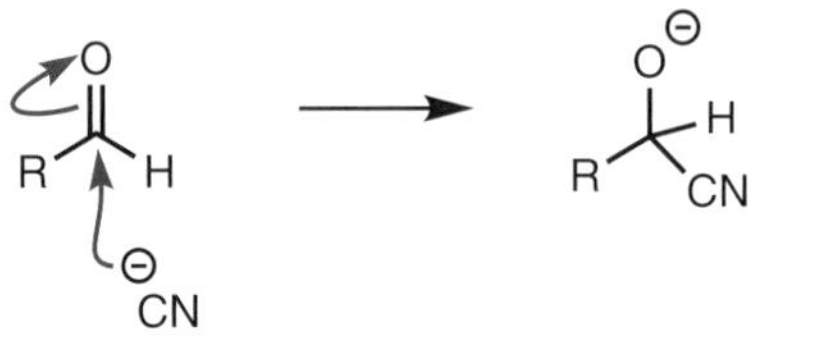

A DIFFERENT TYPE OF CURLY ARROW—FREE-RADICAL REACTIONS

Up to now, we have stated that a curly arrow is used to represent the movement of a **pair** of electrons. This is true for the types of reaction we have been looking at, which we could broadly class as ionic.

However, there are some reactions where we need to think about the movement of single electrons.

When we shine a light[27] on chlorine, Cl_2, the Cl–Cl bond can be broken. However, it doesn't give Cl^+ and Cl^-. Instead, it breaks in such a way that each chlorine atom retains one electron from the Cl–Cl bond.

We can show this as follows:

$$Cl{-}Cl \longrightarrow Cl\cdot \quad \cdot Cl$$

We get two Cl radicals, 'Cl dot'. Note that I haven't shown all the 'lone pair' electrons on Cl. We often focus on the key point rather than showing everything. You should be getting used to this by now.

[27] It needs to be the correct frequency (energy) of light!

Because the curly arrow we have been using so far represents the movement of a pair of electrons, we need a different type of curly arrow to represent the movement of a single electron. We use a 'fishhook arrow' for this purpose, as shown above.

When free-radical chemistry became prominent in the 1950s, books on radical chemistry were banned in some parts of the world! It sounds bonkers but it is true. Some politicians didn't like the idea of radicals of any sort, particularly free ones!

HOMOLYTIC BOND CLEAVAGE

The reaction we saw above (here it is again) is an example of homolytic bond cleavage. The bond is broken in an 'equal' way, so that each atom keeps one electron.

$$Cl{-}Cl \longrightarrow Cl\cdot \quad \cdot Cl$$

*We won't encounter any reactions that feature homolytic cleavage of bonds in this book. However, when we talk about bond dissociation energies, we are referring to the enthalpy change for a homolytic bond cleavage of this type (**Basics 12**).*

HETEROLYTIC BOND CLEAVAGE

This is when one atom in a bond cleavage gets both the electrons. For example,

$$\text{—Br} \longrightarrow \oplus \quad Br^{\ominus}$$

The Br atom ends up with a negative charge because it starts out with an equal share in two electrons (so effectively it has one of the electrons) and it gets both the electrons.

1

FUNDAMENTAL REACTION TYPE 1

Nucleophilic Substitution at Saturated Carbon

WHAT IS A SUBSTITUTION REACTION?

Quite simply, a substitution reaction is where one group on an atom is replaced with another group.

When we are talking about a substitution reaction at saturated (sp^3) carbon, the following process is a substitution. We are replacing the iodine atom with a hydroxyl group to give an alcohol product.

$$HO^{\ominus} + H_3C{-}I \longrightarrow HO{-}CH_3 + I^{\ominus}$$

In this context, the hydroxide anion is acting as a **nucleophile** (we will define the term in Basics 9, but you are already getting the hang of what it means!). We call the iodide anion the **leaving group**. In this case, the 'organic' bit is simply methyl, but it could be a whole host of other alkyl groups with various other substituents.

*We are going to work up to looking at the impact of each of these variables, as well as reaction conditions such as solvent, in **Reaction Detail 1**. For now, though, I want to guide you through a way of thinking about what possible mechanisms you might be able to draw. By considering all possibilities, you can then start to analyse what is good or bad about each.*

ORDERS OF BOND FORMATION[28]

Let's define very clearly what is happening in this reaction. We are forming a new C–O bond and breaking a C–I bond.

[28] Do not confuse these conceptual 'orders of bond formation' with the idea of 'order of reaction'. In the latter case we are talking about the kinetics, and how the concentrations of the various reactants affect the overall rate of reaction. All we are doing here is looking at which bonds break/form first.

From a **purely conceptual** point of view, this could happen in three different ways.

1. The C–O bond forms and the C–I bond breaks at the same time.
2. The C–O bond forms first, and then the C–I bond breaks.
3. The C–I bond breaks, and then the C–O bond forms.

There are no other 'possibilities'. Now we need to determine whether all of these are actually possible in reality.

We can exclude option 2. If the C–O bond was to form first, this is what would have to happen.

$$HO^{\ominus} + H_3C{-}I \not\longrightarrow HO{-}\overset{\ominus}{C}H_3{-}I \longrightarrow HO{-}CH_3 + I^{\ominus}$$

NOT POSSIBLE!

We have just drawn a structure with a carbon atom with a negative charge and five bonds. The negative charge in this case would not represent a non-bonded pair of electrons (as it does in a carbanion—more about these in **Basics 16**). The best way to count the electrons is to start with the iodomethane carbon having a share in eight outer-shell electrons. The curly arrow indicates that we are sharing a pair of electrons from oxygen with this carbon atom. Therefore, by having five bonds, the carbon has a share in ten outer-shell electrons. It cannot have more than eight outer-shell electrons. This mechanism is simply impossible!

We therefore only have two ways in which a nucleophilic substitution reaction can take place at a saturated carbon atom. Either the leaving group leaves first, followed by attack of the nucleophile, or the nucleophile can attack at the same time as the leaving group leaves. You will probably already have seen this before. These are the classic S_N1 and S_N2 reactions.

We will now look at each one in turn.

THE S_N1 MECHANISM

The basic reaction pathway is as follows:

$$R{-}X \xrightarrow{\text{slow loss of } X^{\ominus}} R^{\oplus} + Nu^{\ominus} \xrightarrow{\text{fast}} R{-}Nu$$

We want to avoid equations as much as possible, but we do need to consider the rate equation for the basic reaction types.

An S_N1 reaction has rate = k[RX]

where *k* is the 'rate constant' and [RX] is the concentration of the substrate. The nucleophile does not enter into this equation because it isn't involved in the reaction until after the (slow) rate-determining step. To put this another way, since the second step is faster than the first, a 'better' nucleophile (we will see what this means later) won't make the overall reaction any faster.

The '1' in S_N1 refers to the fact that the rate of reaction is proportional to the concentration of only one of the components. It is a '**first order**' reaction.

You will probably have noticed that I switched from using a specific example at the start of this chapter, to a more general (R, X and Nu) one. There are a number of reasons for this, but let's just focus on one of them for now. If I had stuck with the same example, we would have had CH_3^+ being generated. In **Basics 16** we will see that this is a very unstable carbocation. Therefore, you will never see an S_N1 reaction at a methyl group. It requires too much energy. We will consider the energetics of substitution reactions in detail in **Reaction Detail 1**.

THE S_N2 MECHANISM

In this mechanism, the attack of the nucleophile and the loss of leaving group occur at the same time. This reaction takes place by a 5-coordinate transition state, where the C–X bond is breaking and the C–Nu bond is forming at the same time. We have actually seen this a couple of times while introducing fundamental principles.

Now we can use the example of iodomethane again, because it turns out to be a perfectly good S_N2 reaction.

1.14

$$HO^{\ominus} + H_3C{-}I \longrightarrow \left[HO{\cdots}CH_3{\cdots}I\right]^{\ddagger\ominus} \longrightarrow HO{-}CH_3 + I^{\ominus}$$

At a first glance, this might look a lot like the reaction we said could not happen. However, it is different in subtle but important ways. Because we are breaking the C–I bond **at the same time** as we form the C–O bond, we don't really get a species that has five bonds to carbon.

*The species in square brackets in the middle is a **transition state**, which is given the symbol '‡'. We will look at these in more detail in **Basics 23**.*

Instead of giving the carbon a share in two more electrons (from oxygen) *before* taking a share in two electrons away (to iodine), we are giving and taking away at the same time, so that the carbon always has a share in eight outer electrons.

This may seem like a subtle distinction, but it is an important one. When you are drawing curly arrow representations of reactions, you have to consider the precise sequence of breaking/forming bonds.

For now, we are just focusing on the very basic points that you **must** know. Normally we would not draw the transition state, so the reaction would look like this.

$$HO^{\ominus} + H_3C-I \longrightarrow HO-CH_3 + I^{\ominus}$$

This S_N2 reaction has rate = $k[CH_3I][HO^-]$

There is only one step, so it must be rate-determining. Since it involves both species, changing the concentration of either will alter the rate of the reaction. This is a '**second order**' reaction in terms of kinetics.

CAUTION

It is not uncommon for students to mix up S_N1 and S_N2 mechanisms, because the S_N1 mechanism has **two** steps, whereas the S_N2 mechanism has **one** step. You will probably make this mistake if you read this section once, feel happy that you understand it, and decide that you will have another look at it in the few days before the exam, in order to *refresh* your memory. At this point, you will be trying to remember which reaction is which. It is actually easier, and far more reliable, to keep reinforcing the basic points until you cannot forget or confuse them.

I would say the problem is a little more subtle. If you confuse S_N1 and S_N2, you will probably then try to explain, in your exam, why a given reaction is S_N1, even if it is not! At that point, you will be giving incorrect information about the stability of intermediates and transition states to try to justify something that is simply incorrect.

PRACTICE 6

Electronegativity in Context

THE PURPOSE OF THE EXERCISE

We have looked at substitution reactions, so you are now in a position to consider whether you could attack a particular carbon in a structure and have 'something' leave. That 'something' needs to be able to cope with a negative charge (it needs to be electronegative!). At this point, we are not yet ready to consider whether the reaction will actually take place.

By doing this, you will be drawing more structures. Make sure you keep them tidy. You might need to move parts of the molecule out of the way so that you can fit the new bits in. Make sure your curly arrows are clean and tidy. A nucleophile should always be attacking a carbon atom with a δ+ charge. Make sure the curly arrows reflect this.

THE EXERCISE

Here are those natural product structures again. Now consider which of the δ+ carbon atoms could be attacked by a nucleophile. Don't break carbon–carbon bonds! Which of the δ– atoms could be attacked by an electrophile (use a proton)? Draw sensible curly arrows!

Strychnine

Hemibrevetoxin B

Pseudomonic acid C

Penicillin G

Artemisinin

Vigulariol

Discodermolide

Doxorubicin

CAUTION

Come back to your answers from this when we have covered nucleophilic substitution in more detail, and consider whether any of the reactions you have drawn are particularly good or particularly bad.[29]

An experienced organic chemist will know exactly what we mean by this. A reaction that is particularly bad simply won't happen. A good reaction is more likely. It still doesn't mean it will happen quickly at room temperature.

[29] Have you looked up what these compounds do yet? Why would organic chemists want to spend years developing neat ways to make them, apart from the challenge?

FUNDAMENTAL REACTION TYPE 2

Elimination Reactions

INTRODUCTION

The very first thing we need to do is define what we mean by an elimination reaction. The second thing we need to do is get to grips with the electron flow and curly arrow drawing. We will look at these together.

First of all, when we use the word '**elimination**', we are generally talking about a '1,2-elimination' where two substituents are removed from adjacent carbon atoms. Let's call these groups X and Y for now. All of the elimination reactions which we will be talking about are polar in nature, so one of the groups (let's say X) must be lost as a positive species, X^+, and the other must be lost as a negative species, Y^-. Here is the overall reaction without curly arrows.

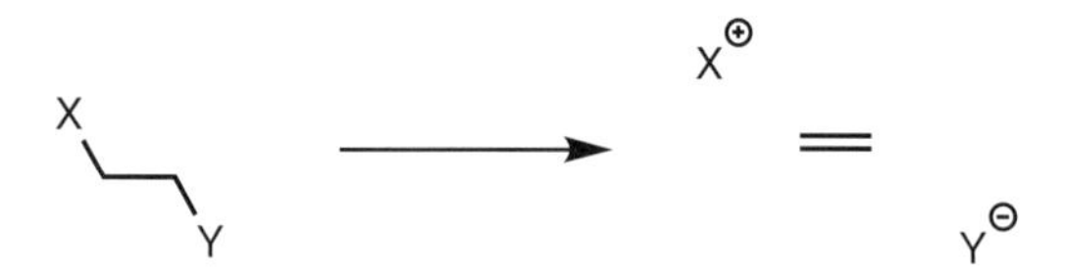

We have formed a new double bond between the two carbon atoms which X and Y were bonded to.

It is important that you are clear about this point. I often find that when I set an exam question stating that it is about elimination, but without drawing the reaction out, some students will draw a substitution reaction, probably because it is 'eliminating' the leaving group. You need to know the basic reaction types. That's why I am covering them with just the absolute basics for now.

Let's not beat about the bush. If you cannot confidently tell the difference between substitution and elimination, you're going to be in a mess. If you make this mistake in an exam, chances are it is because you didn't fully internalize the explanation. Sorry if this offends, but it's the truth!

Now let's think about the curly arrows we need. Since we are forming a new double bond between these carbon atoms, we definitely need a curly arrow which *ends* at the **middle** of this carbon–carbon bond. Where should this arrow start? Well, we are losing the X group as X⁺, so we know that the electrons from the C–X bond are staying with the carbon. These are the electrons which are going to form the new double bond, so the arrow needs to start at the **middle** of the C–X bond.

What about the other arrow? Well, we are losing Y as Y⁻, so we need to take the electrons from the C–Y bond and move them onto Y.

The curly arrows we need are shown below.

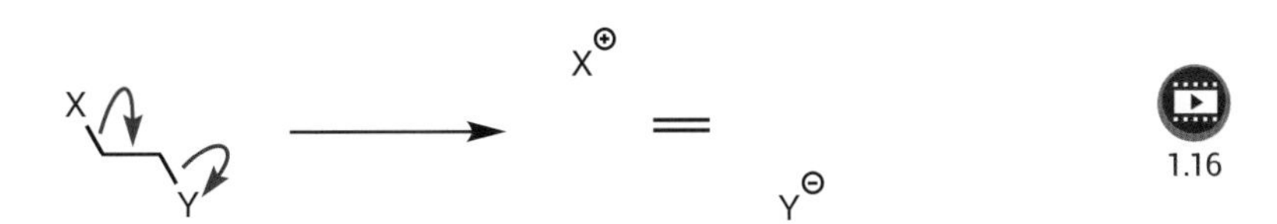

I know you are thinking by now that I am laying this on a bit thick, but I do see a lot of mistakes with these arrows, and your continued success in organic chemistry depends on you being able to draw curly arrows correctly without needing to think about it. Remember how you learned to write in the first place. You did it by endless repetition until you could form the letters and words clearly and correctly. You are now learning to write again, this time using bonds and arrows. Get practising!

WHAT ARE THE POSSIBLE MECHANISMS?

We will do the same as we did with substitution reactions in **Fundamental Reaction Type 1**, and consider the possible order of steps. We have two groups leaving from adjacent carbons. Let's simplify things a bit, and assume that the X group is a hydrogen atom—it usually is. You probably guessed that this was coming. After all, what is the easiest group to remove as a positively charged species? A proton, of course!

Conceptually, either X⁺ or Y⁻ could leave first, or they could both leave at the same time. This gives us three possible mechanisms.

This directly parallels the situation we encountered for substitution reactions. There is just one small but significant difference—this time, all three mechanisms are actually possible.

We will simply classify them according to their mechanisms. Here they are, complete with curly arrows:

THE E1 ELIMINATION MECHANISM

This one is referred to as the E1 mechanism. There is only one species involved in the rate-determining step—the substrate. For this mechanism to operate, the carbocation will need to be stabilized. We will start to see the factors that stabilize carbocations in **Basics 16**, and at that point you will immediately be able to speculate about the types of substrate which will undergo E1 elimination.

The primary carbocation shown above would not form. All we are doing for the moment is drawing curly arrow mechanisms.

THE E1cB ELIMINATION MECHANISM

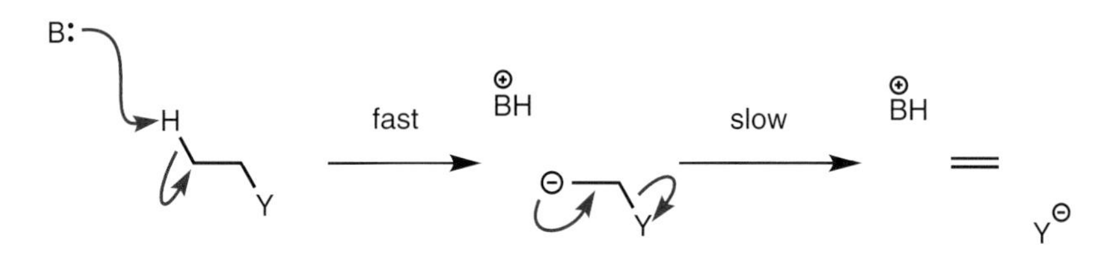

We now need a base to remove this proton, and this occurs rapidly. Then, the slow step is elimination of the Y group. Because the base is involved prior to the rate-determining step, the reaction has second order kinetics.

*This is the least common of the three mechanisms for elimination, and will only be favoured in cases where the intermediate anion is stabilized (see **Basics 18** for the sort of carbanions we would need). After all, why else would a substrate undergo fast deprotonation?*

THE E2 ELIMINATION MECHANISM

These first two mechanisms can be considered extreme cases, where either the carbocation or carbanion intermediates are particularly stable. However, in most cases, we see a middle ground where neither is sufficiently stable to dominate the reaction pathway. In this case, loss of both groups occurs simultaneously, and the reaction only has one step, and proceeds through a transition state rather than an intermediate. We will cover the distinction between transition states and intermediates, along with their energy profiles, in **Basics 14** and **Basics 23**.

You would normally need a base to remove the proton.

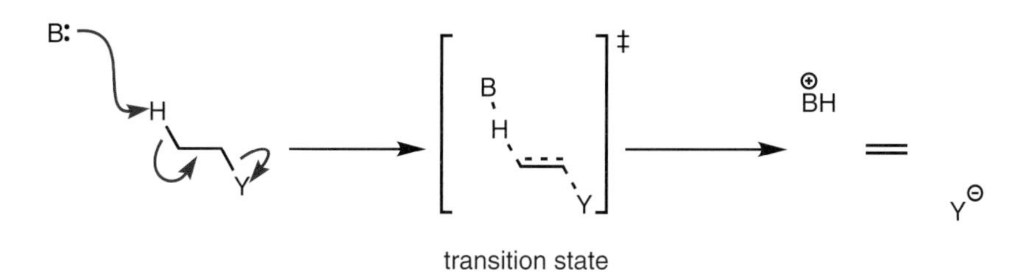

transition state

COMPETING REACTIONS

The E1 elimination shares a first step with the S_N1 substitution. If we add a nucleophile to the carbocation intermediate, we get a substitution reaction overall. If we lose a proton, we get elimination.

The E2 elimination is a one-step process, with a transition state—exactly like the S_N2 substitution. If a nucleophile attacks directly at carbon, we get a substitution reaction. If a base attacks a suitably disposed hydrogen, we get an elimination. We will look at the relationship between nucleophiles and bases in **Basics 9**.

There is often competition between substitution and elimination reactions. We will look at this in a series of worked problems in **Worked Problem 2** and **Worked Problem 3**.

Substrates which undergo substitution by an S_N1 mechanism will undergo elimination by an E1 mechanism. This shouldn't surprise you too much, as the mechanisms only diverge after the rate-determining step. Substrates which undergo substitution by an S_N2 mechanism will undergo elimination by an E2 mechanism.

ONE MORE THING

When we look at the E1cB elimination mechanism in more detail, we can see that this corresponds to the 'impossible' mechanism in substitution reactions—therefore there is no direct parallel.

You should make sure you understand how to draw all three mechanisms before we add further complexity.

 Draw them all out a few times before you go any further!

SECTION 2

BUILDING ON THE FOUNDATIONS

INTRODUCTION

Let's take stock of where we are.

We know how to draw structures. We know what the common errors are, so we can be mindful of them and eventually eliminate them.

You will still make mistakes. It's okay. Just keep checking the structures you draw. Learn to spot the mistakes, and in time you will find that you stop making the mistakes in the first place. But this does need to be an active process. ***You*** *need to keep checking.*

We know how to talk about the bonding in organic compounds. This will allow us to describe how the bonding changes during the course of a reaction.

We know the rules associated with drawing curly arrow representations of mechanisms. We can already apply them to the two fundamental reaction types covered in this book. We recognize that there are different mechanisms for each reaction type, depending on which bonds we break/form first.

There is still a lot we don't know. We don't know what is happening to the bonding during a reaction—when we draw a particular curly arrow, what does it **mean**? We don't know when each particular mechanism (or indeed each particular reaction) will take place. If we don't know which mechanism is operative, how can we get the best outcome from it (highest yield of product)? We need to define parameters that allow us to predict how far and how fast each reaction will go.

The purpose of this section is to build on the initial foundations to provide us with a reasonably detailed understanding of the two key reaction types that we are focusing on in this book.

We will focus on the energy changes that accompany the transformation of starting materials into products. We will start to understand how these energy changes determine which mechanism a particular reaction follows, and what that means for the structures that we will encounter along the way.

We always need to think about the shape of molecules, but we won't talk about the consequences of shape in terms of reactivity too much in this section. That is coming up in Section 3. However, we will consider some aspects of shape. It is impossible not to!

We still need to incorporate all of the aspects we encountered in the previous section.

You cannot progress too far if you are not drawing nice tidy organic structures. Take your time with the drawing, and take every opportunity to practise. It's how you will get good at this!

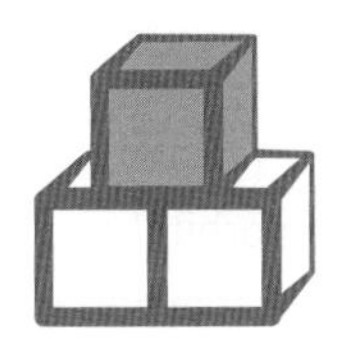

BASICS 9

Breaking Bonds—Linking Curly Arrows and Molecular Orbitals

Now we have examined the nature of bonding in organic compounds, we need to revisit the curly arrows from **Basics 8**, and present them in a more formal way, considering what type of bond is being formed or broken in each case. We should also consider what the curly arrows mean in terms of overlap of molecular orbitals.

2.1

BONDING AND ANTIBONDING ORBITALS

We looked at bonding and antibonding orbitals in **Basics 7**. We saw the following molecular orbital energy diagram for the bonding in the hydrogen molecule, H_2.

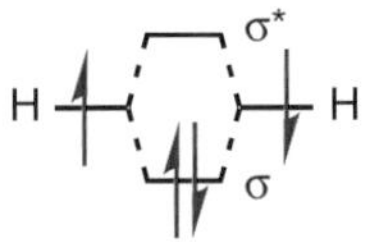

We saw that with two electrons, only the σ orbital is filled, so that energetically this situation is favourable compared to two separate hydrogen radicals.

With helium, the situation was rather different. A helium atom has two electrons. Therefore, a hypothetical He_2 molecule would have four electrons.

Because we now have two electrons in a bonding orbital and two electrons in an antibonding orbital, there is no net bonding.

What does this mean for bonds in organic molecules? Quite a lot! If we are able to add electrons into an antibonding molecular orbital, we will break the bond.

It's not the only way to break a bond, but it is perhaps the least intuitive. Now let's look at all the possibilities.

HOW DO WE BREAK A BOND?

Let's look at the situation of a single bond, comprised of a filled σ bonding orbital and an empty σ* antibonding orbital. There are only three ways that we can break this bond.

1. Take the two electrons and give each atom one of them, so that they are no longer shared.
2. Take the two electrons and give both of them to one of the atoms, so that they are no longer shared.
3. Put some electrons (don't worry about where we get them from just yet) into the antibonding orbital.

There isn't another possibility.

Systematically identifying every possible outcome in a situation is always a good approach. If you can consider all outcomes, you can then evaluate them. We will be doing quite a lot of this, particularly when we discuss reactivity.

Now we can start to think about where the electrons might be coming from, and where they might be going.

DEFINING 'NUCLEOPHILE' AND 'ELECTROPHILE'

We already used the term 'nucleophile' in **Fundamental Reaction Type 1**. Now let's define it. We cannot define the term 'nucleophile' without defining the corresponding term 'electrophile'. Here is a schematic reaction that covers the vast majority of organic reactions.

$$Nu^{\ominus} \quad E^{\oplus} \longrightarrow Nu{-}E$$

A nucleophile is simply something with electrons that it can use to form a new bond. The nucleophile donates electrons from a filled orbital. The electrophile accepts electrons into an empty orbital in order to form the bond.

BACK TO NUCLEOPHILIC SUBSTITUTION

Here is the substitution reaction in which the hydroxide anion is reacting with iodomethane. We saw this reaction in **Fundamental Reaction Type 1**.

$$HO^{\ominus} + H_3C{-}I \longrightarrow HO{-}CH_3 + I^{\ominus}$$

Now we can see that the C–I bond we are breaking is a σ bond. Here are the curly arrows again.

$$HO^{\ominus} + H_3C{-}I \longrightarrow HO{-}CH_3 + I^{\ominus}$$

The hydroxide anion is sharing electrons with the carbon atom as the iodide anion leaves. The nucleophile can attack the carbon because the iodine is electronegative (**Basics 5**). The carbon has a δ+ charge. The electrons from the hydroxide have got to go somewhere. In this case, it is the σ* orbital of the C–I bond. As we have seen, this will break the bond.

In this context, iodomethane is behaving as an electrophile. It is accepting electrons (although it 'gives' them to the iodine pretty quickly!).

We need to define a couple more terms relating to molecular orbitals.

FRONTIER ORBITALS—HOMO AND LUMO

In the above reaction, the hydroxide anion is sharing electrons from a filled orbital. This orbital represents a non-bonded pair of electrons. If we think about the orbitals of the hydroxide anion, we will have bonding electrons in the O–H bond, and three pairs of non-bonding electrons on the oxygen.

If we used the electrons in the O–H bond, we would have to start the curly arrow at the bond. This would imply that we break this bond, which is not the case. We must use the non-bonding electrons, which are higher in energy.

In this case, the non-bonded electrons on oxygen are the highest energy electrons in the system. These are the highest occupied molecular orbital.

__HOMO__ is an abbreviation for highest occupied molecular orbital.

These electrons are being shared with the carbon atom of iodomethane. More specifically, they are being shared with the σ* orbital of the C–I bond, which is unoccupied. More specifically, it is the lowest energy unoccupied molecular orbital in the system.

__LUMO__ is an abbreviation for lowest unoccupied molecular orbital.

Consider a molecular orbital energy diagram such as the following. We have one compound on each side. Each has a number of filled orbitals and a number of empty orbitals.[1] I have labelled the HOMO and LUMO in each case.

[1] This diagram is purely schematic. Real molecules will have various numbers of filled and empty orbitals, including some with the same or similar energy.

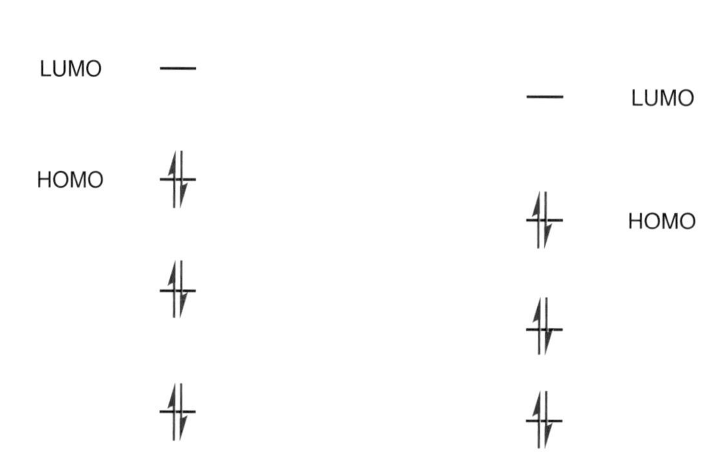

You will get a more favourable interaction between two orbitals that are close in energy.

If we consider the HOMO of the compound on the left, it will interact more favourably with the LUMO of the compound on the right, than it will with the 'next LUMO' (next lowest unoccupied molecular orbital).

The most favourable interaction of the HOMO of the compound on the right will be with the LUMO of the compound on the left.

However, because the HOMO(left)–LUMO(right) gap is smaller than the HOMO(right)–LUMO(left) gap, we would expect the compound on the left to act as the nucleophile and the compound on the right to act as the electrophile.

We can show this as follows.

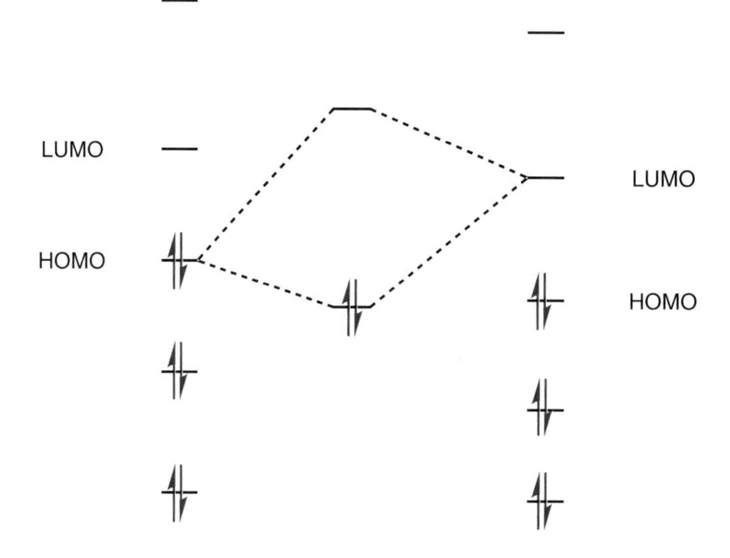

Don't get obsessed by this. Most organic chemists don't try to work out which reagent has the highest HOMO/lowest LUMO.

They will recognize reagents and their properties/patterns of reactivity.

However, it is useful to have some insight into the orbital interactions (even at this rather basic level) so that when you need them, you already know the basics.

At this stage, you should understand what we mean by HOMO and LUMO.

Collectively, the HOMO and the LUMO are known as the Frontier Orbitals.

APPLYING PRINCIPLES AND MAKING PREDICTIONS

Here is a compound we have not encountered before.

Hopefully you can see that if we react this compound with hydroxide, there are two possible substitution reactions.

Give a systematic name for the compound. Draw curly arrows for both possible substitution reactions.

We need to know which will take place preferentially. It turns out that the σ* orbital of the C–I bond is lower in energy than that of the C–Cl bond. Therefore, we will gain more stabilization (better overlap) by displacing iodide rather than chloride.[2] We preferentially displace iodide.

We don't often think of the LUMO energy when we talk about S_N2 substitution reactions. We tend to compare the ability of various leaving groups. Talking only about the orbital overlap assumes that the entire course of the reaction is determined in the initial approach of the nucleophile, which is a bit of a simplification.

There is one aspect that is entirely determined by the initial approach of the hydroxide and the overlap of molecular orbitals. We aren't going to worry too much about

[2] A discussion of why the σ* orbital of the C–I bond is lower in energy is beyond the scope of this book.

the shape of the HOMO of hydroxide. However, we do know the shape of the LUMO of iodomethane. Because we know the shape of this orbital (**Basics 7**), we can see exactly which direction the hydroxide will attack from.

hydroxide attacks here to overlap with the σ* orbital

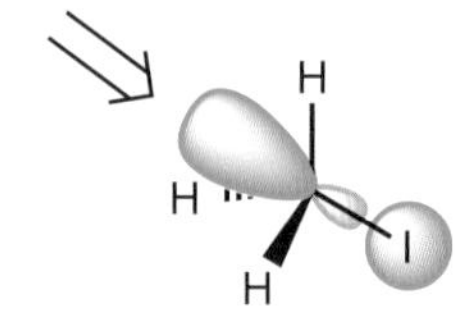

Once we start considering the stereochemistry of reactions, we will see that this is very important indeed!

Now let's see the same treatment for the E2 elimination mechanism. The S_N1 and E1 mechanisms can wait!

Before we do this, there is one more connection we need to make.

ACIDS AND BASES, NUCLEOPHILES AND ELECTROPHILES

In **Basics 8** we saw the attack of hydroxide in the deprotonation of the hydronium ion. It was abstracting a proton, but the curly arrows are exactly the same as in the S_N2 substitution mechanism.

When a nucleophile attacks a hydrogen, we call it a base. When a base attacks a carbon, we call it a nucleophile.

In **Basics 8**, we had a lone pair on water attacking a proton. Does this make the proton an electrophile? Absolutely! The Lewis definition of an acid is an electron acceptor. This sounds like an electrophile to me!

An acid is also an electrophile.

It is really important that you make this connection, because in many reactions, hydroxide has the choice between acting as a base, or as a nucleophile.

In due course, we will look at **chemoselectivity** (**Basics 26**), **regioselectivity** (**Basics 27**), and **stereoselectivity** (**Basics 28**).

In this case, we are talking about chemoselectivity, but don't get confused. Selectivity is all about getting the most favourable outcome from a number of possibilities.

ELIMINATION REACTIONS AND MOLECULAR ORBITALS

We are going to do this again in **Reaction Detail 4**, but let's have it here as well for completeness. Repetition is good!

Here is the curly arrow mechanism for E2 elimination again.

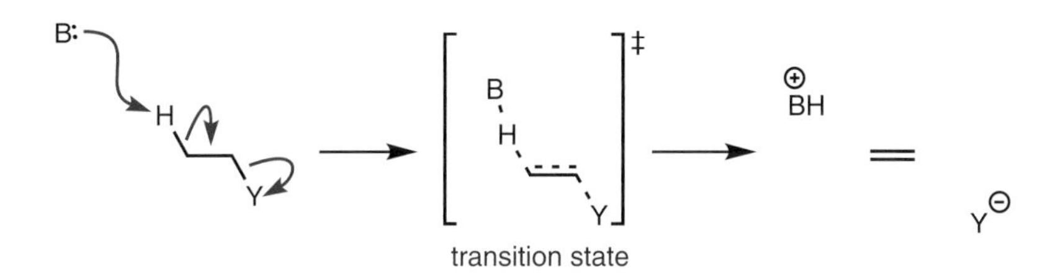

transition state

We are breaking the C–Y bond, where Y is a leaving group—it could even be iodide. If we are breaking the C–Y bond, by sharing electrons from somewhere else, then we must be donating the electrons into the σ^* orbital of the C–Y bond, exactly as we saw above.

The **only** difference, in this case, is that instead of the electrons coming from the nucleophile, they come from the C–H bond. We need the occupied orbital of this bond, and we know what that looks like—it is an sp^3 hybrid orbital.

If you look at the three possibilities we highlighted early in this chapter, we have options 2 and 3 happening here. Option 2 breaks the C–H bond. Option 3 breaks the C–Y bond.

2. Take the two electrons and give both of them to one of the atoms, so that they are no longer shared.
3. Put some electrons (don't worry about where we get them from) into the antibonding orbital.

Because we are forming the new double bond at the same time as we break the C–H and C–Y bonds, we must consider the shape and symmetry of the orbitals involved. The σ orbital of the C–H bond and the σ^* orbital of the C–Y bond must overlap.

When we draw these orbitals, it looks like this.

sp^3 hybridized antibonding orbital

sp^3 hybridized bonding orbital

H

Y

When we look at the orbitals like this, it is already starting to look like a π bond. As the reaction proceeds, these orbitals overlap and distort, and eventually become the π bonding orbital of the alkene. This results in a 'rehybridization' of the carbon atoms to become planar, sp^2.

Let's emphasize the key point again. In order for the reaction to begin, the C–H and C–Y bond orbitals must be able to overlap. In fact, they must be able to be aligned exactly as above, where the H and Y groups are on opposite sides of the central C–C bond. We refer to this arrangement as ***anti*** **periplanar**. This just means that H–C–C–Y are all in the same plane, but H and Y are on opposite sides.

This has important consequences for the outcome of the reaction. We will discuss these consequences in ***Applications 4***.

CONCLUSION

Ideally, you should always understand what the curly arrows mean in terms of the molecular orbitals involved. However, don't obsess about this, or beat yourself up if you cannot understand all of the possibilities yet. You need to be able to draw good curly arrows, so you need to understand the principles. The extra detail is something you can work on alongside the reactions. If you keep coming back to this chapter, it will eventually click. Once you've got it, it will stay with you.

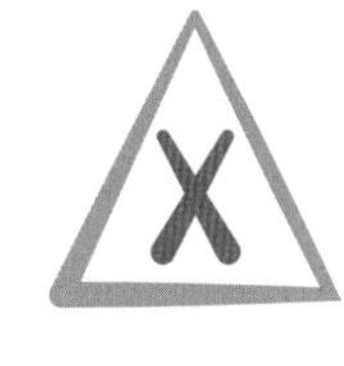

COMMON ERROR 2

Curly Arrows

Curly arrows represent the ‘flow’ of electrons. The start of the curly arrow has to be somewhere that there are electrons. The end of the curly arrow has to be something that can accept electrons.

Here are the main issues. Remember, every scheme in this chapter is **wrong**. Study this very carefully!

2.2

PROTONATION

Many reactions involve adding H^+. This is the **only** thing you will ever encounter that doesn’t have **any** electrons.

Starting a curly arrow at H^+ is ***always*** *going to be wrong.*

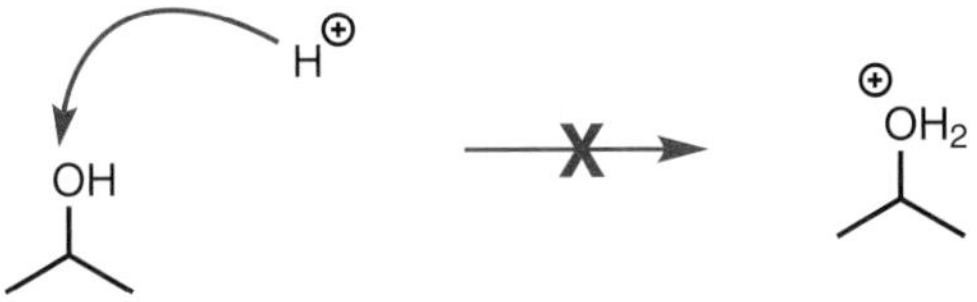

IMPLIED FIVE-VALENT CARBON

We have already looked at the situation where you draw a five-valent carbon, and this will be wrong. This curly arrow mistake is a little bit more subtle, but it ends up in the same place.

Look at the following curly arrows.

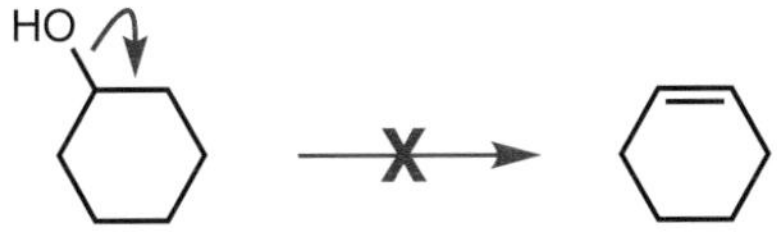

Initially this doesn’t seem too bad. The curly arrow starts at a bond and finishes at a bond. The bond it starts at is broken. The bond it finishes at becomes a double bond.

Let's look a little deeper. If we draw this curly arrow, what would be formed? Here it is.

Yes, we are losing HO^+, not HO^-. That's not good, but it's far from the worst mistake. The carbon atom that the curly arrow is moving towards has two attached hydrogen atoms, and this doesn't change! We are giving it more electrons. It therefore becomes five-valent, and will have a negative charge.

This is a disaster!

I think this is a really good illustration of why organic chemistry is difficult to learn. You have to be so mindful of what you are drawing in the early stages, or you risk learning the wrong 'patterns' and developing bad habits.

Don't rush when you draw a mechanism. Take your time and think about what each curly arrow means. There is plenty of time to draw quickly, once you are drawing everything correctly.

HYDRIDE LEAVING

Let's focus on the rule for now. There are some apparent exceptions, but we won't be seeing them in this book. For now, whenever you break a C–H bond, the H leaves as H^+, not as H^-. While the curly arrow has to start at the bond, it will move **away** from the H, not towards it.

We will look at this in the context of the previous problem.

Recall that the big issue is a five-valent, negatively charged carbon. If there are too many bonds, we could break one of them!

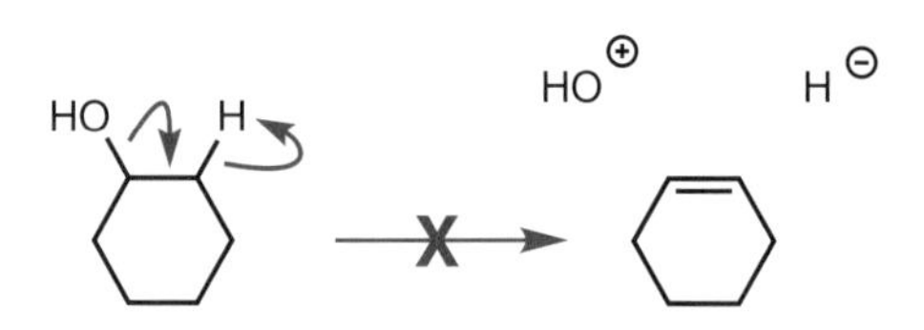

The problem is, this generates a hydride leaving group. This is very unstable, and won't be formed!

It's tempting to combine the HO^+ and H^- to give water, but that doesn't solve the fundamental problem.

The problem is that the curly arrows are going the wrong way. Look back at **Fundamental Reaction Type 2** for the correct curly arrows.

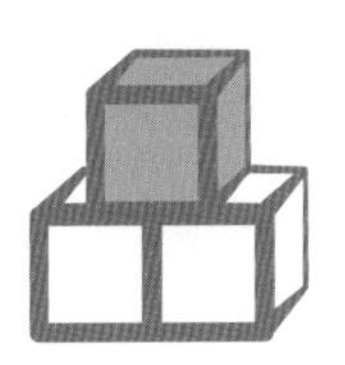

BASICS 10

Conjugation and Resonance

There are few concepts in organic chemistry more important than conjugation.[3] With practice, it is fairly easy to recognize conjugated systems by their alternating double and single bonds. In the two isomers below, the upper isomer is non-conjugated—there is a CH_2 group between the double bonds.

NON-CONJUGATED

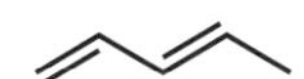

CONJUGATED

An experienced organic chemist will spot a conjugated system at a glance. It is a pattern that you will (and must) get used to.

As a result of the conjugation, you can draw different curly arrows, and hence identify different types of reactivity. In this chapter, I want to emphasize the point that you need to develop your pattern recognition so that you never confuse a conjugated system with a non-conjugated system.

MOLECULAR ORBITALS FOR ETHENE

Ethene is not conjugated. Conjugation requires at least two double bonds. However, we need to start somewhere.

Let's recap the molecular orbitals for ethene, so that we can build on them. Recall that the π bond is made by overlap of p orbitals on adjacent carbon atoms as shown below. Whenever we have a bonding orbital, we also have an antibonding orbital. If we consider the overlap of two p orbitals to make the π bond, there are two possibilities that differ in symmetry.

[3] I said something very similar about curly arrows. That was true as well. It probably won't be the last!

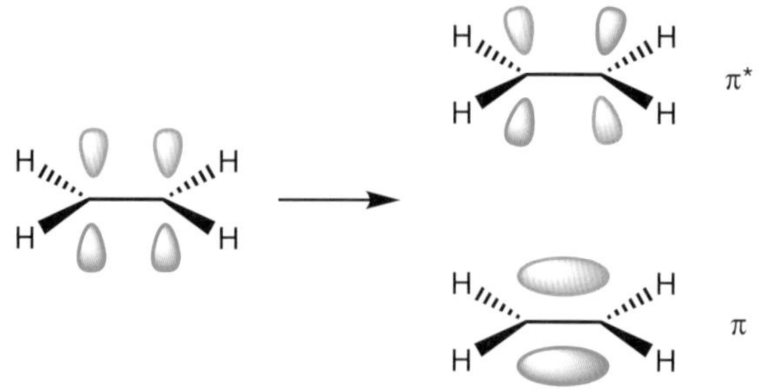

We can represent the energies of the orbitals as shown below. Only the bonding orbital is occupied normally.

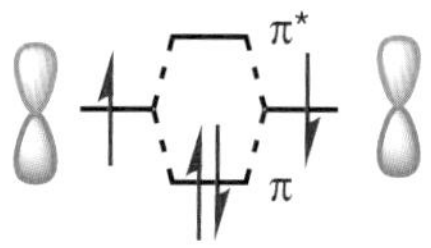

So, here is the key principle. When two atomic orbitals overlap, we form two new molecular orbitals, one bonding and one antibonding.

Now let's add another double bond, in conjugation.

BUTADIENE

This is butadiene. Strictly speaking it is buta-1,3-diene, but we don't usually need to specify.[4]

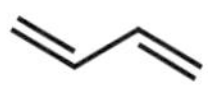

With four p orbitals to overlap, we will get four new molecular orbitals. Two of these will be bonding and two will be antibonding. We have the lowest energy orbital (ψ_1) which has the same symmetry across all carbon atoms. Then we have a second orbital (ψ_2) that has a node between C2 and C3. Since each double bond contributes two electrons, these are the only occupied orbitals.

[4] It is actually possible to have buta-1,2-diene. It looks a bit strange. Think about the molecular orbitals in such a compound!

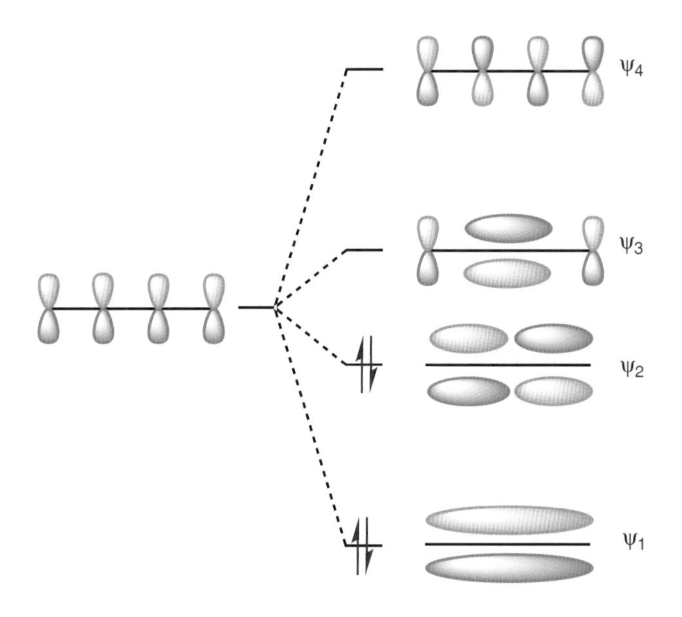

Orbitals ψ_3 and ψ_4 are unoccupied, and they are antibonding. These have two and three nodes respectively. Symmetry of orbitals is really important, but you won't see the full implications of this until much later in your studies.

We haven't added the energies of the orbitals to this diagram, as they would need to be calculated and you won't know how to do this. For now, let's just accept that we get more stabilization from two conjugated double bonds that we would from two non-conjugated double bonds.

WHAT ARE THE IMPLICATIONS OF CONJUGATION?

We started the chapter with the structures of penta-1,3-diene and penta-1,4-diene. Penta-1,3-diene is approximately 20 kJ mol^{-1} more stable.

It's important that you recognize that conjugation makes a structure more stable.

We will explore the evidence for the stability associated with conjugation in **Basics 20**.

HOW DOES CONJUGATION AFFECT BOND LENGTHS?

As a result of the additional orbital overlap between C2 and C3 in buta-1,3-diene, this bond is slightly shorter than a typical σ bond.

Here are some bond lengths in ethane, ethene, buta-1,3-diene, and (2*E*,4*E*)-3,4-dimethylhexa-2,4-diene.

1.57 Å 1.35 Å 1.37 Å 1.47 Å 1.47 Å 1.52 Å 1.35 Å

The C–C bond length in ethane is 1.57 Å. The C5–C6 bond length in 3,4-dimethylhexa-2,4-diene is 1.52 Å. Remember from **Basics 6** that a σ bond to a sp^2 carbon is shorter than one to a sp^3 carbon.

The C2–C3 bond distance in buta-1,3-diene is 1.47 Å. This is significantly shorter than would be expected without conjugation.

BENZENE AND AROMATICITY

Benzene compounds are particularly stable. Let's look at the structure again. Drawn in this way, we might think of benzene as cyclohexa-1,3,5-triene. We haven't looked at energetics yet, so for now I will simply state that benzene is approximately 150 kJ mol^{-1} more stable than it would be if it was three separate double bonds. This is a lot more energy than the stabilization due to the conjugation of two double bonds, and this stabilization dominates the chemistry of benzene and related compounds.

*You simply need to see a benzene ring and think '**more stable, much more stable**' without needing to over-analyse it. It needs to become a 'gut reaction'.*

We can, for now, look at the molecular orbitals of benzene. With six p orbitals on adjacent carbon atoms, we are looking to form six new molecular orbitals. Three of these will be filled, and three will be empty.

The lowest energy orbital has overlap across all six carbon atoms, exactly as we saw with buta-1,3-diene. Then there are two **degenerate** (equal energy) orbitals, each with one node.[5]

Then we move up to the unoccupied orbitals. There are another two degenerate orbitals, each with two nodes. Finally, the highest energy π orbital has three nodes.

[5] This is a top-down view. All of the orbitals have a node in the plane of the benzene ring, since these are derived from p orbitals. It is the additional nodes that we are discussing above.

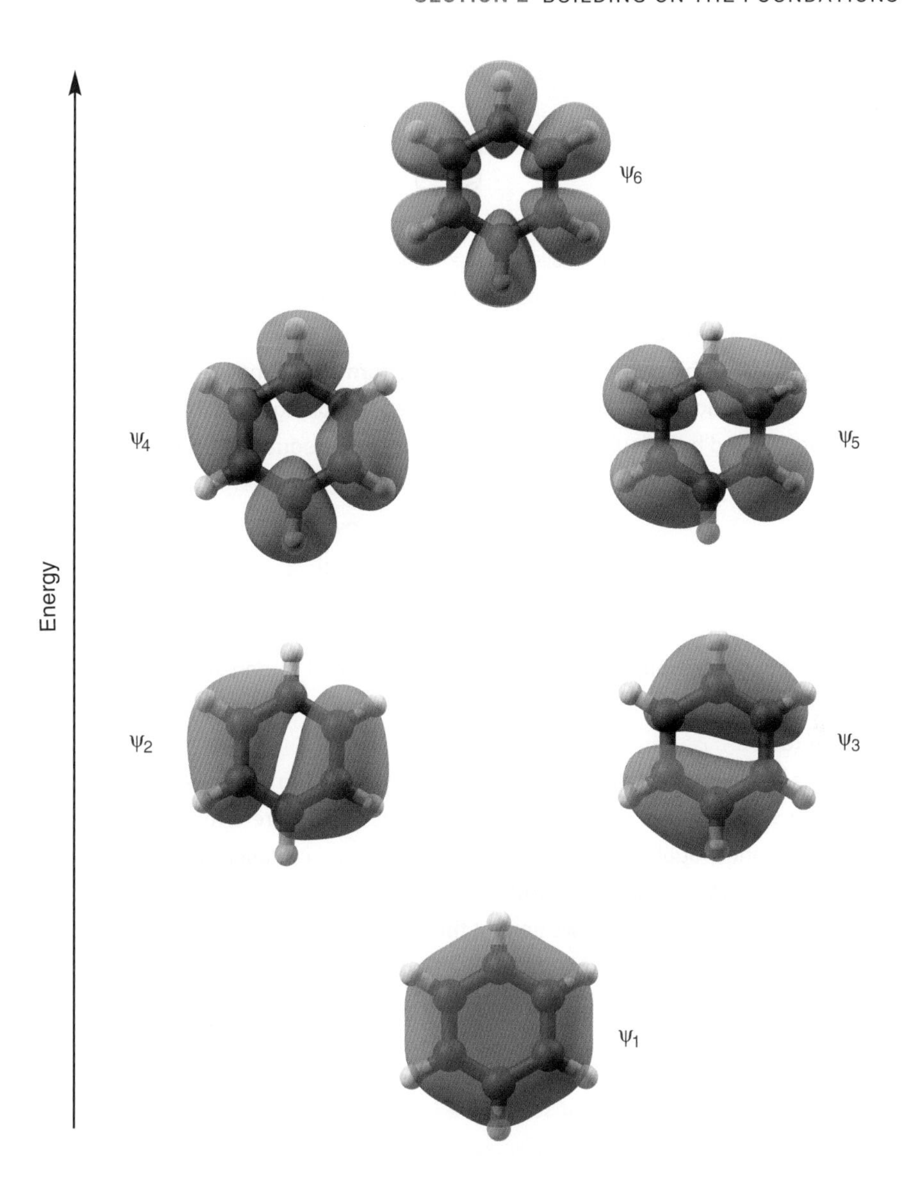

DRAWING RESONANCE CANONICAL FORMS—WHAT DO THEY MEAN?

The structure we drew for benzene above is not a perfect representation. If benzene had alternating single and double bonds, then we would assume that the double bonds would be shorter than the single bonds. The structure above has all the carbon–carbon bonds the same length (which is accurate) but still has alternating single and double bonds.

But which are the double bonds and which are the single bonds? We could just as easily draw this:

Have we moved the double bonds, or have we simply rotated the structure through 60°? Without adding a substituent, we cannot tell. In fact, all the carbon–carbon bonds in benzene are '1.5' bonds.

To help us see this, we can draw **resonance forms** of benzene. These are a curly arrow representation of the stabilization provided by conjugation. These are drawn, by convention, linked by a double-headed arrow and inside square brackets.

We will encounter resonance forms every time we need to consider the stability of a structure with any sort of conjugation. I want to introduce the idea slowly so that you get used to it.

There is one important point here. Neither resonance form fully explains all aspects of the structure. There **is** a real structure for benzene, often referred to as a **resonance hybrid** of the individual resonance forms. As discussed in **Basics 1**, we could take the easy way out, and draw benzene with a circle inside the hexagon. The major problem with this representation is that we cannot then draw curly arrows to represent reactivity.

You simply have to get used to seeing the 'cyclohexa-1,3,5-triene' structure and knowing that all of the bonds are of equal length. After a while, you will be able to look at a structure and know what resonance forms are possible, and which are the most important. Of course, you will only reach this point with practice, so we will do lots of that, starting in ***Practice 8****!*

THE HÜCKEL DEFINITION OF AROMATICITY

We can provide a more formal definition of aromaticity. An aromatic system is fully conjugated (alternating single and double bonds with no CH_2 groups in between!) with **4n + 2** π electrons, where n is an integer.

For benzene, n = 1. Naphthalene has 10 π electrons (5 double bonds). This is aromatic, with n = 2.

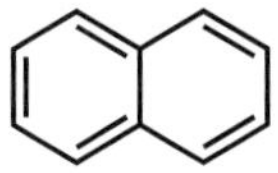

naphthalene

You will see many more aromatic systems during your studies. They are all significantly more stable than they would be if they were not aromatic.

Once you get used to the rule, if you have to consciously count the number of double bonds to determine if a compound is aromatic, you haven't internalized it. Fortunately, the solution is simple if tedious—practise!

MESOMERIC EFFECTS—ELECTRON-DONATING GROUPS

We can think about other types of conjugation. Let's have a look at the structure of methoxyethene.

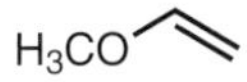

H_3CO

In **Basics 5** we considered inductive effects. Based on this, we might imagine that oxygen is electron-withdrawing, since it is electronegative. This isn't the whole truth. We have lone pair electrons on an oxygen atom, and these can overlap with the π bonding electrons on an adjacent double bond.

We can think about the molecular orbitals in detail, but for now, we will simply draw the following to show that we have non-bonded (lone pair) electrons on oxygen, and p orbitals on carbon. These can overlap, as we have already seen. Note that I am not worrying about whether the oxygen lone pair is a p orbital or an sp^3 hybrid orbital. Either way, there is some 'p orbital symmetry'.

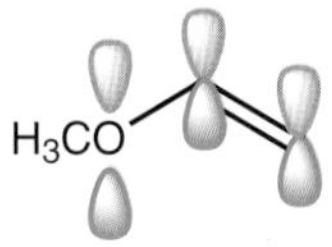

H_3CO

There is another way to represent this orbital overlap, using resonance forms. This is the way most organic chemists would automatically do this, so you need to get used to it. We draw curly arrows as follows.

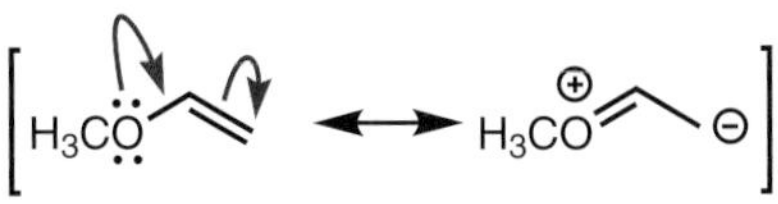

H_3CO ⟷ H_3CO ⊕ ⊖

The oxygen can be considered to be 'moving' electrons to the double bond. We refer to electronic effects which involve the overlap of lone pair and π electrons as **mesomeric** effects. These are distinct from the **inductive** effects we saw in **Basics 5**.

Overall, methoxyethene has inductive and mesomeric effects. The oxygen is electronegative, and will inductively withdraw electron density from the carbon atoms to which it is attached. However, there is also the mesomeric electron-donating effect which needs to be considered. Overall, the carbon atom which the oxygen is directly bonded to has a partial positive charge, but the carbon atom further away has a partial negative charge.

δ+
H_3CO δ−

This mesomeric effect has very profound implications for reactivity. Which carbon atom of the alkene do you think will be attacked by an electrophile?

MESOMERIC EFFECTS—ELECTRON-WITHDRAWING GROUPS

We can consider similar effects with conjugated electron-withdrawing groups. Consider the compound below, commonly known as methyl vinyl ketone. You can push curly arrows as follows, to move a negative charge onto the oxygen atom.

O CH_3 ⟷ ⊕ O⊖ CH_3

As a result of this, the carbon atom furthest from the carbonyl has a partial positive charge.

δ− O δ+ CH_3

If you were only considering inductive effects, you would not have expected the furthest carbon atom to be more affected.

It is very common for students to think that a carbonyl group is electron-donating, because the oxygen atom has lone pairs.

Can you draw a sensible resonance form to show the donation of these lone pair electrons?

Let's just answer the question now—you can't! But did you try? We will come back to this point in **Basics 17**, in the context of carbocation stability.

Whenever you need to evaluate the effect of any substituent in an organic compound, you should consider the following questions:

1. Is it a steric effect (*i.e.* simply due to the size of the substituent)?
2. Is it an electronic effect?
3. If it is an electronic effect, is it inductive, mesomeric, or a combination of the two?

Very often, steric effects and electronic effects, or inductive effects and mesomeric effects, will predict opposite outcomes in reactions. As you build a body of knowledge, you will improve your ability to recognize these situations. For now, it is important that you recognize the principles and ask the right questions every time.

With any effect, if it isn't steric, it must be electronic. If it isn't electronic, it must be steric. Or it could be both!

DELOCALIZATION

There is another word we need to consider—delocalization. It means exactly what it sounds like. In conjugated systems, electrons **are not localized** on one atom. They are spread throughout the conjugated system. In methoxyethene, the oxygen lone pair is delocalized into the double bond.

A REALLY IMPORTANT POINT

Don't lose sight of what resonance means—what conjugation and delocalization mean. It's all about the overlap of orbitals.

Make a molecular model of the following compound.

H_3C CH_3
H_3C CH_3

It is definitely conjugated according to the definition of having alternating single and double bonds. However, there is a problem. With all those methyl groups, it would not be possible for the two double bonds to be co-planar. If they cannot be co-planar, the π bonding orbitals cannot overlap across the entire conjugated system.

2.3

If the orbitals cannot overlap, it cannot ***really*** *be conjugated.*

Here is what it would look like in three dimensions. The two alkene bonds would be almost at right angles.[6]

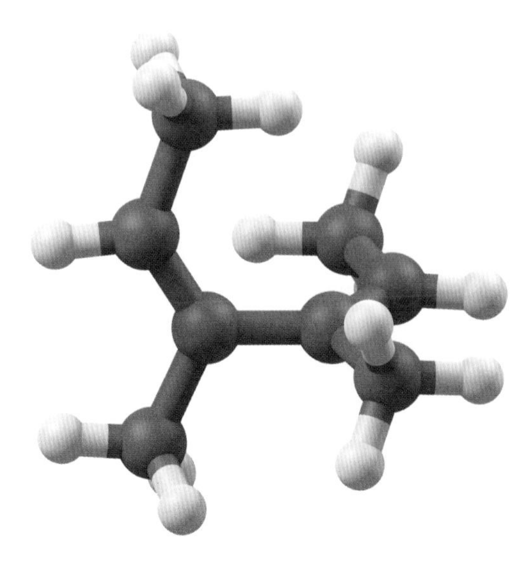

In Section 5 we are going to look at the energy changes as we rotate bonds.

In this case, we would gain approximately 20 kJ mol^{-1} by having the double bonds in the same plane (conjugation) but it turns out we lose much more than that by forcing the methyl groups together.

We can consider this problem on several levels, and each level is important. The first level is seeing the structure and recognizing that the two double bonds are conjugated. It's a pattern, and you are really good at recognizing patterns! The second level is visualizing the overlap of orbitals that conjugation 'means'.

It is perfectly possible to develop the ability to spot conjugation without actually knowing what it means. This is a good thing, because it is an important step along the way!

The third level is spotting that even though a system 'looks conjugated' it cannot be because the double bonds could not be co-planar.

This is reasonably easy to do for simple systems, but as the number of atoms, bonds and rings increases, it becomes harder to do. The solution? Get used to looking at increasingly complex structures![7]

[6] This is an image based on calculated data. I don't even know if this compound has been prepared. It doesn't really matter. The principles are sound!

[7] You weren't expecting anything profound or insightful there were you? I really wish there was a shortcut you could take. It would make my life so much easier!

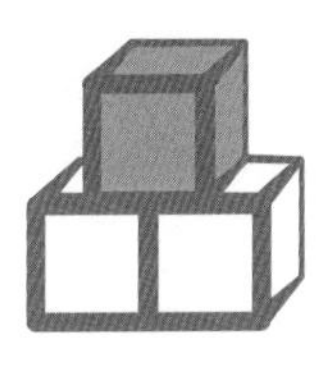

BASICS 11

Thermodynamic Definitions

We now need to start building the framework of how far and how fast a reaction will go. Let's start slowly, and consider the energy differences between starting materials and products.

INTRODUCTION

If you put a ball at the top of a hill, it will tend to roll down. If you put a ball at the bottom of a hill, it never spontaneously rolls up the hill. All systems tend towards their lowest energy state.

In the vast majority of cases, reactants and products will have different energies. If the products are more stable than the starting materials, a reaction will be favourable. If starting materials are more stable than products, the reaction will not be favourable.

There is a little more to it than this, and in order to discuss the energetic of reactions, we need some terminology.

EXOTHERMIC AND ENDOTHERMIC

Consider a hypothetical reaction.

$$\text{A} \longrightarrow \text{B}$$

If B is more stable than A, the reaction will be favoured. By 'more stable', we mean that the product, B, has a lower **enthalpy** than the reactant, A. ΔH, the change in enthalpy, is negative.

We call this an **exothermic** reaction.

If A is more stable than B, the reaction will not be favoured. In this case, the product, A, has a lower **enthalpy** than the reactant, B. ΔH, the change in enthalpy, is positive.

We call this an **endothermic** reaction.

EXERGONIC AND ENDERGONIC

Rather than considering the enthalpy change, we should really be considering the free-energy change.

Let's consider that reaction again.

Assuming that this is in solution, and we have one molecule of reactant giving one molecule of product, then the entropy change, ΔS, for this process will be close to zero.

Since $\Delta G = \Delta H - T\Delta S$, if ΔS is zero then $\Delta G = \Delta H$.

However, if we have a reaction such as the following, then we have one molecule reacting to produce two molecules.

$$A \longrightarrow B + C$$

The 'product' side of the equation is inherently more disordered, and ΔS will be positive. We then need to consider the free-energy change for the reaction, ΔG, rather than the enthalpy change for the reaction, ΔH.

An **exergonic** reaction has a negative ΔG value. An **endergonic** reaction has a positive ΔG value.

A CAUTIONARY NOTE

None of this tells us anything about **how fast** a reaction is. There are many reactions that are *exothermic/exergonic*, but that proceed slowly. Trees in an atmosphere containing oxygen are thermodynamically unstable compared to carbon dioxide and water. However, their rate of oxidation is negligible.

This is not simply because a tree is alive, although this naturally does have some consequences.

Of course, we can give the system enough energy to make the reaction proceed at a faster rate. In general, we increase the temperature of a system in order to increase the rate. In the case of wood, we could set fire to it.

We will look at rates of reaction in Basics 15.

WHERE ARE WE GOING WITH THIS?

After a while, you will learn to look at reactants and products, and instinctively know which are more stable. You will spot reactions with a significant change in entropy—

these are often reactions which produce a gas as one of the products, and they are therefore effectively irreversible.

In order to develop that instinct, we will need to consider the bond breaking and bond forming processes during a reaction in order to determine whether a reaction is *exothermic* or *endothermic*. We will have a closer look at bond dissociation energies in **Basics 12**, and we will see how to use them in **Basics 13**.

We will see that there is an added level of complexity to these calculations in **Basics 35**.

Before we get to that point, we will look at the complete energetic course of a reaction in **Basics 14**, and then we will need to consider the nature of intermediates (**Basics 16**) and transition states (**Basics 23**).

BASICS 12
Bond Dissociation Energy

In order to better understand bond dissociation energy, and the energetics of bond breaking, we need to look at three reactions, all of which simply involve the breaking of a C–H bond.

Make sure you follow the curly arrows in each case, and you understand why the products shown are formed.

ACIDITY—FORMATION OF A CARBANION

The following is an acid–base equilibrium. I am showing you this first, as it will be most familiar (because it generates H^+). It will also generate a carbanion.

$$R_3C{-}H \rightleftharpoons R_3C^{\ominus} + H^{\oplus}$$

We are breaking a bond so that both electrons go to one of the atoms. This is heterolytic bond cleavage, which we encountered in **Basics 8**.

In fact, when we measure the stability of carbanions, we use acidity as the measurement. We are going to look at this in more detail in **Basics 18**.

HYDRIDE ION AFFINITY—FORMATION OF A CARBOCATION

The next reaction shows a curly arrow that I wouldn't draw very often.

I don't want you drawing it too often either! There are (rare) occasions when it is correct, but most of the time when I see this, it is because a student has made a mistake.

We are breaking a C–H bond to form H^-. The reason I don't like it is that H^- is very unstable, so we don't break C–H bonds in this manner. However, as a **hypothetical** reaction, it is fine.

$$R_3C{-}H \rightleftharpoons R_3C^{\oplus} + H^{\ominus}$$

In fact, the reverse reaction would be the 'affinity' of the carbocation for a hydride ion, and we can use hydride ion affinity as a measure of carbocation stability. We are going to look at this in more detail in **Perspective 2**.

Hydride ion affinity isn't commonly taught in undergraduate degree courses. The reason for this is simple. Hydrocarbons don't tend to dissociate to give a hydride ion. If you draw a mechanism in which you break a C–H bond to kick out a hydride, you have generally made a mistake.

This doesn't mean it has less value as a conceptual tool. Don't get confused by this.

This example is another heterolytic bond cleavage.

HOMOLYTIC CLEAVAGE—FORMATION OF A FREE-RADICAL

In the above cases, both electrons from the C–H bond went the same way. In this final example, we break the bond homolytically so that each 'end' gets a single electron—a free-radical.

$$R_3C{-}H \rightleftharpoons R_3\dot{C} + H^{\bullet}$$

WHAT ARE BOND DISSOCIATION ENERGIES?

Quite simply, a bond dissociation energy is the enthalpy change for the **homolytic cleavage** of the corresponding bond to form two free-radicals.

What this means is that the bond dissociation energy is not always a good indication of how easy or difficult it is to break a particular bond.

As an example, we have seen (**Basics 6**) that a C–H bond in an alkyne (R–C≡C–H) is much stronger than a C–H bond in methane (CH_4). However, the alkyne hydrogen atom is quite acidic (we will see why in **Basics 18**), so it is relatively easy to deprotonate an alkyne using a strong base. We would never try to deprotonate methane, despite it having a 'weaker' C–H bond.

Immediately, we can see that bond dissociation energy is a measure we should use with caution. I am afraid it is only going to get more complicated (**Perspective 1**). This doesn't mean bond dissociation energies are not useful. It just means that we should be careful to use the correct values for the correct purpose.

We will explain this, and we will show how we can use bond dissociation energies to determine the overall enthalpy change of a chemical reaction, in **Basics 13**.

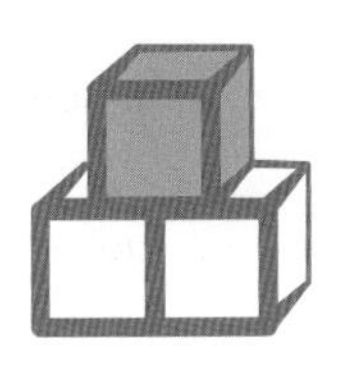

BASICS 13

Calculating Enthalpy of Reaction from Bond Dissociation Energies

INTRODUCTION

If we know how strong each chemical bond in a molecule is, we can calculate the enthalpy change for a reaction.

Here is a table of **average** bond dissociation energies, all in kJ mol^{-1}. These are representative values for the type of bond, except for those specific cases such as H−Cl, for which there can only be one value. There are limitations to the applicability of these numbers, and we will add some more detail in **Perspective 1**. For now, just accept these numbers.

Average Bond Dissociation Energy/kJ mol^{-1}					
H−H	436	**C−C**	350	**C=C**	611
H−Cl	432	**C−H**	410	**C≡C**	835
H−Br	366	**C−O**	350	**C=O**	732
H−I	298	**C−Cl**	330	**C≡N**	898
H−O	460	**C−Br**	270		
H−S	340	**C−I**	240		

We will look at a couple of worked examples, and then give you some more to practise.

We need to be clear about this. Everyone can do this. If you find that you are getting these calculations wrong, it won't be because you don't understand how to do it. You will have missed one or more bonds, or added when you should have subtracted. What is the solution to this (as if you didn't know!)? Practise until you stop making these mistakes—or at least until you can spot the mistakes yourself (which is a step on the way to not making the mistakes).

EXAMPLE 1

Here is a relatively simple substitution reaction.

$$H_3C{-}I \quad + \quad H_2O \longrightarrow H_3C{-}OH \quad + \quad HI$$

Note that we are deliberately avoiding charged species in these examples. We will see how to handle charges in **Basics 35**.

We have broken a C–I bond and an O–H bond. We have formed a C–O bond and a H–I bond.

The bonds we form count as a 'minus' and the bonds we break count as a 'plus'.[8]

So, $\Delta H = (240 + 460) - (350 + 298) = +52\ \text{kJ mol}^{-1}$.

This reaction is *endothermic*.

Perhaps you didn't expect this. After all, the C–O bond is much stronger than the C–I bond. But you are losing an O–H bond and getting back a much weaker H–I bond.

You need to be careful with these numbers.

Not all O–H bonds are the same.

The table has an average number, but an O–H bond in one compound will be stronger than that in another compound. You will learn to look at various structural factors to 'guess' whether the bond dissociation energy in a given compound is likely to be higher or lower than the average.

Remember—we are unlikely to be breaking an O–H bond homolytically!

EXAMPLE 2

This one is an elimination reaction. By now you should recognize the reaction type. Don't worry about this too much though. For now, we are focusing on counting up the bonds formed and broken. These are the key skills we have been developing.

We are breaking a C–H bond and a C–I bond. We are forming an H–I bond. So far, so good.

We are also changing a C–C single bond into a C=C double bond. To put this another way, we are 'breaking' a C–C single bond, and we are forming a C=C double bond. Don't

[8] Think of it this way. If we had a reaction that **only** formed one bond, with bond dissociation energy *x*, ΔH would equal −*x*.

worry about the fact that we aren't actually breaking the C–C single bond. If we replace one bond with another bond, treat it as breaking one and forming another.

$$\text{So, } \Delta H = (410 + 240 + 350) - (611 + 298) = +91 \text{ kJ mol}^{-1}.$$

It's another *endothermic* reaction.

In reality, when we do an elimination reaction, we tend to use a base to remove the hydrogen atom, so we aren't actually forming HI. As you progress, you will learn to consider reaction conditions and determine what you are really forming.

Of course, you need to learn the basics first. Everything you need in terms of factual information has been covered in this chapter. You could apply this to a reaction that forms one bond, or twenty bonds.

The skill you need is to keep track of exactly what has changed.

To do this, we need to also consider what has not changed. We looked at this important idea in **Habit 3**. Then we will get to practising calculating enthalpy changes for reactions.

I get confused with these calculations. I add when I should subtract, and I subtract when I should add. But I know I do this, so I do the calculation graphically by drawing 'energy levels' going up/down by the bond dissociation energy for the bond I am breaking/forming.

That way, I reduce the chance that I will make a mistake! It's all about having a 'plan'.

PERSPECTIVE 1
A Closer Look at Bond Dissociation Energies

This is a new type of chapter. The purpose is to give you a bit more detail. The downside of this is that some of the ideas are quite complicated. I wanted to keep this material with the rest of the discussion of bond dissociation energies, but the reality is you might want to skim this and then come back to it later.

You may have noticed that some of the bond dissociation energies in the table in **Basics 13** are very different to those in **Basics 6**.

In **Basics 6** we established that a C–H bond to an alkene is stronger than a C–H bond to an alkane. Here are the data again.

377 kJ mol^{-1} (1.57 Å)
420 kJ mol^{-1} (1.10 Å)
1

728 kJ mol^{-1} (1.35 Å)
458 kJ mol^{-1} (1.07 Å)
2

954 kJ mol^{-1} (1.21 Å)
H–C≡C–H
549 kJ mol^{-1} (1.06 Å)
3

Here is the table from **Basics 13** again.

Average Bond Dissociation Energy/kJ mol^{-1}					
H–H	436	**C–C**	350	**C=C**	611
H–Cl	432	**C–H**	410	**C≡C**	835
H–Br	366	**C–O**	350	**C=O**	732
H–I	298	**C–Cl**	330	**C≡N**	898
H–O	460	**C–Br**	270		
H–S	340	**C–I**	240		

WHAT'S THE PROBLEM?

We have a bond dissociation energy for the double bond in ethene of 728 kJ mol^{-1}, and the average bond dissociation energy for this type of bond is 611 kJ mol^{-1}.

What on earth are we taking the average of to get a number so much lower than that of the simplest possible alkene?

The short answer is that it's not quite that simple. In explaining this, we are going to see some very odd structures. This chapter is one to read quickly, get the key 'take-home message', and then come back to later.

*The take-home message is '**it's complicated**'.*

Let's look at the bond dissociation energies for methane, ethene, and acetylene. We will consider the sequential dissociation of each C–H bond.

$CH_4 \xrightarrow{+439\ kJ\ mol^{-1}} CH_3^{\bullet} + H^{\bullet} \xrightarrow{+462\ kJ\ mol^{-1}} CH_2 + 2\ H^{\bullet} \xrightarrow{+424\ kJ\ mol^{-1}} CH + 3\ H^{\bullet} \xrightarrow{+339\ kJ\ mol^{-1}} C + 4\ H^{\bullet}$

How strong is a C–H bond in methane? Is it the average of these four numbers (**416 kJ** $\mathbf{mol^{-1}}$) or is it the first dissociation energy? First let's have a look at ethene and acetylene. We can do these together, because we get acetylene on the way from ethene.

$C_2H_4 \xrightarrow{+458\ kJ\ mol^{-1}} C_2H_3^{\bullet} + H^{\bullet} \xrightarrow{+141\ kJ\ mol^{-1}} H{-}C{\equiv}C{-}H + 2\ H^{\bullet}$

$C_2H_3^{\bullet} \xrightarrow{+339\ kJ\ mol^{-1}} H_2C{=}C{:} + 2H^{\bullet}$

$H{-}C{\equiv}C{-}H + 2\ H^{\bullet} \xrightarrow{+549\ kJ\ mol^{-1}} H{-}C{\equiv}C^{\bullet} + 3\ H^{\bullet} \xrightarrow{+487\ kJ\ mol^{-1}} {}^{\bullet}C{\equiv}C^{\bullet} + 4\ H^{\bullet}$

$H{-}C{\equiv}C^{\bullet} + 3\ H^{\bullet} \xrightarrow{+351\ kJ\ mol^{-1}} H_2C{=}C{:} + 2H^{\bullet}$

The first C–H bond dissociation energy for ethene is 458 kJ mol^{-1}. For acetylene it is 549 kJ mol^{-1}. The average bond dissociation energy for ethene is **409 kJ mol^{-1}**. The average for acetylene is **518 kJ mol^{-1}**.

If we use the average value, we find that the dissociation energy for ethene is lower than that for methane. This is true **as an average** but not true for the dissociation of one bond.

This is a result of the stability of the various intermediates along the pathway for the complete dissociation of ethene.

If we use average bond dissociation energies to compare the strengths of individual bonds, we would not see the correct trends. If we use 'first' bond dissociation energies to calculate enthalpy changes for reactions, we would be miles out!

WORKED PROBLEM

Let's use a specific calculation to highlight the nature of the problem.

$$H_2C{=}CH_2 + H_2 \longrightarrow H_3C{-}CH_3$$

Calculate the enthalpy change for this reaction using the individual bond dissociation energies at the start of the chapter, and the average bond dissociation energies in the table. You need to use the value for H–H in the table in both cases.

Here are the solutions. Using the specific bond dissociation energies, we get the following.

Losses	Gains
Four sp^2 C–H bonds (4 × 458 kJ mol^{-1})	Six sp^3 C–H bonds (6 × 420 kJ mol^{-1})
One H–H bond (436 kJ mol^{-1})	One C–C bond (377 kJ mol^{-1})
One C=C bond (728 kJ mol^{-1})	
2996 kJ mol^{-1}	**2897 kJ mol^{-1}**

On this basis the reaction is calculated to be *endothermic* by 99 kJ mol^{-1}.

If we use the average bond dissociation energies, we get the following result.

Losses	Gains
Four sp^2 C–H bonds (4 × 410 kJ mol^{-1}) One H–H bond (436 kJ mol^{-1}) One C=C bond (611 kJ mol^{-1})	Six sp^3 C–H bonds (6 × 410 kJ mol^{-1}) One C–C bond (350 kJ mol^{-1})
2687 kJ mol^{-1}	**2810 kJ mol^{-1}**

1

On this basis the reaction is calculated to be *exothermic* by 123 kJ mol^{-1}.

The accepted enthalpy change for hydrogenation of ethene is –137 kJ mol^{-1} (*exothermic*). The calculation using average bond dissociation energies agrees reasonably well with this. The calculation using the specific bond dissociation energies doesn't even get the sign of the enthalpy change correct, never mind the magnitude!

SUMMARY

It's complicated! We have specific individual bond dissociation energies and we have average bond dissociation energies. The former give information about the strengths of individual bonds within a molecule. The latter are representative numbers that can be used additively to work out the expected enthalpy change of a given reaction.

Furthermore, if we wanted to compare the enthalpy change for the hydrogenation reactions of two different alkenes, we would have to locate good average bond dissociation energy data for each system. These data are not always easy to find.

We should use these numbers with some caution. We will practise these calculations, but remember that not all C–H bonds (or any other type of bond) are the same. Treat your answers as a guide, and nothing more!

PRACTICE 7
Calculating Enthalpy of Reaction from Bond Dissociation Energy

7

THE PURPOSE OF THE EXERCISE

In my experience, it is pretty rare for an organic chemist to carry out one of these calculations. However, they will look at a reaction and intuitively know whether it is likely to be *endothermic* or *exothermic*.

They can do this because they have put in the leg work!

THE EXERCISE

Here is the table of average bond dissociation energies again. We are going to use this to practise calculating the enthalpy changes for reactions. There are a few extra numbers in this table. You will need some of them, but not all!

Average Bond Dissociation Energy/kJ mol^{-1}					
H–H	436	**C–C**	350	**C≡N**	898
H–Cl	432	**C–H**	410	**C≡C**	835
H–Br	366	**C–O**	350	**C=C**	611
H–I	298	**C–N**	300	**C=O**	732
H–O	460	**C–Cl**	330	**C=N**	615
H–N	390	**C–Br**	270	**N–N**	240
H–S	340	**C–I**	240		

We will look at some of the reaction types covered in this book, and other reaction types that are not. It doesn't matter if we haven't covered the reaction type. This is all about counting bonds and adding and subtracting numbers. Don't over-think it!

In this case, I am going to give you the answers at the end. Just the numbers though. You still have to work out how to get to them!

QUESTION 1

Let's start with a substitution reaction. In a substitution reaction, we are breaking one bond and forming one bond. In the example below, this happens **twice**.

This doesn't change what you have to be able to do.

Note that the NC group isn't a cyano group. It's an isonitrile. This is a rather odd functional group, but it is drawn in a very specific way so that you know that the nitrogen atom is directly bonded to the carbon on its lower left.

Of course, since we are not making or breaking any bonds involving this atom, it doesn't matter for the problem!

QUESTION 2

Now a problem that looks more complicated!

There is an eight-membered ring and two three-membered rings in one of the starting materials. Spotting the connectivity in the product is not trivial when it is drawn rather differently.

Why do you think I included this problem?

Perhaps you should make a model of the product if you cannot see how it relates to the starting material.

When you get to Reaction Detail 2 you will appreciate that substitution reactions have particular stereochemical consequences. Come back to this question at that point, and make sure you understand why only this stereoisomer is formed in the reaction.

? Which other regioisomeric (**Basics 27**) product might you expect to see formed?

7

Have a go at this now—draw some curly arrows and product structures. Don't worry about whether your answer makes too much sense yet. You are still getting used to drawing structures. The second time through the book, you will find you have a lot more perspective on the problem.

QUESTION 3

Now an elimination reaction. Once again, I am choosing an example that looks quite complicated, so that you can develop the key skill of focusing on the parts of the structure that actually matter.

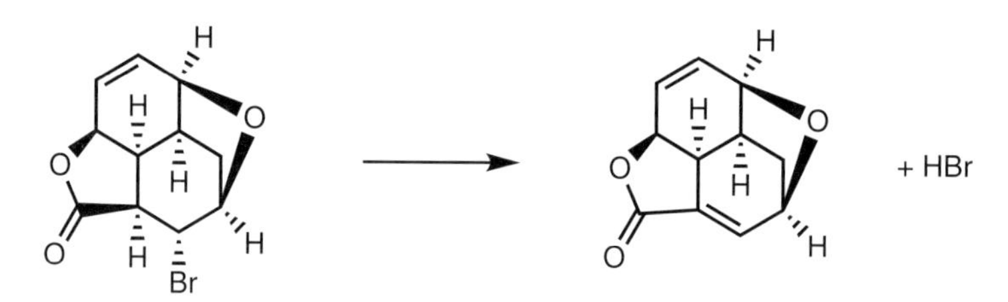

There are C–O bonds that look very long. You just cannot draw this on paper and have all of the bonds approximately the same length.

Make a model of the compound. Come back to this once you have worked through **Section 6**. At that point, what must the mechanism be, and what reaction conditions would you use?

For now, don't worry about the shape. Learn to look past this to focus on the bonds being formed and broken.

QUESTION 4

Here is a reaction that we won't be covering in this book, but that doesn't matter. There are substituents with abbreviations that you may not know, but they don't change, so that doesn't matter either.

What matters is that if a double bond becomes a single bond, then you treat it as completely breaking the double bond and forming the single bond.

QUESTION 5

Here is quite a tricky one to finish with. This is the first step in a classic total synthesis of strychnine. There are quite a few bonds being formed and broken. Of course, you don't need to know **how** that happens. You just need to keep track of the bonds.

$+ NH_3 + H_2O$

*Perhaps in this case, it is easier to just add up **all** the bonds on each side of the reaction arrow and then 'cancel them out' to see what you are left with. Ultimately, how you get there is up to you. My way might not work for you. Find your own way!*

FINDING MORE PROBLEMS

In principle you can do this for any reaction, but you may need to consider bond dissociation energies for some quite complicated reagents. It tends to be easier (and more useful) if you only have organic molecules reacting with no change in oxidation state. We have not defined oxidation state in this book, as none of the reactions covered feature oxidation or reduction.

Stick to examples that only have the following elements in all compounds involved: C, H, N, O, Cl, Br, I. Make sure you use examples with balanced equations.

THE ANSWERS

I am not going to make this too easy for you. The numerical answers are −178 kJ mol^{-1}, −72 kJ mol^{-1}, −40 kJ mol^{-1}, −19 kJ mol^{-1}, and +53 kJ mol^{-1}. I am not going to tell you which one is which.

BASICS 14

Energetics and Reaction Profiles

INTRODUCTION—REACTION PROFILES

In Basics 11 we looked at the energy differences between starting materials and products. We might be tempted to think that if a reaction product is more stable than the starting material, the reaction would proceed. It's a lot more complicated than this. We need to think about what happens in between starting materials and products. We need to consider not just whether a reaction can proceed, but whether it will proceed at a meaningful rate. We can use **reaction profiles** for this.

*Fundamentally, a **reaction profile** (or reaction coordinate diagram) is a graph of energy versus the progress of a reaction.*

THE S_N2 SUBSTITUTION MECHANISM—REACTION PROFILE WITH NO INTERMEDIATE

Here is a very simple reaction profile. The starting materials are on the left. The products are on the right.

In this case, the products are more stable than the starting materials. This reaction is therefore *exothermic*. This is good. It doesn't tell us how fast the reaction will be, but if the reaction could reach equilibrium then there would be more product than starting material.

There is an energy barrier, called the **activation energy**, to be overcome before the reaction can proceed. If a large enough proportion of molecules have enough energy, they will undergo the reaction.

*The activation energy determines the rate of reaction. We will quantify this with an equation in **Basics 15**.*

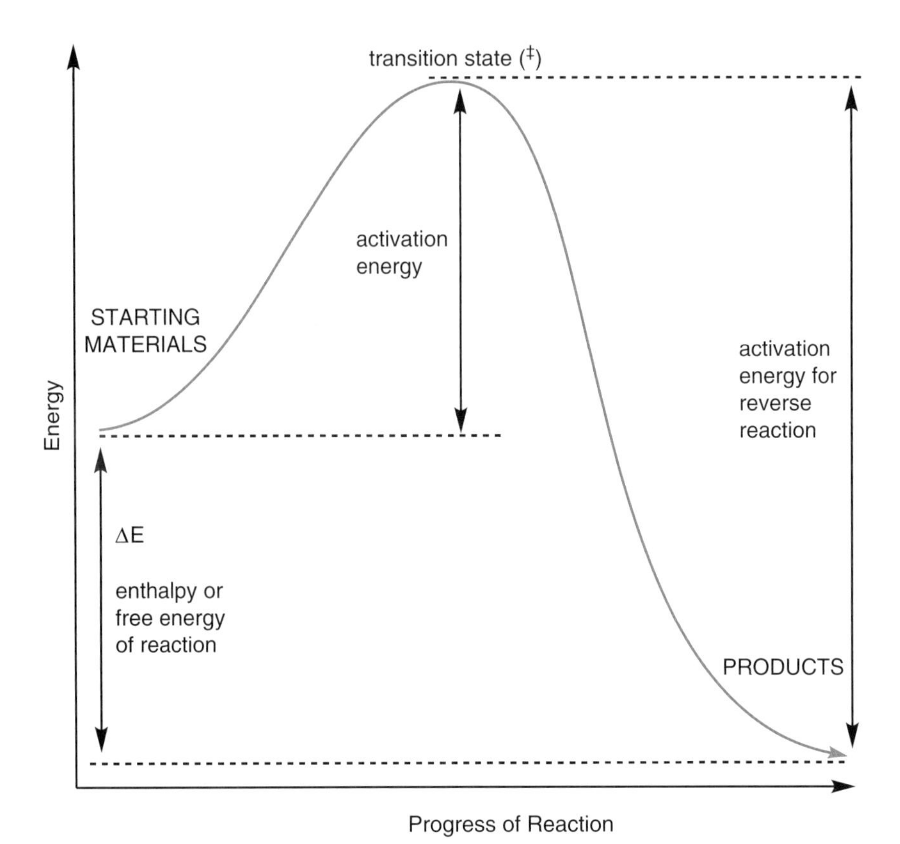

The higher the barrier, the fewer molecules will have enough energy at any given time, and the slower the reaction will be. If you want more of the molecules to have the requisite energy, you could increase the temperature of the reaction.

Now we have the tools to talk about 'how far' and 'how fast'.

There is just one more point along this reaction profile that we need to talk about—the transition state.

Here is the S_N2 reaction of iodomethane with hydroxide.

$$HO^{\ominus} + H_3C{-}I \longrightarrow \left[HO\cdots C(H)(H)(H)\cdots I\right]^{\ddagger\ominus} \longrightarrow HO{-}CH_3 + I^{\ominus}$$

The structure in the middle, in square brackets, is a transition state. We use the '‡' symbol to indicate a transition state. We will look at these in detail in **Basics 23**.

For now, just accept that as the nucleophile attacks the carbon atom, the iodide starts to leave. When the C–O bond is partly formed, and the C–I bond is partly broken, we will move through a maximum energy before it all gets more stable as we move towards the product.

There are no intermediates, with a discrete existence, along this reaction profile. We will look at a reaction profile with an intermediate in a moment.

*In **Fundamental Reaction Type I**, we saw that there are two possible mechanisms for substitution reactions. What we need to be able to determine is whether a given substrate will undergo a substitution reaction at all, and which mechanism will be operative if it does.*

THE S_N1 SUBSTITUTION MECHANISM—REACTION PROFILE WITH ONE INTERMEDIATE

Let's look at a similar process that proceeds via a different mechanism. In this case, it is the S_N1 substitution of an alkyl halide. The first step is formation of a carbocation. We are going to discuss carbocations starting in **Basics 16**. For now, accept that this is a charged species formed by breaking a bond, and as such it will have higher energy (be less stable) than either the starting material or the product.

$$R-X \xrightarrow{\text{slow loss of } X^{\ominus}} R^{\oplus} + Nu^{\ominus} \xrightarrow{\text{fast}} R-Nu \qquad (+\ X^{\ominus})$$

The reaction profile is shown over the page.

Instead of having a single 'hump', this one has a dip after the first hump, and then it has a second hump. Between the two is the carbocation **intermediate** (or more precisely, it is R^+ and X^-, which have been completely separated). Think of the R–X bond as being like a spring. You stretch it and stretch it, until it breaks. You will need to keep putting energy into it while you are stretching it, right up until the point at which it breaks. Then you will get some of that energy back. The 'highest energy' point along the reaction coordinate is a transition state (TS1‡). Again, we use the '‡' symbol to indicate a transition state.

The opposite happens as you form a bond between the carbocation and the nucleophile. The energy increases at first, and then you start to gain energy as the bond forms. There is another transition state (TS2‡) between the intermediate and the product.

We need to look in more detail at transition states and intermediates. It is really important that you understand what these are. We will also need to understand the reaction profile in more detail.

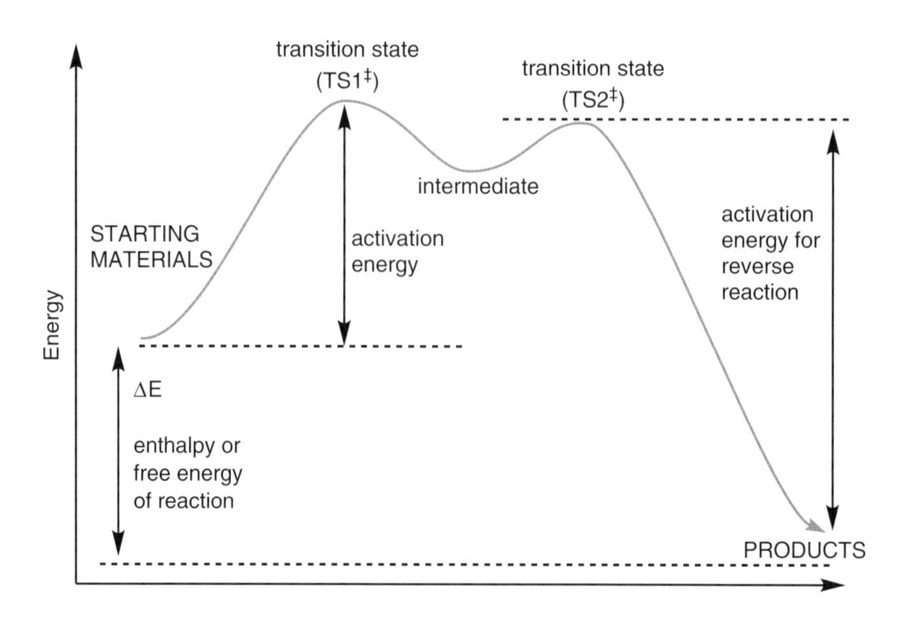

WHAT IS AN INTERMEDIATE?

Let's think of a simple general reaction.

The thing in the middle, 'B', is something that is passed through on the way to the final product 'C'. Depending on the reaction, 'B' could be many things. Fundamentally though, we have two different situations.

1. 'A' is converted into 'B' faster than 'B' is converted into 'C'.
2. 'A' is converted into 'B' more slowly than 'B' is converted into 'C'.

In the first case, in principle the reaction could be stopped at 'B' and 'B' is a potential product from the reaction. In effect 'B' is 'too stable'.

We are more interested in the latter case, where 'B' has some level of stability, but still reacts rapidly to give the product.

For now, the most important intermediates we will consider are **carbocations** and **carbanions**. We are going to start looking at the factors that affect the stability of these species in Basics 16. If an intermediate is more stable, this places it at lower energy on the reaction coordinate. Once we introduce the Hammond postulate (Basics 19) to make the link between the energy of a transition state and that of an intermediate, we will see that this can also tell us quite a lot about how quickly it will be formed. Bear with me on that one. We will get there in the end. At that point, you should come back to this chapter and read it one more time!

WHAT IS A TRANSITION STATE?

This is a complicated question. At the simplest level, it is the entity at the top of the energy curve. Consider an S_N2 substitution. You start to form the bond from the nucleophile to the carbon at the same time as you break the bond from carbon to the leaving group. At some point, where the energy is a maximum along the direction of the reaction, you could think of the two bonds as being 'half bonds'.

This explanation will serve our purpose for now. Transition states are explained in more detail in Basics 23.

A FUNDAMENTAL POINT—SELECTIVITY IN REACTIONS

We will encounter several types of selectivity—chemoselectivity, regioselectivity, stereoselectivity.

Don't let this confuse you. Where you have two (or more) possible outcomes for a given reaction, the one that happens fastest will be the one with the lower barrier.

This is a slight simplification, as we are assuming that the products of the various processes cannot be interconverted.

Chemoselectivity is discussed in Basics 26. Regioselectivity is discussed in Basics 27. Stereoselectivity is discussed in Basics 28.

Make sure you know exactly what each type of selectivity is. It is very easy to read an exam question about one type of selectivity, and give an answer all about a different type of selectivity!

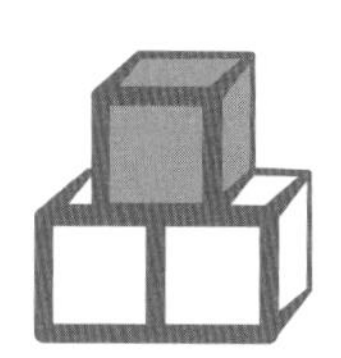

BASICS 15

How Fast Are Reactions?

We have seen (**Basics 13**) how to determine whether a reaction is energetically favourable (*exothermic*). This is important.

We have also looked at reaction profiles (**Basics 14**), and we have seen that we have an activation 'barrier' to reactions. If the barrier is too high, it doesn't matter whether the reaction is *exothermic* (or exergonic) or not. It will simply be too slow.

Trees, in an atmosphere of oxygen, are unstable relative to carbon dioxide and water. However, trees don't spontaneously oxidize. They might not be thermodynamically stable, but they are kinetically stable. They are oxidized very slowly.

A CHALLENGING CALCULATION

Here is a reaction scheme for the oxidation of glucose. I have omitted the stereochemistry, as it won't change anything. Use the table of bond dissociation energies below to calculate the enthalpy change for the reaction.[9]

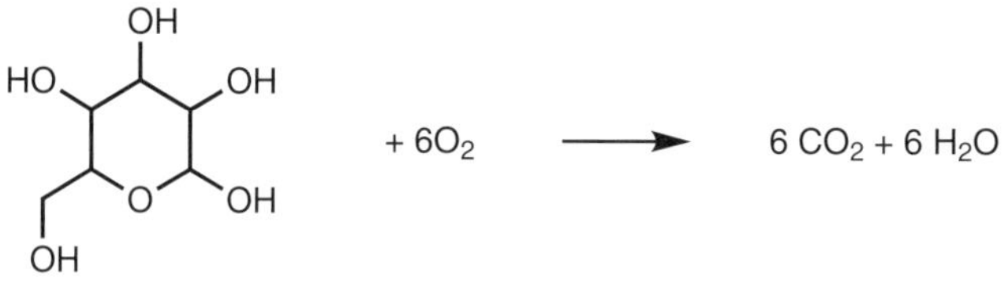

OH, HO, OH, O, OH, OH

$$+ 6O_2 \longrightarrow 6\,CO_2 + 6\,H_2O$$

Average Bond Dissociation Energy/kJ mol^{-1}					
H–H	436	**C–C**	350	**C=C**	611
H–Cl	432	**C–H**	410	**C≡C**	835
H–Br	366	**C–O**	350	**C=O**	732
H–I	298	**C–Cl**	330	**O=O**	498
H–O	460	**C–Br**	270	**C≡N**	898
H–S	340	**C–I**	240		

HINT You are completely breaking up the glucose. Break all the bonds. Don't forget the C–H bonds that are not explicitly shown!

[9] How good do you think the answer is? Do you think CO_2 is a typical carbonyl compound? Are all the C–O bonds in glucose the same?

NOW A LOOK AT REACTION RATES

We need to look at an equation. This is basic physical chemistry kinetics.

$$k = Ae^{\frac{-E_a}{RT}}$$

k is the rate constant for the reaction. E_a is the activation energy (J mol^{-1}). R is the gas constant (8.314 J mol^{-1} K^{-1}). T is the temperature in K. A is a pre-exponential factor related to the frequency of collisions.

I often talk about the 'shape of an equation'. What I mean by this is the development of an understanding of how the output changes according to changes in the input. In this case, the relationship is exponential.

There are a couple of key questions that we need to consider.

1. How does the rate of reaction change with a change in temperature?
2. What activation barriers are reasonable?

We can address both issues by applying a bit of reasoning, and by putting numbers into the equation. I've done this for a few numbers.

RELATIVE RATES OF REACTION FOR GIVEN ACTIVATION ENERGIES AND TEMPERATURES

Here is a table of some relative rates of reactions with different activation energies and at different temperatures.

Temperature/K	Activation Energy/kJ mol^{-1}		
	10	**20**	**50**
273	45,000,000	550,000	1
293	60,000,000	1,000,000	4.5
313	79,500,000	1,700,000	16.8

What do these numbers mean? Let's say that a reaction with an activation energy of 20 kJ mol^{-1} takes **1 minute** to reach completion at 293 K.[10] If that reaction had an activation energy of 50 kJ mol^{-1}, the reaction would take almost **22 weeks** to reach completion.

[10] We have to be hypothetical, as we don't know the other important factors, such as the concentration of reagents.

You could do a similar comparison for any other pair of numbers in the table. Try a few, just to get a better idea of what the differences between these numbers really mean.

The rate of a reaction increases as temperature increases. It is an exponential relationship rather than a linear relationship, so it isn't possible to make a simple statement such as 'increasing the temperature of a reaction by *x* degrees will double the rate'.

A reaction that has a barrier above 100 kJ mol^{-1} will be very slow. You'll have to heat it to get it to proceed at an acceptable rate.

Many reactions proceed rapidly at temperatures as low as −78 °C. In other cases, it is okay to heat a reaction to 150 °C (or sometimes even higher). This does give us a substantial temperature 'window' in which to work. We can use catalysts to lower the activation barrier, increasing the rate of a reaction. But for now, we should remember that relatively small increases in activation energy result in significant decreases in reaction rate.

Slowing one reaction down can mean that an alternative reaction becomes the more favourable process. We will look at many instances of reactions where there is more than one possible outcome.

BASICS 16

Introduction to Carbocations, Carbanions, and Free-Radicals

We are going to start this with some facts and explanations, and then start discussing 'how to succeed'. We will discuss aspects of carbocation and carbanion stability together in order to show the common themes.

SIMPLE CARBOCATIONS—STRUCTURE AND STABILITY

A carbocation is a species containing a positively charged carbon atom with three bonds (R_3C^+). Textbooks often refer to such a species as a carbonium ion,[11] although the correct term is a carbenium ion. We will avoid issues of nomenclature by simply referring to these structures as carbocations.

The electronic structure of a carbocation consists of a sp^2 hybridized carbon atom, with the positive charge localized in a p orbital (*i.e.* an empty p orbital). Carbocations are therefore **planar**, similar to alkenes.

empty p orbital

$R_3C^{\oplus}$ (planar, with empty p orbital perpendicular to the plane)

We need to consider the factors which lead to stabilization of carbocations. The basic trend for carbocations with methyl substituents is:

tertiary		*secondary*		*primary*		*methyl*
$(CH_3)_3C^{\oplus}$	>	$(CH_3)_2CH^{\oplus}$	>	$CH_3CH_2^{\oplus}$	>	$CH_3^{\oplus}$
Most stable						Least stable

[11] A carbonium ion is R_5C^+. Since R_3C^+ and R_5C^+ both contain positively charged carbon, it is correct to refer to both of them as carbocations.

*The nomenclature—tertiary, secondary, primary—needs to become part of your language. If you remember that a tertiary carbocation is more stable, but cannot remember which one **is** the tertiary one, there isn't much point. You need to understand the stability trend and learn the language. Write them out—repeatedly—until you cannot forget!*

16

A more stabilized carbocation is one in which the positive charge is 'reduced' by something releasing electron density towards it. In effect, with this trend we are saying that in order to stabilize the positive charge, a methyl group must be electron-donating! It must be able to release electron density towards the positive charge. When you see something like this, you need to ask a very fundamental question—**which electrons**?

This needs to become your habit. Don't just count the alkyl groups. Ask the right questions! It is only a small adjustment, but it will make a massive difference to you!

Now we get to a fundamental principle. Whatever electrons are being released by the methyl groups **must** have the correct symmetry to overlap with a p orbital.

In the diagram below, we can see the vacant p orbital of the carbocation and the filled sp^3 orbital of the C–H bond. They have the correct symmetry to overlap. (Remember that the shape of the sp^3 orbital derives from the p orbital contribution, so they would have the correct symmetry!)

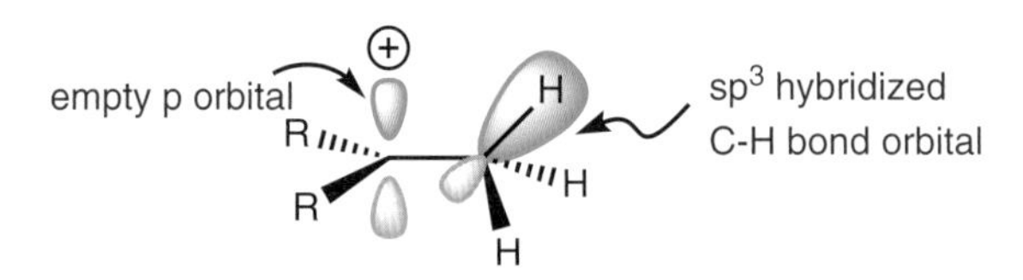

2.6

How can we represent this stabilization? We have two orbitals that can overlap: one filled, one empty. The two orbitals overlap to form two new orbitals, one that is lower in energy and one that is higher in energy. Since there are only two electrons, only the lower energy orbital is occupied so that the net energy of the system is lowered.

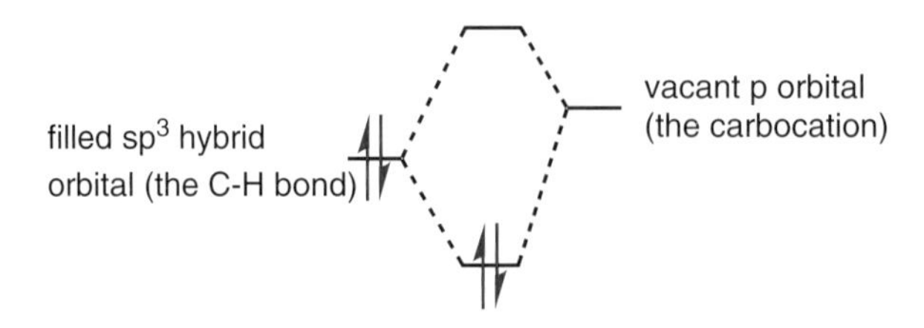

*The sp^2 hybridization of carbocations is quite natural. We only have three substituents attached to carbon, and six outer electrons. We saw in **Basics 6** that a bond to sp^2 carbon is stronger than a bond to sp^3 carbon. We maximize the bonding, as well as allowing the electrons and substituents to be as far apart as possible.*

Is this unique to C–H bonds? No! However, it turns out that the relative energies of the carbocation p orbital and sp^3 C–H or C–C bond orbitals lead to a larger stabilization than for most other bonds. The reasons for this are not needed at this stage. The more C–H or C–C bonds available, the greater the stabilization, so that $(CH_3)_3C^+$ is much more stable than CH_3^+. Remember, though, that neither of these are very stable. We will return to this point.

HYPERCONJUGATION

We refer to this type of stabilization as 'hyperconjugation'. We can draw resonance forms for this. We have the carbocation structure on the left with the positive charge on the central carbon atom. We have one C–H bond explicitly drawn in. We are using a curly arrow to show the movement of electrons. The arrow starts in the middle of the C–H bond, and goes to the middle of the C–C bond. This breaks the C–H bond and makes the C–C bond into a double bond.

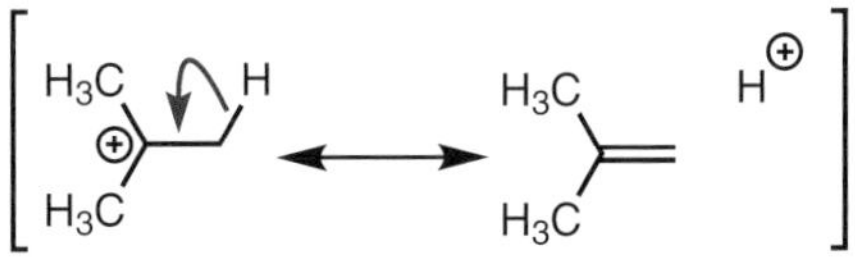

When I first learned this topic, I was not shown the molecular orbital representation, so I was very confused by being shown resonance forms where the C–H bond is broken, and then being told that it isn't being broken. Thankfully, we are in a position to link the resonance forms with the molecular orbital representation. The resonance forms are showing you that there is donation of electron density from the C–H bond to stabilize the carbocation.

You absolutely need to become comfortable with drawing resonance forms and to understand what they tell you about stability. After a while you will be able to visualize all possible resonance forms without drawing them. You'll know which one to draw to show a particular aspect of stability.

The only way you will reach this level is with practice. You won't get to this stage without going through the 'pain' of drawing them out over and over again. It's how you learn.

WHAT DO WE MEAN BY STABILITY?

We have looked at how the substituents attached to a carbocation affect its stability. But can we directly compare the stability of a methyl cation and a *t*-butyl cation?

We cannot make a direct comparison because they are not isomers. We ***can*** *compare the intrinsic stability (heat of formation) of a t-butyl cation and an n-butyl cation.*

That doesn't devalue the above discussion. We just need to add a frame of reference. We must compare the stability of carbocations to the stability of a suitable precursor. For now, we will consider carbocation formation from a generic alkyl halide, R–X.

$$R{-}X \longrightarrow R^{\oplus} + X^{\ominus}$$

Have a look at the reaction profile below, taken almost directly from **Basics 14**. I have removed some of the information, so we can focus on what really matters for now.

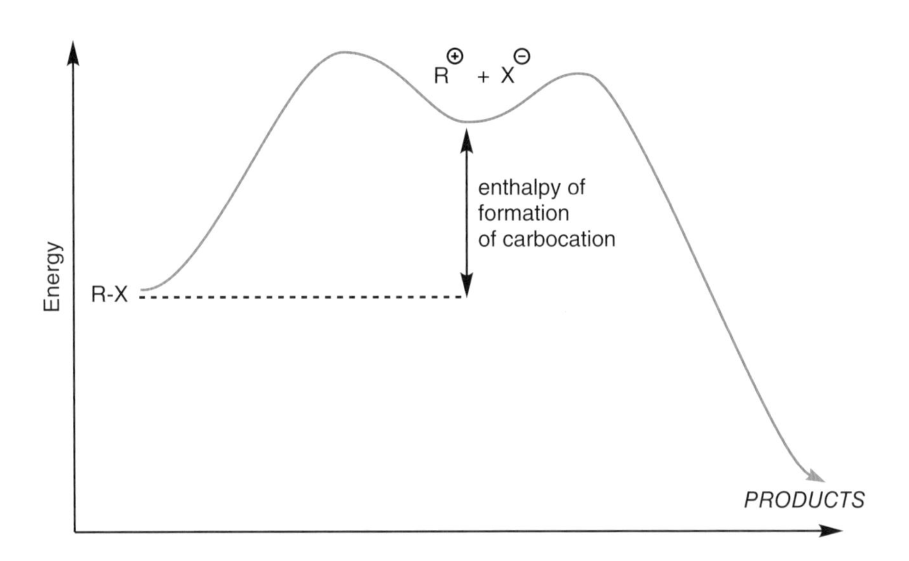

As long as we use the same 'X' group in all cases, then the stability of X^- won't change. Although I have included it, we can simply consider that we are looking at the stability of R^+. So, the energy difference I have highlighted above is the stability of the carbocation relative to the precursor.

The smaller the energy difference, the more stable the carbocation. But it's always uphill!

STERIC FACTORS IN CARBOCATION STABILITY

It's always nice to start with a basic fact.

Carbocations are planar. Alkyl halides are tetrahedral.

Here is a reaction. I have included wedges and dashes to give it a bit of shape. It is good to start getting used to these representations.

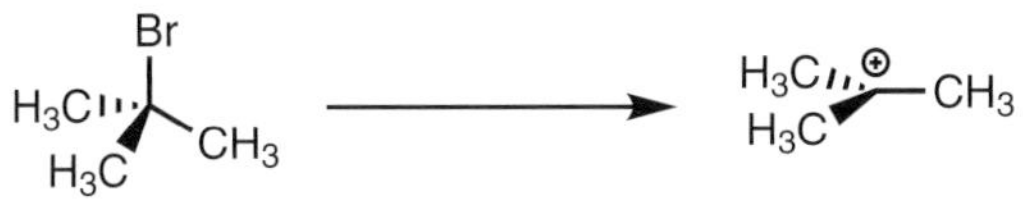

As we form the carbocation, we are breaking the C–Br bond, so we are going uphill energetically. Look at the reaction profile above.

At the same time, the methyl groups of the *t*-butyl group are moving further apart. This is good—we are relieving steric strain. We are therefore not moving uphill quite as far as we would without the steric effect. In *t*-BuBr, the methyl groups are compressed against one another. In terms of the reaction profile on the previous page, R–X is destabilized (higher in energy) as a result of steric strain. Therefore, the enthalpy of carbocation formation is lower.

So, we have an electronic effect in which the methyl groups stabilize the positive charge in the carbocation, and we also have a steric effect that destabilizes the alkyl halide precursor.[12]

SIMPLE CARBANIONS—STRUCTURE AND STABILITY

A carbanion is a species containing a carbon atom bearing a negative charge. The stability of carbanions with simple alkyl substituents follows the opposite trend to that seen with carbocations.

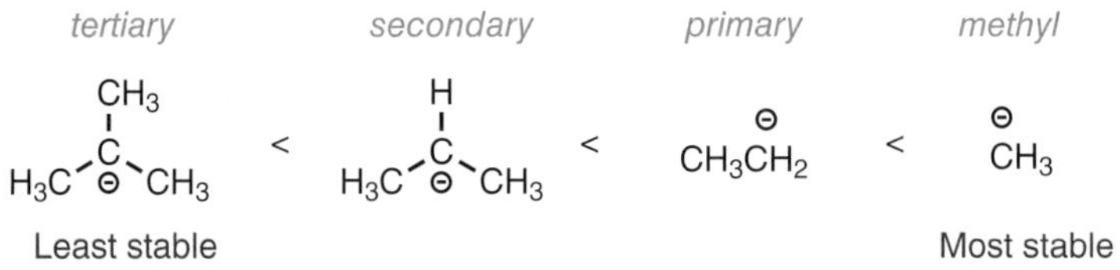

tertiary < *secondary* < *primary* < *methyl*

Least stable ... Most stable

[12] What if we make a carbocation by protonation of an alkene? Would we have the same steric effect?

This makes sense. If the extra methyl groups will stabilize a positive charge, they will destabilize a negative charge. We can draw a MO diagram for this, and now we see that we don't get the stabilization that we saw for carbocations. It isn't so easy to see from this diagram, but we actually get destabilization.

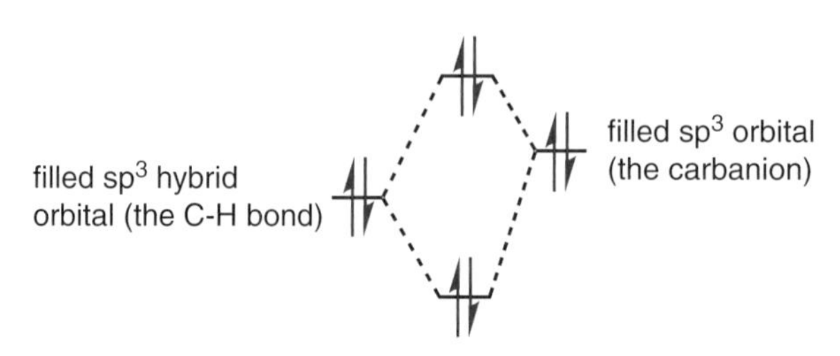

Simple carbanions are tetrahedral in shape. We have eight outer electrons, which are accommodated in four molecular orbitals. Two of these electrons are non-bonded. It's a bit like having a lone pair on carbon. These carbanions are tetrahedral in shape, but one of the positions is occupied by the non-bonded electrons, so it is more correct to say that they are trigonal pyramidal. In general, a carbanion will undergo rapid 'inversion' as shown below. For some types of carbanion, this is important, but for now we will simply note that this methyl carbanion is isoelectronic with ammonia (NH_3), and ammonia does the same thing.

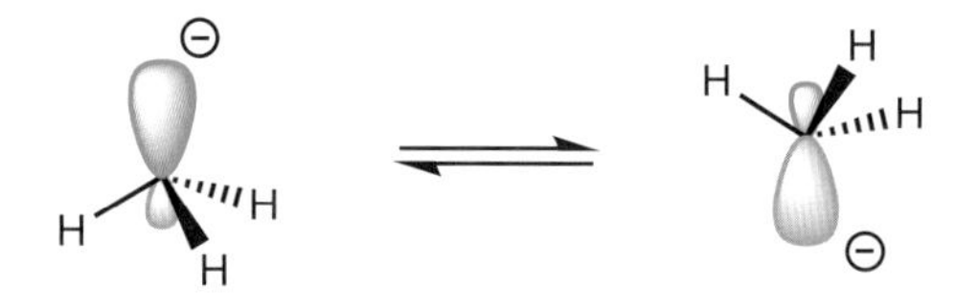

This is shape and shape is stereochemistry. We will come back to stereochemistry in Section 3.

STERIC FACTORS IN CARBANION STABILITY

We have just looked at steric effects in carbocation stability/formation. Carbanions are tetrahedral.

 Will increasing the number of alkyl substituents increase or decrease the stability of a carbanion **purely on steric grounds**?

Hopefully you concluded that increasing alkyl substitution will increase steric strain in the tetrahedral carbanion, so that a more substituted (with simple alkyl groups) carbanion is less stable on steric grounds as well as on electronic grounds.

FREE-RADICALS

Free-radicals are species with a single electron. They are useful synthetic intermediates and very important in biological systems. We aren't going to say too much about them at this point, other than to point out that for simple free-radicals, any feature that stabilizes a carbocation will also stabilize a free-radical, so that the trends in stability are identical.

tertiary		*secondary*		*primary*		*methyl*
$(CH_3)_3\dot{C}$	>	$(CH_3)_2\dot{C}H$	>	$CH_3\dot{C}H_2$	>	$\dot{C}H_3$
Most stable						Least stable

We represent a free-radical, as above, with a dot to indicate a single electron. Free-radicals are neutral, although it is possible to have radical anions and radical cations. Don't worry about these for now.

If we look at an orbital energy diagram that is similar to the one we drew for carbocations, we can see why the same trend would apply.

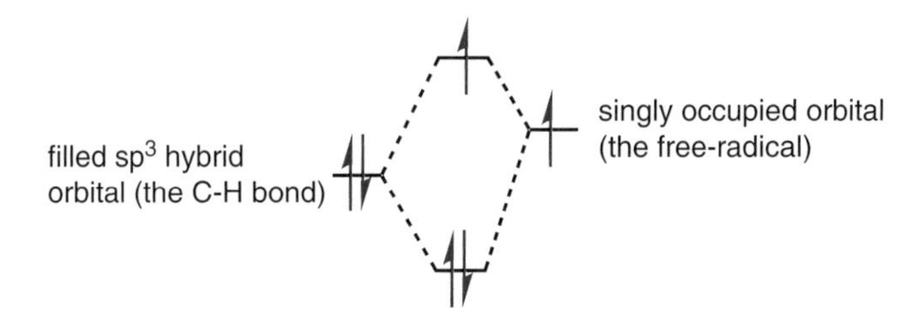

Here we have two electrons moving down in energy and one moving up. There is a net stabilization, so that the C–H bond in this case stabilizes the free-radical.

WHAT DO YOU NEED TO BE ABLE TO DO WITH THIS INFORMATION?

You need to be absolutely confident that you know and understand the stability trends. We have only considered the basics so far. We will add more different types of substituent and consider their effect on carbocation and carbanion stability. You need to have a framework into which to place the new information. When we start talking about reactivity, we will very often compare the rate of reaction of two different compounds, or of two different functional groups in one compound. These aspects of selectivity are fundamental to organic synthesis, and indeed to life.

We will come back to some of the problems in Common Error 5. However, it is worth getting some of the issues in now as well.

The trends for stability are different for carbocations and for carbanions. I tend to cover carbocation stability first, and students learn that 'more substituted = more stable'. This becomes a habit, so that when they see a series of carbanions, they claim that the more substituted carbanion is more stable. While the reasons for this error are understandable, the consequences are significant.

Let's add two more pieces of information. The first one is rather subtle.

ALL OF THIS APPLIES TO PARTIAL CHARGES

You won't fully understand the implications of this yet, but I want to sow the seeds of this idea. Almost every reaction you encounter will involve a redistribution of charge as we progress along the reaction coordinate. There will be a build-up of either positive or negative charge which will be stabilized or destabilized depending on the detailed structure. This level of (de)stabilization can often determine whether a reaction takes place at all, or which of several possible reactions will occur. If you don't understand carbocation and carbanion stability, you certainly won't recognize when a partial charge is stabilized.

WHEN WE SAY STABLE, WE DON'T REALLY MEAN STABLE!

In general, all carbocations, carbanions, and free-radicals are unstable. However, some of them are **less unstable** than others.

*When you see the word 'stable' you might assume that these species **really are** stable. This then leads to the following argument: 'If a particular carbocation is more stable, it will be less reactive. Therefore, the reaction that involves the most stable carbocation will be slowest.' This logic is flawed.*

Look back at the reaction profile we used to discuss carbocation stability. The final product is more stable than the reactant. If it wasn't, the reaction would not go in the direction we want.

There is an activation barrier for the formation of the carbocation, and an activation barrier for reaction of the carbocation. It would be very difficult to envisage a situation in which the barrier for carbocation reaction is higher than that for carbocation formation.

Whenever we are looking at the rate of a reaction involving a carbocation, the slow step is the carbocation formation. Therefore, if we compare two reactions, the fastest of these will be the one that involves the more stable carbocation, because this carbocation will form more quickly. Reaction of the carbocation, no matter how stable, will still be faster than its formation.

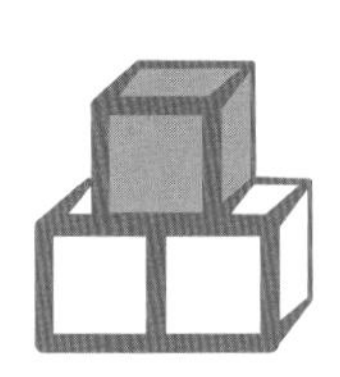

BASICS 17

Carbocations 2—More Factors Affecting Stability

In **Basics 16**, we introduced the idea that the number of alkyl groups attached to a carbocation carbon atom will significantly affect its stability. Of course, there are many different groups that we can attach to a carbocation carbon atom, and they will affect the stability of the carbocation in different ways.

We are now in a position to consider this aspect in more detail. Let's start with one absolute.

Carbocations are stabilized by electron-donating groups, and are destabilized by electron-withdrawing groups.

The important thing at this stage is to be able to look at any given carbocation and to systematically examine the substituents attached and determine whether each of them will stabilize or destabilize the carbocation.

Very often, students in the early stages develop a knee-jerk reaction such as 'more substituents = more stable', or 'that group has electrons so it must be electron-donating'. Both of these statements ***can*** *be true, but are* ***not necessarily*** *true. We need to replace this knee-jerk reaction with an equally quick and instinctive determination based on practice.*

PLANAR CARBOCATIONS ARE MORE STABLE THAN TETRAHEDRAL CARBOCATIONS

In **Basics 16**, we established that carbocations are planar. Let's explore this in more detail. You will notice in the heading to this section that I have made direct comparison with tetrahedral carbocations.

If you were to compare the energy of a planar carbocation with that of the otherwise identical tetrahedral carbocation, you would find that the planar carbocation is about 100 kJ mol^{-1} more stable. I'm not going to explain how we do this now,[13] and I don't think you should learn this number.

[13] We will look at this in more detail in **Perspective 2**.

*What you should learn is that **if** a carbocation cannot be planar, it will be considerably less stable.*

We have looked at curly arrows already, so it is appropriate to consider reactions that could lead to carbocation formation, as well as considering the carbocation itself. Here is a reaction that could hypothetically be used to form a *tertiary* carbocation. It turns out that this reaction is extremely slow, to the point that we can simply say 'it does not happen'. Why is this?

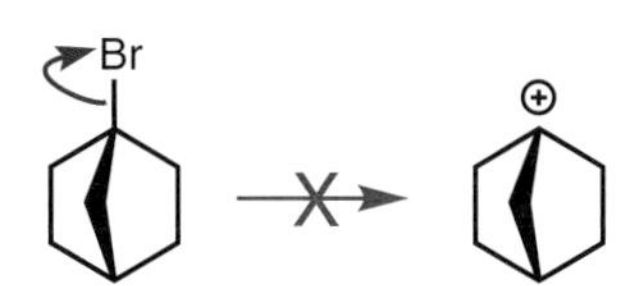

The reason is very simple. There is no way this carbocation could be planar. Therefore, it is less stable, to the point where it is not formed.

2.7

The simplest way to work this out is to make a molecular model. Make sure you have a set of plastic molecular models, and that you use them to help you see what shape things are. After a while, you'll find you no longer need to use them to work out the shape of things.

In this case, if you make a model of the above carbocation, and try to force the carbocation carbon to become planar, you will probably break the molecular model. It takes a considerable amount of force to make this carbon planar. It is a direct parallel with strain in the actual molecular structure. Once you convince yourself of this, you won't forget it as easily.

Let's look at some more substituents that we could attach to carbocations, and see what effect they have.

DOUBLE BONDS STABILIZING CARBOCATIONS

There are a number of other factors which lead to the stabilization of carbocations. However, when we examine them, we will see that they all have their origin in a similar effect. Fundamentally, anything that can release electron density towards the carbocation carbon atom will stabilize it.

Or we could put this another way—anything with a filled orbital that is of the correct symmetry and geometrically able to align with the empty p orbital will provide stabilization.

One place we could get electrons from is an alkene. The allyl cation shown below has two (identical) resonance forms. We can show them in square brackets with a double-headed arrow between them. The term 'allyl' refers to CH_2=CH−CH_2.

allyl bromide allyl cation

We can take this a stage further by using a benzene ring instead of just one double bond. The benzyl cation is shown below. This is stabilized by the electrons in the π orbitals of the benzene ring. We can represent this simply using resonance forms as shown below.

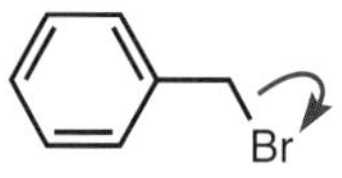

benzyl bromide

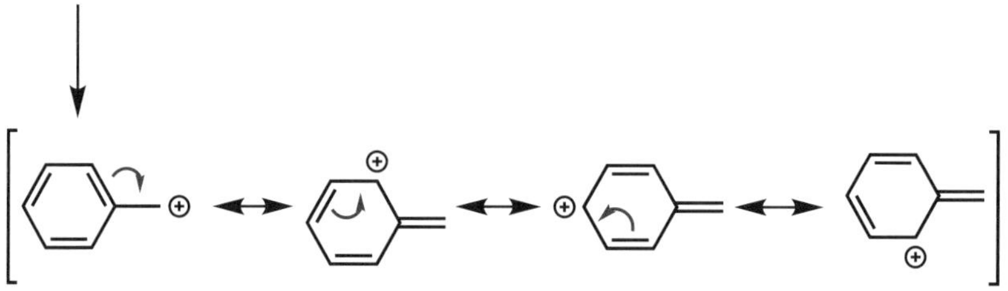

benzyl cation

There are a number of points which you should pay attention to when drawing resonance forms.

1. We use the double-headed arrow between the canonical forms to indicate resonance, and we place all the resonance forms in square brackets.
2. You cannot put the positive charge on any carbon atom—just some of them. Being able to visualize which without drawing them is a good indicator that you are making progress.
3. We don't talk about the cation 'resonating between the different structures'. It's important that you talk like a professional chemist. We use these simplified representations to indicate the distribution of charges, degree of double bonds, etc. The true structure, which in the above case has a partial positive charge on each of four carbon atoms, is described as a resonance hybrid of the above canonical forms.

We can show the stabilization as an orbital overlap, as seen below. This is directly analogous to the MO representation of hyperconjugation (**Basics 16**). We could draw something similar for the allyl cation.

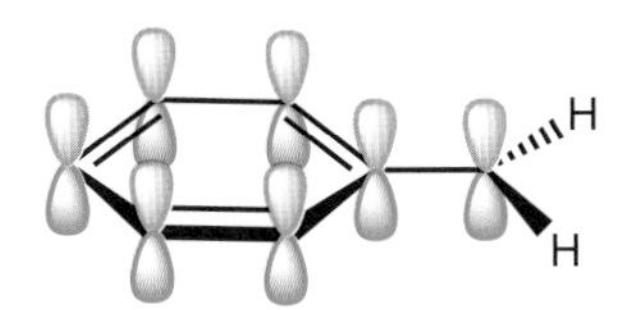

The net outcome of this is that the benzyl cation is almost as stable as $(CH_3)_3C^+$. One aromatic ring as a substituent provides comparable stabilization to nine C–H bonds (remember that not all nine C–H bonds in $(CH_3)_3C^+$ can be aligned to overlap with the empty p orbital at the same time).

You would be tempted to assume that the following two carbocations are even more stable, having two and three benzene rings respectively.

2.7

This is true. However, the third **phenyl** substituent provides considerably less stabilization than the first two. We can understand this when we look at the shape of this carbocation. Here is a 3D space-filling model of this carbocation. You can hopefully see that the **phenyl** rings are twisted slightly, so that the structure resembles a propeller. This twist makes the orbital overlap shown above less effective, so that you don't get the full stabilization from each **phenyl** substituent.[14]

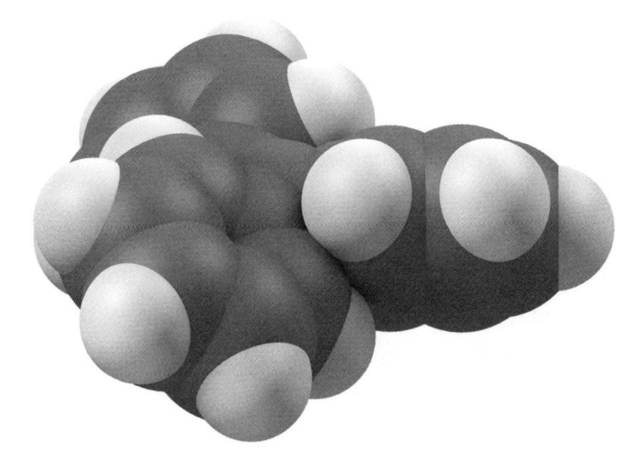

We will quantify this in Perspective 2.

[14] Reminding you that it is phenyl, not phenol. Just saying!

DOUBLE BONDS DESTABILIZING CARBOCATIONS

We mustn't get blinkered into thinking that a double bond will **always** stabilize a carbocation. Consider the following structure.

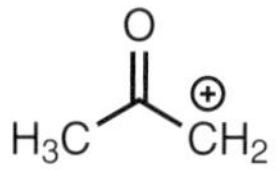

*I have an absolutely visceral reaction to seeing a positive charge on a carbon atom directly attached to a carbonyl group. It is **so wrong** that I actually find it hard to draw. The fact that I feel this way tells you that I have internalized my understanding of structure and stability. Let's look at **why** this is unstable, so that you also develop and internalize your understanding.*

Let's draw some resonance forms. We will start with the typical resonance forms for the carbonyl group.

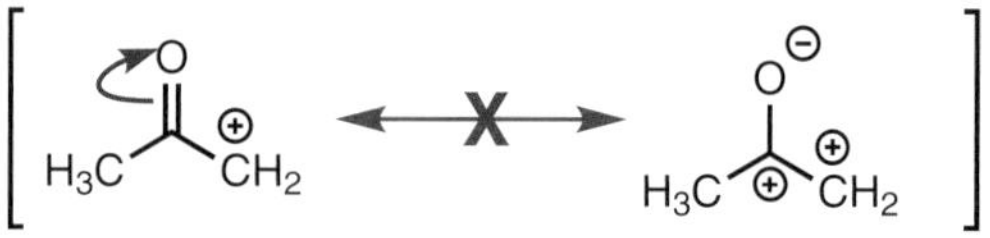

Here, we have a positive charge on two adjacent carbon atoms. This cannot be good!

Now let's consider resonance forms similar to those we drew for the allyl cation.

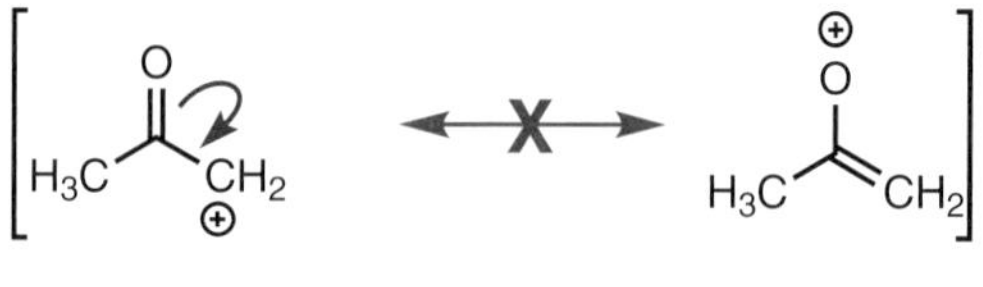

First of all, there is absolutely nothing wrong with this curly arrow. The problem is what it means.

Now we have a positive charge on oxygen. But it's not the sort of positive charge on oxygen that we are used to! Maybe that sounds a little strange. Usually, when we see a positive charge on an oxygen atom, the oxygen has three bonds and one lone pair, so it has a share in eight outer-shell electrons.

That is okay.

Here, we have an oxygen atom with only one bond and a positive charge. The positive charge is due to the 'removal' of two electrons, so that it only has a share in six outer-shell electrons. This is a disaster!

*You need to understand why this carbocation is unstable, so that you can identify other similar systems. This really does highlight the challenges associated with getting good at organic chemistry. You need to see **everything** in a structure.*

STABILIZATION OF CARBOCATIONS BY LONE PAIR ELECTRONS ON NEIGHBOURING ATOMS

Although chlorine is not a great leaving group, methoxymethyl chloride reacts very quickly in S_N1 substitution reactions (more on that later, but it is difficult to discuss carbocations without referring to their reactions). The stabilization of the cation is shown below.

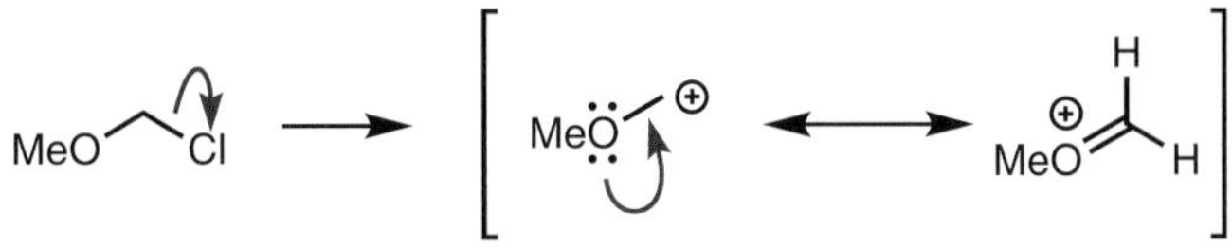

The resonance form on the right is closer to the truth (doesn't it look a bit like an aldehyde?). In terms of the curly arrows for the formation of this cation, it is probably best to draw the reaction as shown below. Think of it as the oxygen lone-pair electrons *pushing* the chlorine out of the molecule. The orbital interaction is similar to that with the C–H bond. Filled orbital overlaps with vacant orbital to give a net stabilization. The electrons are different, but the idea is the same.

How stable is this particular carbocation? It is a lot more stable than a *secondary* alkyl carbocation, but not quite as stable as a *tertiary* alkyl carbocation. We will quantify all of this in **Perspective 2**.

Oxygen isn't the only element with lone pair electrons. We could consider a similar situation with nitrogen.

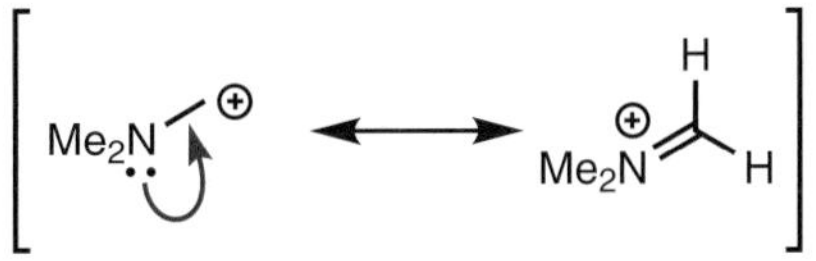

When assessing the stability of the carbocation, we need to ask a fundamental question—what is the difference between oxygen and nitrogen? Nitrogen is less electronegative than oxygen, so it will be happier sharing its lone pair electrons in this way. This carbocation is considerably more stable than a *tertiary* alkyl carbocation.

STABILIZATION OF CARBOCATIONS BY AROMATICITY

The final carbocation is a bit unusual, but relatively easy to understand. Tropylium bromide is a crystalline solid with a melting point of 203 °C. This is higher than might have been expected,[15] and it is because this compound is ionic. The tropylium cation is so stable that the bromide dissociates as shown below. The cation is a delocalized seven-membered ring system with three double bonds and a positive charge.

tropylium bromide
(1-bromocyclohepta-2,4,6-triene)

Draw the curly arrows and resonance forms.

We have already seen aromaticity and the molecular orbitals for benzene. Following from the definition of aromaticity in **Basics 10**, this cation is a $4n + 2\ \pi$ system with $n = 1$ (six π electrons). This cation is aromatic and is therefore very stable.

CAUTION

Have a look at the following four carbocations.

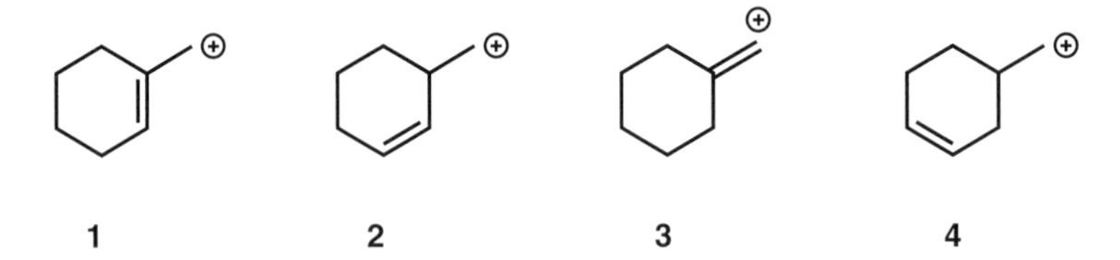

It is tempting to state that **1** is more stable than **2** because the double bond is **closer** to the positive charge. On this basis, you might then conclude that **3** will be even more stable because the double bond is even closer!

This reasoning is flawed.

[15] Maybe not yet, but after a while you get used to 'guessing' whether a compound will be a solid or a liquid. With practice, you will find you get 'luckier'!

Carbocation **1** is the most stable in this entire series because it is **allylic**—the molecular orbitals that make up the double bond can overlap with the carbocation empty p orbital to provide stabilization. This cannot happen with the other three carbocations, and in fact **3** is the least stable. Equally well, you might be tempted to suggest that **2** is more stable than **4** because the double bond is closer to the carbocation. In neither case can you get the orbital overlap mentioned above, so there is essentially no difference in stability between these carbocations.

When discussing stabilizing/destabilizing effects of substituents, don't use vague terms such as 'closer'. Instead, focus on the fundamental reason why the substituent stabilizes or destabilizes the system.

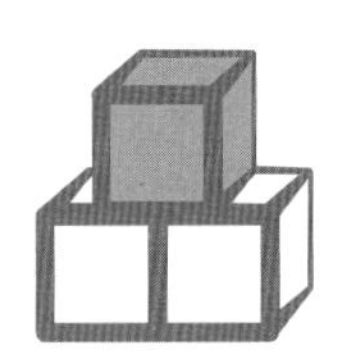

BASICS 18

Carbanions 2—Stability and pK_a

We are still at the stage of building patterns and habits, so I want to focus primarily on relating carbanions to other systems that you know, and build on the theme of drawing and understanding resonance forms.

We are going to discuss the stability of carbanions in the context of the acidity of the corresponding carbon atom. We use a scale called 'pK_a' to quantify this acidity/stability, so we will look at this first. Then we will look at some examples.

DEFINITION OF pK_a

Consider the following equilibrium.

$$HA \rightleftharpoons H^{\oplus} + A^{\ominus}$$

This equilibrium has equilibrium constant K_a which is defined as follows.

$$K_a = \frac{[H^+][A^-]}{[HA]}$$

The more acidic HA is, the larger the equilibrium constant, K_a, will be.

It turns out that the range of possible values is large, so it is convenient to define pK_a as

$$pK_a = -\log_{10} K_a$$

The 'negative' here means that a stronger acid has a lower pK_a than a weaker acid. Because this is a logarithmic scale, a compound with pK_a 19 is ten times as acidic as a compound with pK_a 20.

We need a reference point for pK_a values. Water has a pK_a of 15.7. This should be your reference point. Anything lower than this is acidic. Anything higher than this is not acidic.

*Don't confuse 'not acidic' with 'basic'. Methane has a very high pK_a, which we will look at in a moment. Methane is definitely **not** a base.*

WHY ARE CARBOXYLIC ACIDS ACIDIC?

Bear with me on this. A carboxylic acid isn't a carbanion, but there is no fundamental difference between a C minus and an O minus. The difference is the extent of stability.

Let's compare acetic acid and ethanol. Here are the equilibria for the two compounds. I have included pK_a values. Ethanol has very similar acidity to water. Acetic acid is 10^{11} times more acidic. Of course, acetic acid is not the strongest acid you will ever encounter.

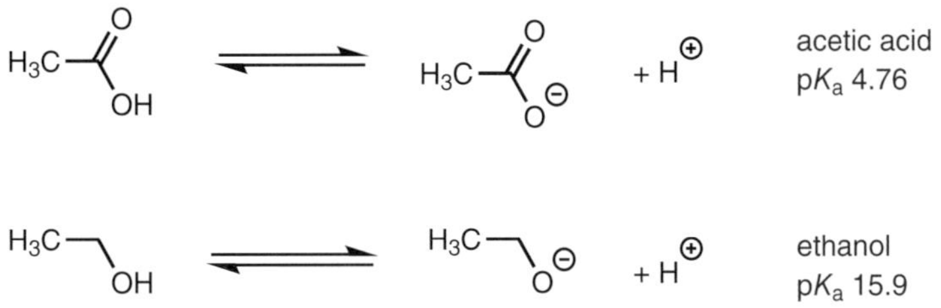

H_3C–COOH ⇌ H_3C–$COO^{\ominus}$ + $H^{\oplus}$ acetic acid pK_a 4.76

H_3C–CH_2OH ⇌ H_3C–$CH_2O^{\ominus}$ + $H^{\oplus}$ ethanol pK_a 15.9

The acidity of acetic acid can be explained by using the delocalized structure below right, showing that the negative charge is equally distributed between the two oxygen atoms, and there is effectively a 1.5 bond between the carbon and each oxygen atom. The resonance structures on the left show exactly the same thing.

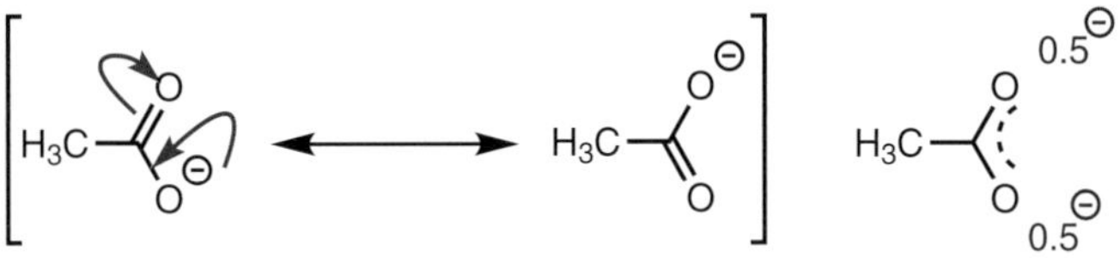

H_3C H_3C H_3C 0.5⊖ 0.5⊖

ANIONS α- TO CARBONYL GROUPS

It is routine in carbonyl chemistry to call the carbon atom adjacent to the carbonyl the α- (alpha) carbon.

Anions α- to carbonyl groups are stabilized. We have seen this above for an anion on oxygen.

We can compare propanone and propane in the same way as we compared acetic acid and ethanol.

propanone
pK_a 19

propane
pK_a 51

Propanone is a **lot** more acidic than propane—10^{32} times more acidic to be precise. To understand why this is, draw the resonance form for the anion. This is the same as we drew above, but let's just run through it one step at a time. The carbon atom will always retain a share in the pair of electrons that correspond to the anion. We are **sharing** them, not **giving** them. Therefore, the curly arrow goes from the negative charge to the C–C bond. This is sharing electrons with the carbonyl carbon. This carbon atom cannot accept any more electrons, unless it is able to get rid of a share in two electrons. It does this by moving the electrons in the C=O π bond towards the oxygen atom.

2.8

Why does this make sense? Because the oxygen atom is electronegative. It 'wants' those electrons! The structure on the right is formed by these two curly arrows. You are moving the negative charge into the C–C bond, so this becomes a double bond. You are moving two electrons from the C=O to the oxygen so this bond becomes a single bond, and the oxygen becomes negative.

Remember what we mean by these structures. We are using a double-headed arrow to indicate that this is resonance. The structure on the left isn't adequate on its own. The structure on the right also isn't adequate on its own. We need both of these structures, because what we are trying to do is to represent an arrangement of nuclei and electrons using letters and lines. It is never going to be perfect, so the important thing is to understand the limitations of structural representations.

What does this mean from a molecular orbital perspective?

The pair of electrons that represents the negative charge must be overlapping with an empty molecular orbital. The curly arrows tell us we are breaking the π bond of the carbonyl. It must be the π* orbital. Job done! Here is a suitable orbital energy diagram, and a representation of the orbital overlap.

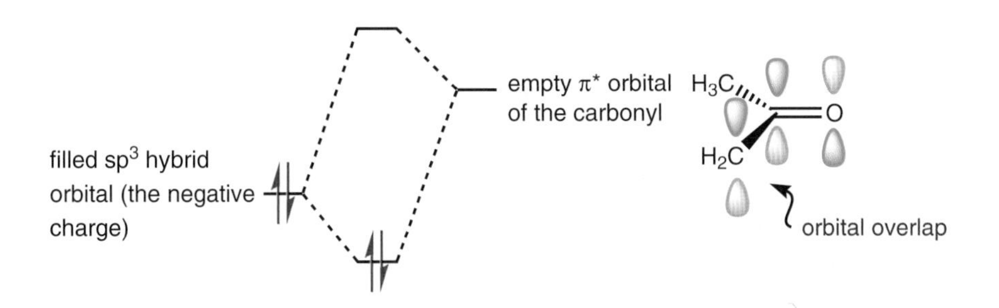

The difference in acidity between propane and propanone is much more than the difference between ethanol and acetic acid. With ethanol and acetic acid, every structure you draw for the anion will have a negative charge on oxygen. Oxygen is electronegative, so it is happy to bear a negative charge. With propanone, only one of the structures you draw has a negative charge on oxygen. Propane does not have an oxygen atom, so it cannot have a structure with a negative charge on oxygen!

The relatively low pK_a of propanone is due to the stability of the corresponding anion. The difference in pK_a between propane and propanone is due to the spectacular lack of acidity of propane.

ALTERNATIVE VIEW OF THE ACIDITY OF PROPANONE—INTRODUCTION TO ENOLS AND ENOLATES

Since we are able to draw the anion formed by deprotonation of propanone as an O minus with a C=C double bond, we could actually think of this being formed from the corresponding alcohol as shown below.

OH, H_3C, CH_2 ⇌ $O^{\ominus}$, H_3C, CH_2 + $H^{\oplus}$

an enol

If we do this, we have to think about it being formed by the deprotonation of the structure that I have labelled above as an **enol**. It's an **alcohol** on an **alkene**. It turns out that enols are really important, although we don't cover their chemistry in this book.[16]

I just want to add one more term—enolate. We have actually seen an enolate above. Here it is again.

[16] This is a very specific term. You cannot apply the term 'enol' to any compound that contains an alcohol and an alkene. The alcohol has to actually be attached to the alkene.

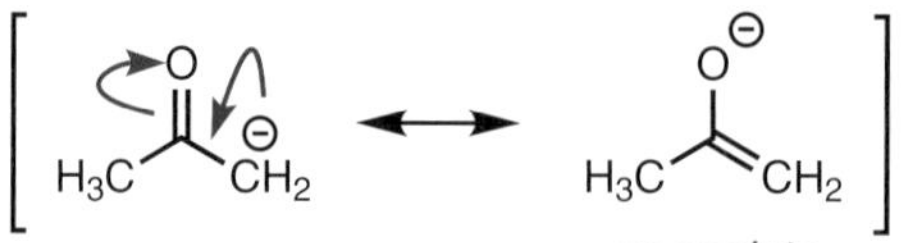

an enolate

Let's look at the term. The '**en**' bit refers to the alkene. The '**ol**' bit refers to an alcohol, although in this case we have removed the alcohol proton to leave a negative charge. Hence the '**ate**' bit of the name. Although I've written 'an enolate' below the structure on the right, because it more obviously applies to this structure, it applies equally to the structure on the left. After all, they are both representations of the same thing.

STRUCTURE AND SHAPE OF CARBANIONS

When we talked about carbocations, we got the shape of these out of the way at the start. This is because they are easier. The shape of carbanions depends on 'where the electrons are'. We have already seen that 'simple' alkyl carbanions are tetrahedral.

For the more stable carbanions, particularly where π bonding is involved, these can become planar. For example, with

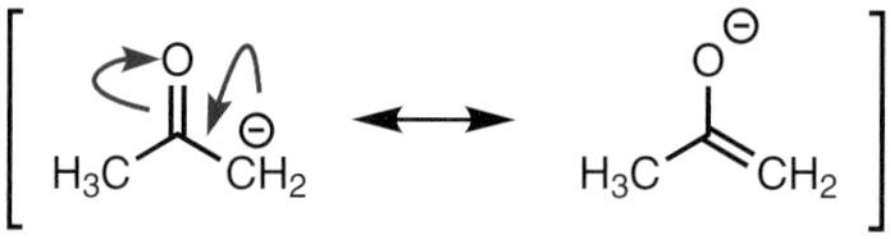

The overlap of the negative charge on C with the carbonyl bond (after all, this is what the resonance forms mean!) means that the structure on the right is closer to the truth. This structure has the carbon atom on the right looking like an alkene, and you would expect this to be planar.

At this point, the carbanion carbon (if it can be called that) has become sp^2 hybridized. There are two choices.

1. The carbanion remains sp^3 hybridized and the sp^3 orbital that constitutes the carbanion doesn't overlap effectively with the π^* orbital of the alkene.
2. The carbanion becomes sp^2 hybridized, and the p orbital that now constitutes the carbanion overlaps very effectively with the π^* orbital of the alkene.

Put like that, hopefully the decision is easy!

PUTTING pK_a VALUES INTO CONTEXT—HOW ACIDIC ARE HYDROCARBONS?

The pK_a value for methane is 48. This means that K_a is 10^{-48}. To put this another way, at equilibrium, one molecule of methane in 10^{48} will be ionized to give the methyl anion.

$$CH_4 \rightleftharpoons CH_3^{\ominus} + H^{\oplus}$$

One molecule of methane in 10^{48} is one molecule in 160 million million million tonnes.

To put this another way, if the Earth was made entirely of methane, this would be ***37 molecules*** *in the weight of the Earth.*

We have already seen that the methyl anion is the most stable of the 'simple' anions, so that methane is the most acidic 'simple' hydrocarbon. However, it really isn't very acidic at all!

A SELECTION OF pK_a VALUES

Here is a small selection of compounds with their respective approximate pK_a values. Remember that the lower the pK_a, the more acidic the hydrogen (indicated in red if there are more than one chemically distinct hydrogen in the molecule), and therefore the more stable the corresponding carbanion.

We have already looked at methane and propanone in detail. We can see that propanone (acetone) is 10^{29} times more acidic than methane. That is what a difference of 29 pK_a units tells us. Let's be really clear about this. That is a 1 followed by 29 zeroes. Write it out. I really want to be sure you understand the sort of numbers we are talking about. Big! Very big! Did you write it out? You have to join in!

As a slight digression, there is no point obsessing about whether the pK_a of methane is 46 or 48 or 50, even though these numbers span four pK_a units, corresponding to a difference in acidity of a factor of 10,000. Similarly, I have seen the pK_a of water quoted as 14 or 15 as well as the generally-accepted 15.7. It really doesn't matter! As long as you are in the right ball park, things will be fine!

Compound	pK_a
CH_4	48
$CH_2=CH-CH_2-CH=CH_2$ (H H)	32
$H_3C-C\equiv C-H$	25
$H_3C-C(=O)-OCH_2CH_3$	25
$H_3C-C(=O)-CH_3$	19

Compound	pK_a
H_2O	15.7
cyclopentadiene (CH_2, H H)	15
$H_3C-C(=O)-CH_2-C(=O)-CH_3$	13
H_3C-NO_2	10

Before we proceed, let's emphasize a point about strategy. Rather than asking 'why does compound X have pK_a Y?', it is always better to find something to compare it to, and ask 'why does compound X have a pK_a that is Y units lower than that of compound Z?'

So, let's compare ethyl ethanoate (**always** called ethyl acetate) with propanone (always called acetone). Ethyl acetate is 1,000,000 times less acidic. Actually, this isn't really a lot when we think of the scale we are using.

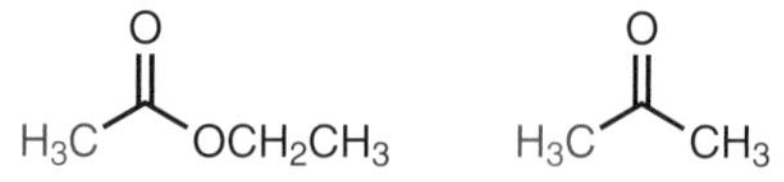

pK_a 25 pK_a 19

The key question is 'what does an OEt group do compared to a methyl group?' We should also remember that pK_a is related to an equilibrium constant for a reaction, shown below.

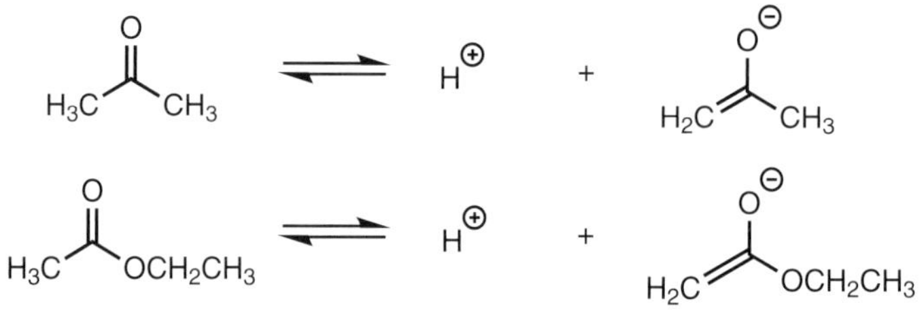

We have a smaller equilibrium constant for the second equilibrium than for the first. This means that the difference in stability between the species on the left-hand side of the equilibrium arrow and those on the right-hand side is greater in the lower case (ethyl acetate).

What we do not know (yet) is whether this is because the anion is less stable, or the ethyl acetate is more stable. It turns out that we can draw one additional resonance form for the carbanion as a result of the lone-pair electrons on the ester oxygen atom. This is shown below. However, a resonance form in which there are two negative charges and one positive charge is unlikely to contribute to stabilization. If anything, there might be some destabilization.

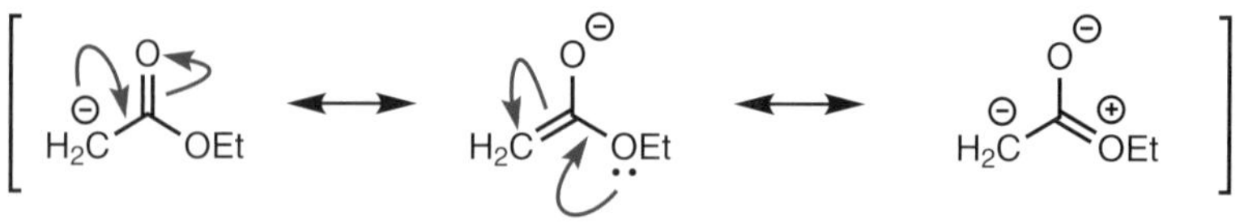

It is generally accepted that resonance stabilization of the ester, shown below, is the most important factor here. You cannot get this stabilization in the anion.

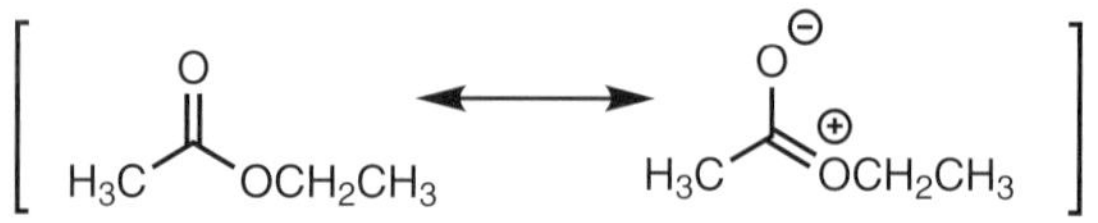

As always, what we have just done is much more important than the conclusion we were able to draw from it. Asking the right questions leads to the right answers.

Now let's consider pentane-1,3-dione. The pK_a of the hydrogen atoms shown in red below is about 13. This is 10 million times more acidic than the hydrogen atom in propanone. We have two carbonyl groups doing the same as the above. A pK_a of 13 is pretty acidic for a hydrogen atom attached to carbon. After all, water has a pK_a of 15.7, so this compound is more acidic than water.

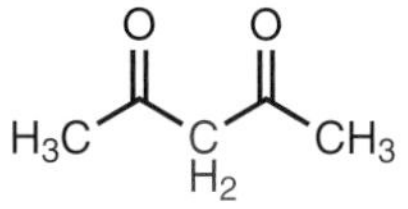

We can see an extreme case for stabilization in nitromethane, pK_a 10.

$$H_3C-NO_2$$

We need to see how to draw the nitro group fully.

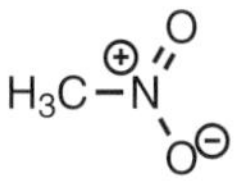

Now, when we form the corresponding anion, we can see that the carbon atom has three electronegative elements attached to it, which are going to stabilize the negative charge through an electron-withdrawing inductive effect (**Basics 5**).

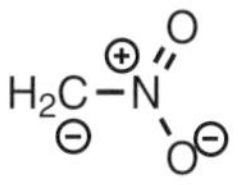

However, once again the mesomeric (resonance) effect is more significant.

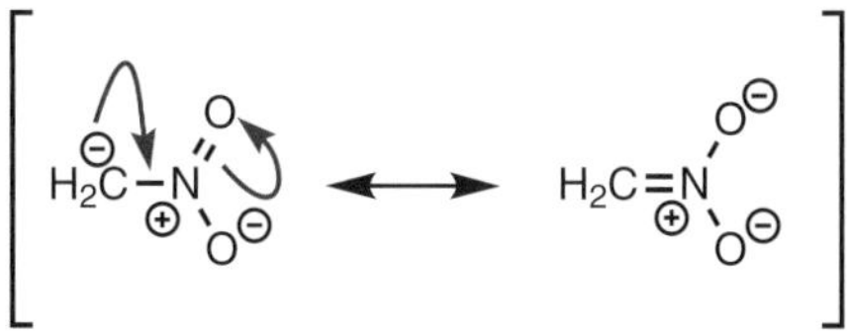

Here's the next one. It's a bit different.

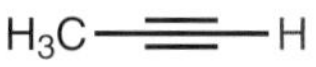

$$H_3C-C\equiv C-H$$

The hydrogen atom directly attached to the alkyne has a pK_a of about 25. This isn't very acidic compared to water, but it is very acidic compared to methane. In **Basics 6** we established that the C–H bond in an alkyne is stronger (about 550 kJ mol^{-1}) than the one in an alkane. In Basics 12 we hinted that it doesn't necessarily follow that a stronger bond is harder to break. Here is the proof.

We must consider the structure and the bonding in the carbanion, shown below. We cannot draw resonance forms to explain the stabilization of this anion.

H_3C—≡⊖

We can discuss the stability in terms of hybridization. The carbon atom is sp hybridized. Therefore, the C–H bond, and the corresponding anion, are both 50 per cent s in character. The s electrons are very tightly held by the nucleus. Therefore, the same factors that make the C–H bond stronger also make the carbanion more stable.

We have two left, and we are going to look at them side by side. We definitely need to make a comparison. Cyclopentadiene has a pK_a of 15. Penta-1,4-diene has a pK_a of 32. And yet both of these compounds have a CH_2 between two C=C double bonds. What is the difference?

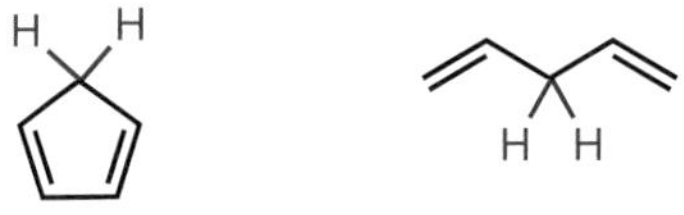

H H

H H

pK_a 15

pK_a 32

Well, let's follow our system and draw the anions. First of all, note that we have only removed one hydrogen atom, even though I explicitly drew two hydrogen atoms in the structures above and none in the structures below.

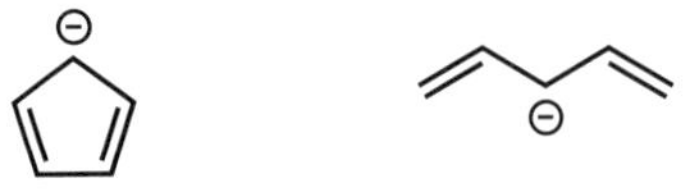

The one on the left is cyclic. It is also fully delocalized. This means we can draw resonance forms that place the negative charge on any of the carbon atoms. These are shown below.

Now let's count up the electrons. Not counting the σ bonds, we have two electrons in each π bond and two electrons in the negative charge (this might seem strange, but if we had one electron on this carbon atom, it would be neutral). So, we have a total of six electrons in a fully delocalized π system (Hückel rule, n = 1). It's **aromatic**. As we have seen before, this makes it **much** more stable.

We can still draw resonance forms for the **acyclic** (not cyclic) pentadienyl anion.

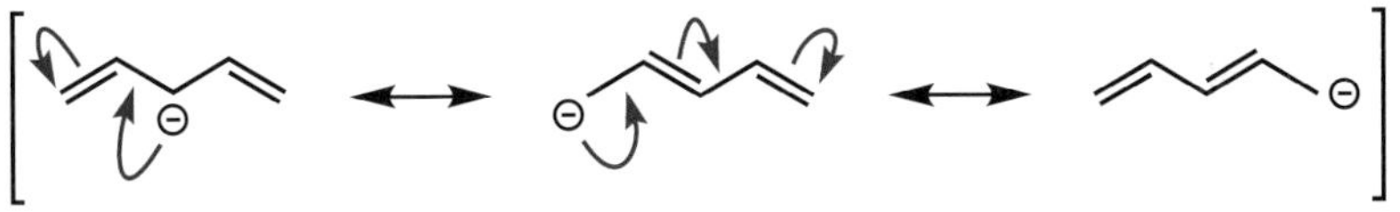

The amount of stabilization this provides is significant. After all, the pK_a is 32, which is **much** lower than that of methane (48). However, the aromatic stabilization provides a further 10^{17} (one hundred thousand million million!) times **more** stabilization! So, to conclude, the **cyclo**pentadienyl anion is much more stable than the **acyclic** pentadienyl anion, because the former is aromatic. This provides further stabilization.

When we looked at cyclohexanes, we established that not all six-membered rings are aromatic. Now we are seeing that not all aromatic rings are six-membered.

There is one further point that we need to make here. Some groups can stabilize positive **and** negative charges. The alkene is one such group.

We sometimes talk about the pK_a of the compound (*e.g.* methane) and we sometimes talk about the pK_a of the anion (*e.g.* methyl anion). We are generally talking about the same property—the acidity of the hydrocarbon and the stability of the anion are two aspects of the same thing.

We said that methane, with a pK_a of 48, is not a base. However, the methyl anion formed by dissociation of methane would be a spectacularly strong base.

PRACTICAL POINT—WHAT BASE TO USE IN REACTIONS?

Carbanions are typically produced by removal of a proton from a neutral organic molecule. This would use a base, so we need to consider which base would be appropriate for any given reaction. In order to do this, we need to have a quantifiable 'scale' of acidity/basicity.

Consider the following equilibrium, where 'HA' is the compound we are trying to deprotonate and 'B^-' is an anionic base.

$$HA + B^{\ominus} \rightleftharpoons A^{\ominus} + HB$$

If HA is more acidic than HB, then the equilibrium will be towards the right. If HA is less acidic than HB, the equilibrium will be towards the left.

Sometimes we need to fully deprotonate a compound—we need a base with a pK_a higher than that of the hydrogen we are trying to remove. Sometimes we only need to generate a small equilibrium concentration of the anion, in which case we can use a base with a lower pK_a, but not too much lower.

pK_a DEPENDS ON SOLVENT

There is one more thing you need to be aware of, although we won't get too bogged down by this. The pK_a value relates to a chemical equilibrium. This would be affected by solvent. Therefore, you may see different pK_a values quoted for the same compound in different solvents. Measurement of pK_a can be very complex, particularly for very weak acids.

PERSPECTIVE 2
A Scale for Carbocation Stability

Here is another 'Perspective' chapter. They tend to go a little bit beyond what is usually covered in textbooks at this level, so you can dip in and out. I like you to have the information, as long as it doesn't confuse you.

Let's identify the problem. We know how to compare the stability of methyl, ethyl, isopropyl, and t-butyl carbocations, as they all have the same substituent. But what provides more substitution, one phenyl or three methyls? We need to be able to quantify stability across a range of substituents.

HYDRIDE ION AFFINITY

When we wanted to quantify the stability of carbanions, we used the pK_a scale to measure the acidity of the corresponding hydrocarbon.

We can do something similar for carbocations. We use a scale called **hydride ion affinity**. We introduced this in Basics 12. To recap, this is defined as ΔG for the following reaction, and it can be measured by various methods. We are not going to worry about how we do the measurement.

$$\text{R-H} \rightleftharpoons \text{R}^{\oplus} + \text{H}^{\ominus}$$

Not surprisingly, the value obtained for a given carbocation depends on a number of parameters, with the main variable being solvent. We are going to avoid any such problems by using calculated data.[17] The data in the table show values which are normalized for the *t*-butyl cation having a value of zero. This means that any cation with a negative value is calculated to be more stable than the *t*-butyl cation, whereas a positive value indicates lower stability.

PLANAR OR TETRAHEDRAL CARBOCATIONS

First of all, comparing entries 1 and 2, or 4 and 10 shows that a tetrahedral carbocation is approximately 120 kJ mol^{-1} less stable than the otherwise identical planar structure. This calculation should be viewed with a little caution because of the methodology used

[17] For the interested reader, the calculations were done using Density Functional Theory (DFT) with the B3LYP hybrid functional and the 6-31+G* basis set. Calculations were carried out by the author using Spartan '10. While it is possible to use data already present in the chemical literature, the difficulty of comparing numbers from different sources makes it safer to calculate all data under identical conditions.

(constraining the H–C–H or C–C–C bond angles to the tetrahedral angle of 109.5°). Any deviation from this would produce slightly different numbers. The larger value for the tetrahedral *t*-Bu cation compared to the methyl cation is almost certainly due to steric reasons. The methyl groups are bulky, and making the *t*-Bu cation tetrahedral means pushing the methyl groups closer together. As always, we should question how much of the stability of the planar structure is steric and how much is electronic in nature.

We looked at a bridged bicyclic carbocation in **Basics 17**. Perhaps surprisingly, this cation (entry 7) is not as unstable as the hypothetical **tetrahedral** *t*-butyl cation because the bonds can distort a little.

Entry	Carbocation	Hydride Ion Affinity Relative to t-Bu$^+$ (kJ mol^{-1})
1	(tetrahedral) $\overset{\oplus}{C}H_3$	459
2	$\overset{\oplus}{C}H_3$	343
3	$H_3C-\overset{\oplus}{C}H_2$	163
4	(tetrahedral) $(CH_3)_3C^{\oplus}$	134
5	$H_2C{=}CH-\overset{\oplus}{C}H_2$	91
6	$H_3C-\overset{\oplus}{C}H-CH_3$	62
7	 bicyclic carbocation	58
8	$H_3CO-\overset{\oplus}{C}H_2$	32
9	$Ph-\overset{\oplus}{C}H_2$	11
10	$(CH_3)_3C^{\oplus}$	0 (reference point)
11	 bicyclic carbocation	–10

Entry	Carbocation	Hydride Ion Affinity Relative to t-Bu$^+$ (kJ mol^{-1})
12	$Ph_2CH^{\oplus}$	−93
13	$(H_3C)_2N{-}CH_2^{\oplus}$	−130
14	$Ph_3C^{\oplus}$	−158
15	$C_7H_7^{\oplus}$ (tropylium)	−172

AROMATIC CARBOCATIONS

We can quickly look at one carbocation at the other end of the scale. In **Basics 17** we encountered the tropylium cation (entry 15). Now we can see just how much more stable it is than the most stable 'simple' carbocation (*t*-Bu). It is 172 kJ mol^{-1} more stable! This is massive!

BENZYLIC CARBOCATIONS

Addition of one benzene ring (entry 2 vs 9) provides a great deal of stabilization, shown schematically below. Addition of a second benzene ring provides significant additional stabilization, and a third ring provides additional stabilization. However, there is a very clear trend that each additional benzene ring provides less stabilization than the one before.

$CH_3^{\oplus}$ (Entry 2) $\xrightarrow{-332\ kJ\ mol^{-1}}$ $PhCH_2^{\oplus}$ (Entry 9) $\xrightarrow{-104\ kJ\ mol^{-1}}$ $Ph_2CH^{\oplus}$ (Entry 12) $\xrightarrow{-65\ kJ\ mol^{-1}}$ $Ph_3C^{\oplus}$ (Entry 14)

It is relatively straightforward to explain this observation if we think about the structure of the cations. In order for the benzene rings to provide stabilization, the π orbitals of the benzene ring must be aligned with the vacant p orbital of the carbocation. This can happen quite easily for one benzene ring, but not so well for two or three benzene rings.

This is an **electronic effect**. We can draw resonance forms (see Basics 17 for an example of this, then practise it!) for the carbocation, which should convince you that this is a positive mesomeric effect.

When we have two or three benzene rings, they will be very crowded. This is a **steric effect**. However, the consequence of the steric effect is that the benzene rings will be twisted slightly (see below) which reduces the ability of the π electrons to overlap with the p orbital.

We can see this in a 3D representation of the triphenylmethyl carbocation. Each benzene ring is twisted approximately 34° from planarity. In this instance, a space-filling model makes it easier to see.

The diphenylmethyl carbocation is essentially planar, although there is a steric interaction which results in the two benzene rings being pushed apart, as shown below. It is apparently better to have some distortion of the bond angles but to retain π orbital overlap.

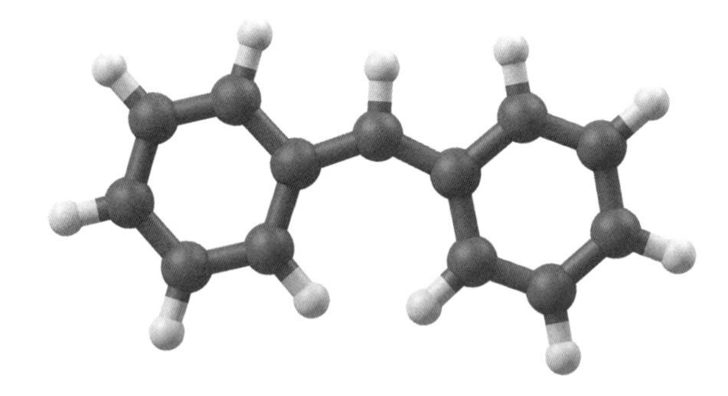

What we have here is a steric effect that has implications for an electronic effect.

Does this seem complicated? After a while, this will become your default way of thinking. Of course, underpinning all of this is the ability to visualize what a particular structure will look like.

We can also discuss the allyl cation (entry 5) at this point, and we note that it is not as stable as the benzyl cation (entry 9). A double bond provides considerably more stabilization than one methyl group (entry 3), but not as much as two methyl groups (entry 6).

You might like to extrapolate to the stability of a secondary or tertiary allylic carbocation. Make sure you understand what these terms refer to!

STABILIZATION BY LONE PAIRS

The following three compounds (entries 2, 8, and 13) allow us to compare **inductive** and **mesomeric** electronic effects.

$\overset{\oplus}{C}H_3$ —(−311 kJ mol^{-1})⟶ $H_3CO{-}\overset{\oplus}{C}H_2$ —(−162 kJ mol^{-1})⟶ $(H_3C)_2N{-}\overset{\oplus}{C}H_2$

Entry 2 Entry 8 Entry 13

For entries 8 and 13 we have an electronegative element attached to a carbon bearing a positive charge. The **inductive** effect is electron-withdrawing, and it will make the positive carbon even more positive. This would destabilize the carbocation.

However, the **mesomeric** effect will donate/share[18] lone pair electrons towards the positively charged carbon. This stabilizes the carbocation.

$[H_3C\ddot{O}{-}\overset{\oplus}{C}H_2 \longleftrightarrow H_3C\overset{\oplus}{O}{=}CH_2]$ $[(H_3C)_2\ddot{N}{-}\overset{\oplus}{C}H_2 \longleftrightarrow (H_3C)_2\overset{\oplus}{N}{=}CH_2]$

If we compare either of these substituents with a hydrogen atom (entry 2 versus entries 8 and 13), we find that the mesomeric effect wins.

*It usually does! The amount of stabilization is **huge**.*

It is also worth comparing the two substituents with each other. The dimethylamino group (entry 13) provides a massive amount of stabilization to the carbocation—more than three methyl groups and more than two benzene rings. The methoxy group provides quite a lot of stabilization, but not as much as a benzene ring. Nitrogen is less electronegative than oxygen, so provides more stabilization via this mesomeric effect.

We can conclude that the dimethylamino group has a stronger mesomeric electron-donating effect and a weaker inductive electron-withdrawing effect compared to the methoxy group. This makes perfect sense, since nitrogen is less electronegative than oxygen.

[18] Remember, the electrons are being shared, but it is common to 'slip up' and talk about the electrons being donated. It is important that you understand the deeper meaning!

NOW A LITTLE SURPRISE!

One very respected, older, textbook[19] quotes the relative rates of solvolysis of the following three compounds in 80 per cent aqueous ethanol (classic S_N1 conditions) as follows:

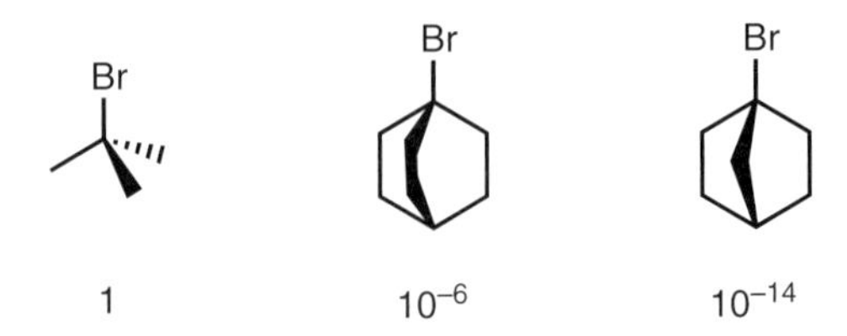

These data were interpreted as being due to the relative stabilities of the carbocations that needed to be formed. There are two aspects to this if we look at the structures of the cyclic carbocations in entries 7 and 11. Here they are again.

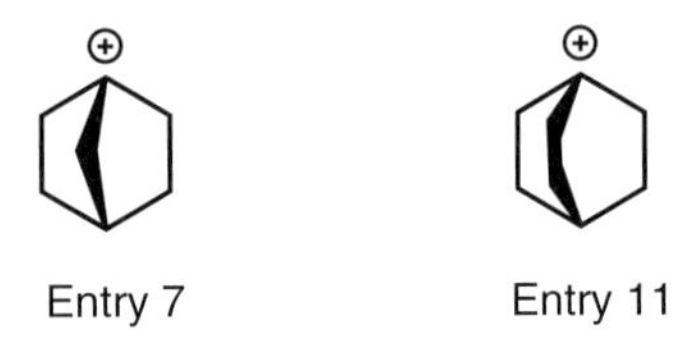

It is clear that neither of these carbocations can be planar, and we have already seen that a planar carbocation is significantly more stable than an otherwise-identical tetrahedral carbocation. The carbocation in entry 7 is considerably more strained, and less planar, than that in entry 11, so we shouldn't be surprised that it is quite a bit (68 kJ mol^{-1}) less stable.

Considering hyperconjugation as a major stabilizing factor, it is absolutely clear that there is no way that any of the C–H bonds in either of these carbocations can properly align with the carbocation p orbital.

Make a model! You'll only learn to visualize this from structures if you have seen it.

So, what's the problem?

The carbocation in entry 11 is actually slightly more stable than the planar t*-butyl cation.*

[19] I won't name the textbook, and I mean no criticism of the author. The data are correct, and the interpretation was considered to be accepted wisdom when the book was written. I show this example to indicate that interpretations **do** change when new data becomes available.

$-58\ \text{kJ mol}^{-1}$ $\Longrightarrow$ CH_3, H_3C, CH_3 $-10\ \text{kJ mol}^{-1}$ $\Longrightarrow$

Entry 7 Entry 10 Entry 11

These data are for stability of the carbocation in the gas phase relative to the corresponding alkane. It isn't easy to make a direct comparison of the stability of the carbocation in entry 10 with that in entry 11, since they have different molecular formulae. This is why we need the hydride ion affinity scale.

Why is the carbocation in entry 11 more stable than the t-butyl carbocation?

First of all, 10 kJ mol^{-1} is a small difference. We mustn't over-interpret it. The carbocation is considerably flatter than the above structure indicates. Hyperconjugation isn't restricted to C–H bonds. Three C–C bond orbitals in entry 11 are constrained to be perfectly aligned with the empty p orbital of the carbocation, and they provide stabilization in the same way.

So why is the solvolysis reaction to produce the carbocation in entry 11 so much slower?

We would be happy that entry 7 is less stable than the *t*-butyl carbocation, and therefore it should be formed more slowly. The carbocation in entry 11 is more stable, so you would expect solvolysis of the alkyl bromide above to proceed more rapidly than that of *t*-BuBr.

So, we have identified the key question, and the problem with the 'simple' answer—the carbocation is less stable (a reasonable conclusion, except it is **not**!).

We have three data points on which to base our understanding. If the rate of solvolysis is determined only by the stability of the carbocation, then formation of the carbocation in entry 11 would be faster than that of the *t*-butyl carbocation (entry 10). This is perfectly reasonable, as long as the intermediates are being formed by the same mechanism, and they are comparable.

*It is **not** faster—experimental results beat everything else!*

So, which of the three data points is the outlier?

Now we are getting somewhere! Much of the point of becoming a good scientist is to learn to ask the right questions.

It turns out that formation of the *t*-butyl carbocation is faster than expected. This is the outlier! The apparently slow formation of the carbocation in entry 11 is a red herring!

The current established wisdom (which actually appears not to be very well known, and it has been challenged) is that the formation of the *t*-butyl cation appears to be assisted by the nucleophilic solvent. This would suggest (at least to me)[20] to be something like the following.

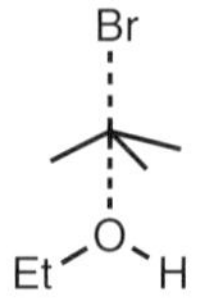

On this basis, it looks like there might be an element of S_N2-type reactivity even in (at least some!) S_N1 reactions.[21]

Of course, this will also apply to most other relatively stable tertiary carbocations, so you can see why the compound in entry 7 appeared, at first, to be the anomaly.

This example is complicated! If you didn't follow all of it, don't worry about it. You can come back to it later. I wanted to include it for two very specific reasons. First of all, this example, along with the incorrect explanation, is in a textbook that most chemists of my generation studied—I wanted to set the record straight. Secondly, and more personally, once I became aware of the problem (more recently than I am happy to admit!) I spent a lot of time finding out what was going on. I don't want that time to be wasted, and you can learn from my efforts.

[20] The paper in which this is first proposed did not give any structures to explain, so I have to make my own interpretation.

[21] Do **NOT** take this to mean that everything you will learn about the S_N1 mechanism is incorrect. Once the carbocation has formed, it can become planar and solvated prior to nucleophilic attack. Alternatively, just because this explanation is in peer-reviewed literature, it might be wrong. Don't forget that a scientific theory is a framework into which to put all the observed data. Another framework might ultimately be better, but we have to know what has gone before. We will see, in **Perspective 4**, that S_N1 and S_N2 are limiting cases in a continuum of mechanisms.

COMMON ERROR 3

Methyl Groups Are Electron-Releasing

WHAT'S THE PROBLEM?

So, here is the thinking.

A methyl group stabilizes a carbocation and destabilizes a carbanion. Therefore, methyl is electron-releasing.

Actually, it's fine up to that point, as long as we are talking about carbocations and carbanions. It just doesn't apply to stable neutral molecules. Let's see some data and an explanation.

We are going to discuss some data from NMR spectroscopy. Don't worry if you aren't too familiar with the technique yet.

Each hydrogen atom in a molecule can be represented by a chemical shift—the position of the peak in the ^{1}H NMR spectrum. The chemical shifts (in ppm) for the hydrogen atoms in red on the structures are indicated below the structure.

CH_4	H_3C-CH_3	$(H_3C)_2CH_2$	$(H_3C)_2CH-CH_3$	H_3C-OH
0.23	0.86	1.33	1.68	3.39

Let's start with a fact. We have methane on the left and methanol on the right. The reason why the chemical shift in methanol is so much higher is because the oxygen is electron-withdrawing.

Oxygen is electronegative. That's a fact that we cannot dispute.

Going from left to right in the figure above, we are sequentially replacing hydrogen atoms with methyl groups. If we replace one hydrogen with a methyl group, we get ethane, with a chemical shift of 0.86.

This implies that a methyl group is electron-withdrawing!

If we replace another hydrogen with a methyl group, and then a third hydrogen with a methyl group, we get two more increases. This reinforces the idea that a methyl group is electron-withdrawing.

Here is a selection of the electronegativity values that we saw in **Basics 5**.

Element	Electronegativity
C	2.55
H	2.20
N	3.04
O	3.44

Carbon is indeed more electronegative than hydrogen. We should expect a methyl group to be electron-withdrawing compared to hydrogen. The chemical shifts above should not be a surprise.

But a methyl group **can** donate electrons towards a carbocation. It's just not an inductive effect!

Of course, this isn't just about methyl groups. Nothing ever is!

Let's remind ourselves what an electronegative oxygen atom does to an adjacent carbocation.

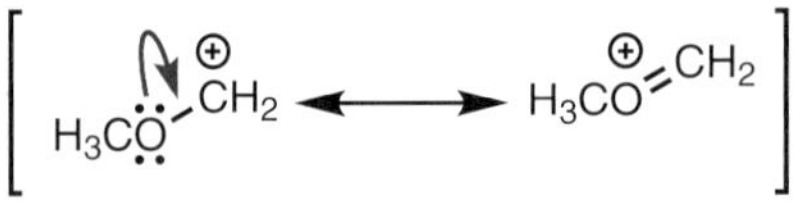

There is definitely an electronegative inductive effect. However, this is offset by the mesomeric electron-donating effect. We quantified this in **Perspective 2**.

2.9

SO, WHAT'S THE STORY?

A methyl group **is** electron-releasing, **if** it has somewhere to release the electrons to! The key question is '**which electrons?**' It is the electrons in the C–H bonds, and we are talking about orbital overlap.

*Go back to **Basics 16** and check that you understand hyperconjugation.*

But, if we are talking about a stable neutral molecule, the inductive effect is more important, and then a methyl group is electron-withdrawing.

Context is everything!

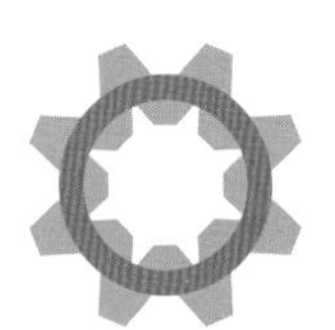

PRACTICE 8
Drawing Resonance Forms for Carbocations and Carbanions

THE PURPOSE OF THE EXERCISE

We have two very specific goals.

1. You should be able to draw correct resonance forms that show how a given substituent will 'interact' with a given charge.
2. With sufficient practice, you will then be able to visualize the most important resonance form of a structure instinctively.

__NOTE__ I am not going to 'test' you on 'simple' carbocations and carbanions with only alkyl groups. You __just__ need to know these inside out. Don't try to distinguish between a methyl group and an ethyl group. If you do this at this level, you will be over-thinking the problem.

We have seen resonance forms for carbocations and carbanions in Basics 17, Basics 18, and Perspective 2. We looked in quite a lot of detail at what various substituents do to the stability of an adjacent charge. Now we need to work on the use of resonance forms to emphasize what you should look for.

Resonance stabilization now needs to become just as natural to you. If you try to move forward without getting the hang of this, you will find that you are memorizing the effect of a substituent, or are using terms such as 'closer to' to describe the relative positioning of substituents (refer back to the end of __Basics 17__).

I am going to include solutions for some of these problems. Let's start with a bit of drawing, and then run through the answers. We will then finish with a bit more drawing.

QUESTION

For each of the following carbocations and carbanions, draw resonance forms, and hence consider whether the carbocation or carbanion is stabilized or destabilized.

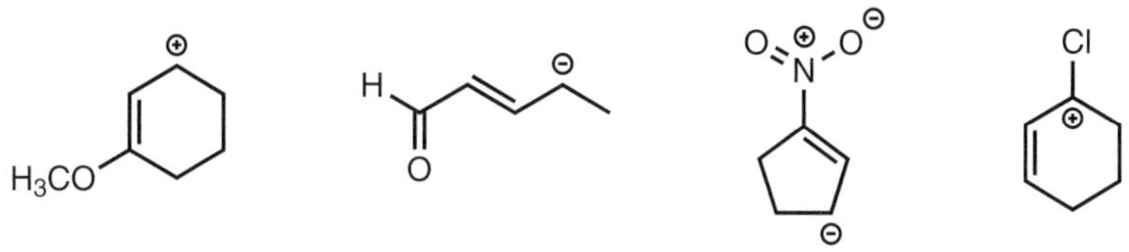

This is quite a 'vague' problem, and yet every good organic chemist will get the point straight away.

Stabilized or destabilized relative to what? I would compare to a carbocation/carbanion with the same number of hydrogen atoms, and every other substituent being a simple alkyl group.

WORKED SOLUTION 1

Let's work through the first one, and then we will add some context.

Make sure you have had a go at this before you continue reading. Perhaps you will make some mistakes and then understand what you did wrong. If you simply continue to the correct answer, all you will do is convince yourself that you would have got it right!

As drawn, the structure is a *secondary* carbocation. It has two substituents and one hydrogen atom attached.

The positive charge can be stabilized by the alkene double bond. We saw this effect in Basics 17. We looked at the resonance forms and also the molecular orbitals. Here are the resonance forms.

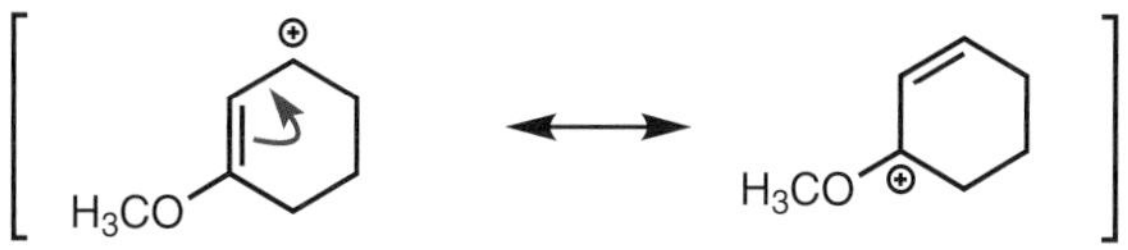

Let's just remind ourselves that the curly arrow starts at the double bond **because that's where the electrons are!** It goes to the adjacent single bond, **not** to the positive charge. We are **sharing** electrons with the carbon, not **giving** electrons to the carbon.

Lots of people make this mistake at first. It may seem like only a small difference—pushing the curly arrow a bit too far. But if you make the effort to get this right, the other stuff will be right as well. Don't think I am being overly pedantic or tell yourself that it doesn't matter. It does!

Now draw the orbitals involved. It is important to connect the resonance forms you draw with the molecular orbitals involved in providing the stabilization.

Look back at ***Basics 17****. You will find the answers there, not here.*

Now let's look for more resonance forms. We have a methoxy substituent. The oxygen atom has a lone pair. We can use it as follows.

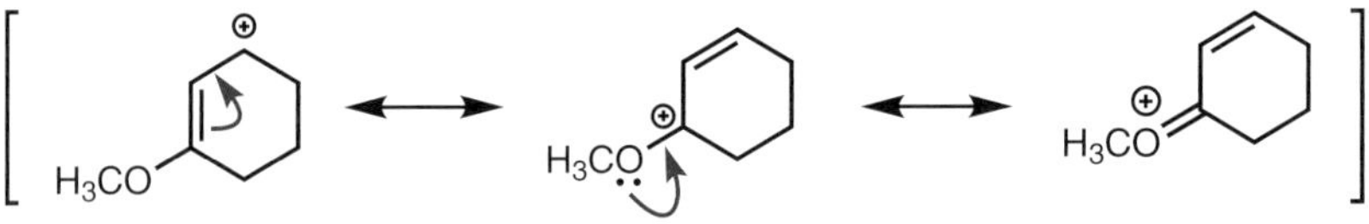

We have already seen that this is a significant stabilizing effect. I would go straight to the resonance form on the right. This is the best representation of this particular carbocation.

When we draw it like this, it isn't even a carbocation. But it is definitely stabilized!

WORKED SOLUTION 2

We aren't going to do this one. Look back at the previous problem and at **Basics 18** and ensure you make the connections. If you struggle, come back to this after looking at the next one.

WORKED SOLUTION 3

Much of this comes down to knowing how to draw a nitro group[22] so that you can draw the resonance forms. Let's start with a relatively simple resonance form for this struc-

[22] We are taking it as read that you know the functional groups, or that you will look them up when you need to.

ture. This shows that the nitro group itself is delocalized. But it does nothing for the stability of the carbanion. It's not that it is wrong. It just doesn't help.

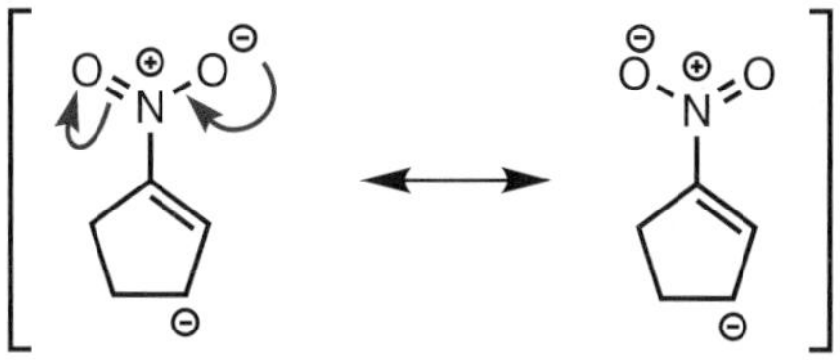

Now let's take a little leap and show you where you need to get to.

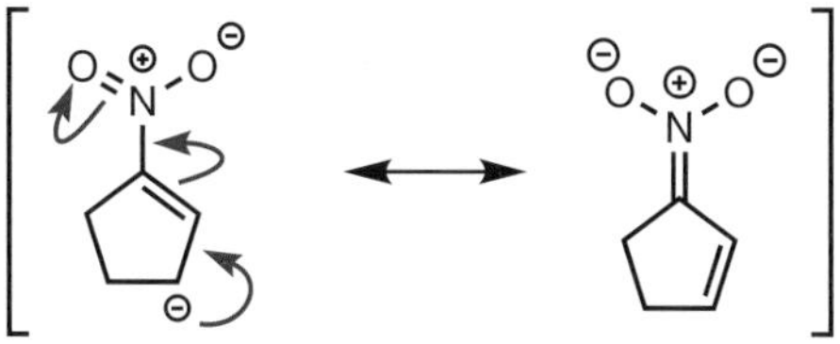

In this case, we share the negative charge on carbon all the way out to the oxygen atom.

Note that only one curly arrow goes to an atom. The others go to bonds. Make sure you can understand the difference.

There is an intermediate stage between the two structures above. Draw it out.

Do you need to draw this? Probably not. Does it add anything to the discussion? Again, probably not!

This leads us to a key question. How do you know when you have found all of the resonance forms?

For now, just start with a curly arrow at the (in this case) negative charge and do the minimum. Then take the resulting structure and again do the minimum. Keep going until you 'run out of molecule'. The current example is complicated by the fact that we have two negative charges that we could start with. Here is the full answer.

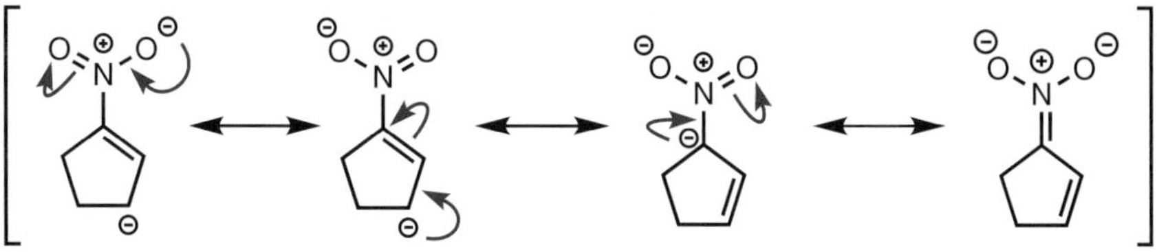

There is another answer to this key question. With enough practice, you will 'just know'. You will probably think this is a rubbish answer right up until the point where you 'get it'. Trust me on this, and keep practising.

WORKED SOLUTION 4

Here, we are only going to look at one aspect. We will take the double bond as read! Let's focus on chlorine.

Chlorine is in group VII of the periodic table. It is electronegative, and it has three lone pairs. We don't often draw the lone pairs on chlorine.

There is a good reason for that, which is where this is heading!

We can draw exactly the same resonance forms for chlorine stabilizing a positive charge as we draw for oxygen. Here it is.

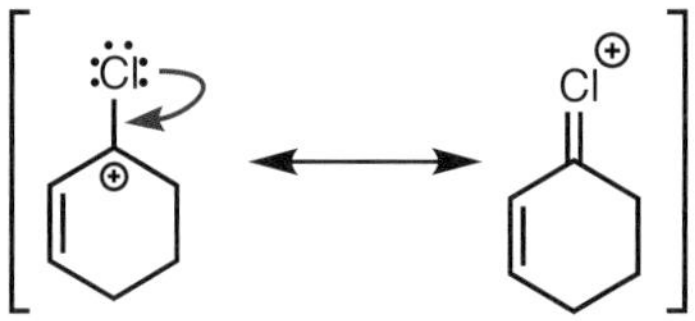

There is nothing fundamentally wrong with the resonance structure on the right. Yet somehow it 'looks' less correct compared to the similar structure with oxygen (**WORKED SOLUTION 1** above). Let's explore the reasons for this.

First of all, it 'looks' less correct because you will see it far less frequently. You will rapidly get used to seeing the similar structures with oxygen, and as a result, you will become more comfortable with them.

This is a good thing. You should learn to trust your instincts![23]

The reason why we see this less frequently is due to the orbitals involved. The positive charge on carbon is a 2p orbital. Lone pairs on oxygen are in the '2' shell. Lone pairs on chlorine are in the '3' shell. We get much more effective orbital overlap from an oxygen lone pair.

Remember, this is what the resonance forms are representing. Never lose sight of this. It is very easy to draw resonance forms without really thinking about it.

[23] Oh no! I'm turning into Yoda!

Using the same measure as in **Perspective 2**, calculated hydride ion affinity, the chloromethyl carbocation ($ClCH_2^+$) is over 200 kJ mol^{-1} less stable that the *t*-butyl carbocation. This places it between the methyl carbocation and the ethyl carbocation in terms of stability.

The Cl substituent is destabilizing compared to a methyl group. Overall, the inductive effect wins over the mesomeric effect and Cl is electron-withdrawing.

The other good comparison to make is with a methoxy group. Chlorine and oxygen have almost identical electronegativity, so any difference has to be due to the **mesomeric** effect. The chloromethyl carbocation ($ClCH_2^+$) is approximately 170 kJ mol^{-1} less stable than the methoxymethyl carbocation ($MeOCH_2^+$).

THE MOST COMMON MISTAKE

Five valent carbon! You draw a curly arrow going towards a carbon atom, but the carbon cannot accept any more electrons. We looked at this in **Common Error 2**. Learn to identify this. The good news is that once you get the hang of this, you'll stop making this mistake.

MORE PROBLEMS TO PRACTISE

Here are some more. After a while they get boring. This is a good thing! But it shouldn't stop you finding more examples to practise with.

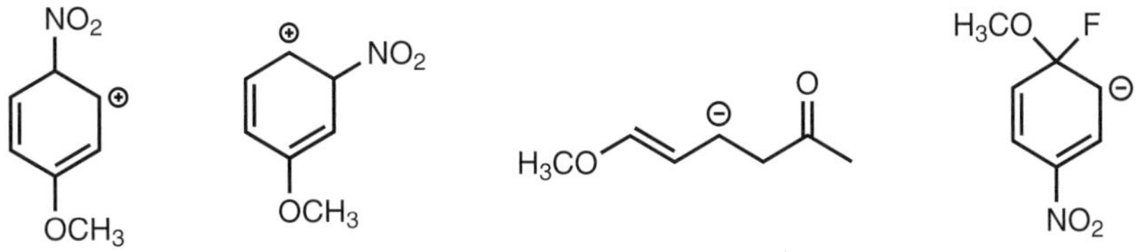

A couple of these problems every week and you will have this cracked!

If you don't practise these regularly, you'll never reach the point where you can just do it without thinking!

2.10

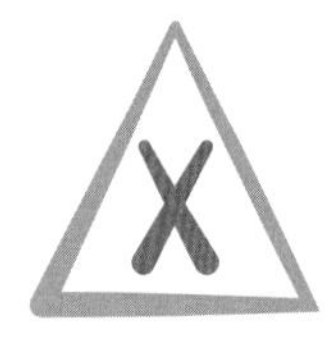

COMMON ERROR 4

Resonance

4

WHAT DOES IT MEAN?

The first common error is a fundamental misunderstanding of what resonance structures mean. I sometimes hear students talk about a compound 'resonating between two structures'.

This is understandable. The word 'resonance' already has meaning for you.

Have another look at the structure of benzene. Each individual structure doesn't show that all of the ring bonds are equivalent. We need both resonance forms to make this clear. **There is definitely no rapid interconversion between the two structures. The structure is the structure!**

Here is an interesting carbocation, shown as the two resonance forms **1** and **2**. Where would a nucleophile attack?
The (flawed) logic goes something like this: '*Structure* **1** *is a primary allylic carbocation.*

Structure **2** *is a tertiary allylic carbocation. Tertiary carbocations are more stable. Therefore, the positive charge is on carbon 3. This is where the nucleophile attacks.'*

Let's unpick this. First of all, the positive charge is delocalized between carbons 1 and 3. You would certainly expect carbon 3 to have more positive charge than carbon 1, because it is indeed better able to stabilize that charge.

Can you see the problem? The correct reasoning is only very slightly different to the flawed reasoning.

Even if you draw structure **2**, you can still draw perfectly good curly arrows for attack at carbon 1.

2

It turns out that this product is more stable than the alternative.

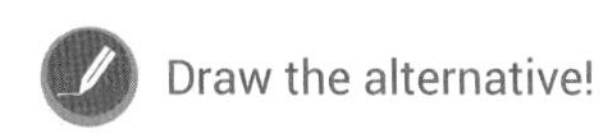

Draw the alternative!

The double bond is more highly substituted (read ahead to **Basics 34**, or wait until you get there), and the nucleophile experiences less crowding when it attacks carbon 1.

You have to adjust your thinking so that resonance forms represent a single structure that reacts in a defined way. It takes time to do this, but after a while, you won't know any other way to think about resonance.

ERROR: A METHOXY GROUP IS ONLY ELECTRON-WITHDRAWING

We already did this in **Basics 10**, but it is worth repeating. Have another look at **Common Error 3** as the principles are the same.

When we have a structure where we can draw resonance forms, for example **3** or **4**, then a methoxy group is electron-donating.

3

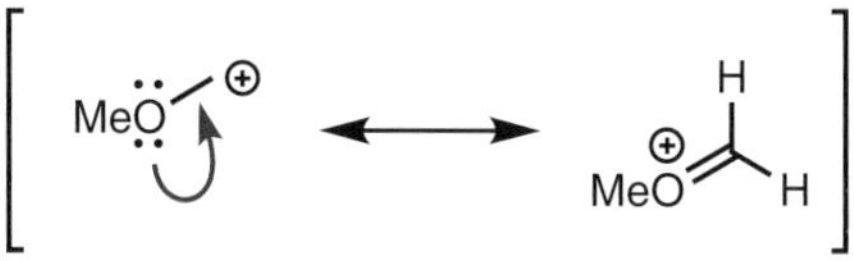

4

Of course, if there is nowhere for the electrons to go, then the methoxy group is electron-withdrawing—you can only get an **inductive** effect.

There is still an inductive electron-withdrawing effect in **3** and in **4**, but the mesomeric electron-donating effect wins!

ERROR: A CARBONYL GROUP IS ELECTRON-DONATING

We did this as well in Basics 10. Following from the logic above, an oxygen atom has a lone pair, and lone pairs can be shared to stabilize positive charges. **Therefore, anything with a lone pair is electron-donating!**

Of course, this is flawed reasoning. If you are going to argue that a substituent is mesomerically electron-donating, you have to be able to do one of the following two things:

1. Draw resonance forms that show the donation of electrons.
2. Draw diagrams that show the overlap of the relevant filled orbital with a suitable empty orbital.

In effect, these are the same thing.

Try to draw any resonance form for structure **5** that has correct curly arrows and shows donation of oxygen lone-pair electrons to the alkene double bond.

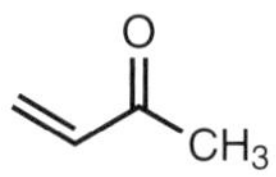

O
CH_3

5

Trying to do this is the best way to convince yourself that it cannot be done.

Of course, you could draw the following to represent sharing of the carbonyl bond electrons, but this leaves an oxygen atom with six outer-shell electrons and a positive charge. **It's just wrong!**

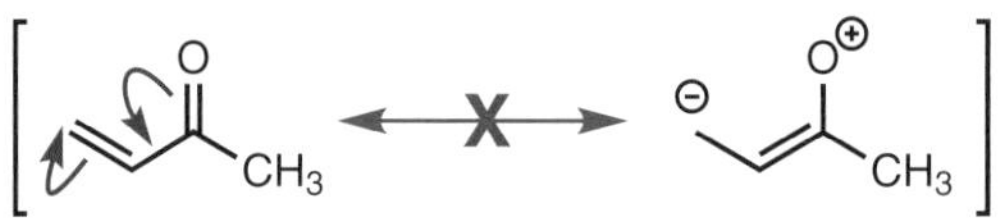

O CH_3 ←X→ ⊖ O⊕ CH_3

BASICS 19

The Hammond Postulate

THE HAMMOND POSTULATE—AN EXPLANATION

The Hammond postulate is a truly elegant and simple idea. When I was taught it, I was taught the words. I don't think I really understood the implications of them. Let's start with the words anyway.

> *'If two states, as, for example, a transition state and an unstable intermediate, occur consecutively during a reaction process and have nearly the same energy content, their interconversion will involve only a small reorganization of the molecular structures.'*

Now let's look at what this really means. Here is a reaction profile with an intermediate. It could be an S_N1 substitution (we saw this in **Fundamental Reaction Type 1**) but it could be one of any number of other processes. It doesn't matter for now.

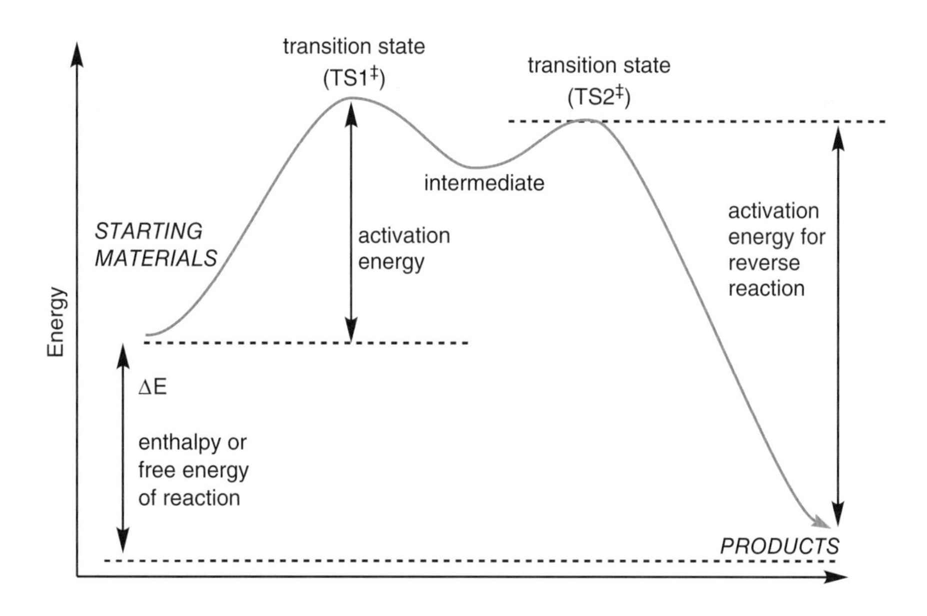

Both transition states TS1‡ and TS2‡ are much closer in energy to the intermediate than they are to starting materials or products. So, there will only be a small reorganization of the molecular structures in going from TS1‡ to the intermediate to TS2‡.

But what does that **really** mean?

If we know the structure of the intermediate, we can examine the factors that will stabilize it. For example, in an S_N1 substitution the intermediate is a carbocation, and we have seen the factors that lead to their stabilization or destabilization.

Now here's the beautiful bit. If there is only a small reorganization of the molecular structure between the intermediate and either TS1‡ or TS2‡, then anything that stabilizes the intermediate will also stabilize TS1‡ and TS2‡. Now we are getting somewhere!

There's more! If we stabilize the transition state relative to the starting materials, we will lower the activation energy. Therefore, the reaction will be faster.

So, the Hammond postulate is a really neat idea that allows us to look at the structure of an intermediate, and the reaction energy profile, and make meaningful predictions about the likely rate of reaction.

Let's compare two hypothetical S_N1 reactions. We haven't seen this in detail, but the key point is that the rate-determining (slow) step is formation of a carbocation.[24]

A $(CH_3)_3C{-}I \longrightarrow (CH_3)_3C^+ \longrightarrow (CH_3)_3C{-}OCH_3$

B $(CH_3)_2CH{-}I \longrightarrow (CH_3)_2CH^+ \longrightarrow (CH_3)_2CH{-}OCH_3$

Reaction **A** and reaction **B** differ only by one methyl group. With one fewer methyl group in reaction **B**, the carbocation intermediate in this case is less stable. But according to the Hammond postulate, the transition state will also be less stable (higher in energy).

[24] There is a bit of a problem with this example from a practical sense. Under the conditions where you would form the carbocation, you would not have methoxide present. The actual nucleophile would need to be methanol, and you would then need to lose a proton. I am simplifying the example so that the point will be clearer.

Therefore, reaction **B** has a higher activation energy than reaction **A**. Since the activation energy determines the rate of reaction, reaction **A** will be faster than reaction **B**.

A GENERALIZATION

The formation of the carbocation intermediate from reactants is **endothermic**. Since the transition state is higher in energy than either the starting materials or the carbocation intermediate, it will be closer in energy to the carbocation intermediate. The Hammond postulate therefore tells us that it will be closer in structure to the carbocation intermediate.

Endothermic reactions tend to have 'late' (product-like) transition states. Any factor that stabilizes the product (which is in this case the carbocation) will stabilize the transition state, and it will therefore accelerate the reaction.

The formation of the product from the carbocation intermediate is **exothermic**. Since the transition state is higher in energy than either the carbocation intermediate or the products, it will be closer in energy to the carbocation intermediate. The Hammond postulate therefore tells us that it will be closer in structure to the carbocation intermediate.

Exothermic reactions tend to have 'early' (reactant-like) transition states. Any factor that stabilizes the reactant (which in this case is the carbocation intermediate) will stabilize the transition state, and it will therefore accelerate the reaction.

The Hammond postulate is our link between the energies of stable structures and reactive intermediates (which we understand) and transition states (which are harder to understand). Intuitively applying the Hammond postulate to all reactions will greatly aid your understanding of the rates of reactions.

CAUTIONARY NOTE

It is possible to get the same outcome from a reaction by different mechanisms. We can only apply the Hammond postulate in this way to compare the rates of two reactions proceeding through the same mechanism. Reaction **B** would not take place via this mechanism because the carbocation intermediate is too unstable. In this case there is another possible mechanism (S_N2) that is more favourable. Be careful you don't try to compare different processes.

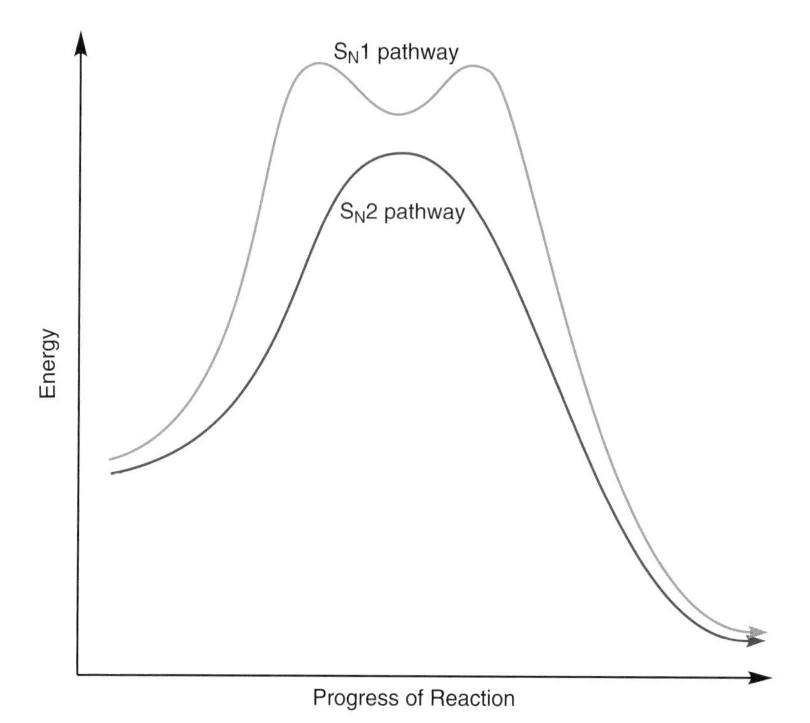

2.11

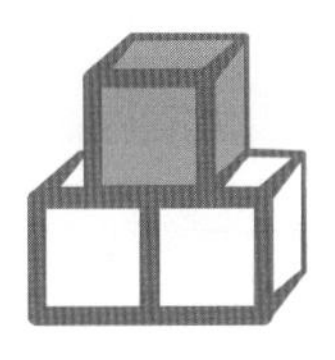

BASICS 20

Conjugation and Stability—The Evidence

In **Basics 10** we presented conjugation and resonance. One of the key points was that conjugation of (overlap between) double bonds imparts additional stability to compounds.

As a scientist, it is important that you understand the facts and theories, but it is also important that you understand the data that was used to reach the theories. The further along the journey we get, the more data you will encounter. If you have followed the discussion so far, you should be ready for this.

We need to present some data, but you also need to do some work here. This is a combination chapter!

ENTHALPY OF HYDROGENATION OF 1,3-BUTADIENE

Consider the following two reactions. It doesn't matter that we aren't discussing this reaction in this book. You can still do everything you need.

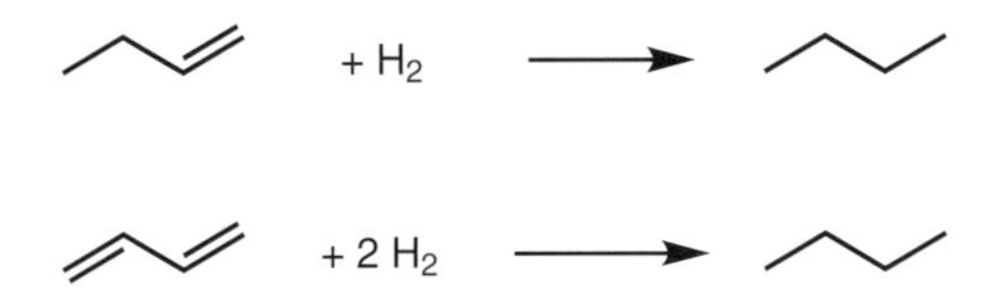

Using the data in the table below, calculate the enthalpy change for these two reactions.

You have more data than you need. This is deliberate. It is important to learn to select the relevant data.

Average Bond Dissociation Energy/kJ mol^{-1}					
H–H	436	**C–C**	350	**C=C**	611
H–Cl	432	**C–H**	410	**C≡C**	835
H–Br	366	**C–O**	350	**C=O**	732
H–I	298	**C–Cl**	330	**C≡N**	898
H–O	460	**C–Br**	270		
H–S	340	**C–I**	240		

WORKED SOLUTION

You did the calculation, right? If you didn't, stop reading, and go back and do it. You may think you will get everything you need by reading the solution and convincing yourself that you understand it. That's not how it works!

For the first reaction, you are breaking a H–H bond and a C=C bond, and we are forming a C–C bond and two C–H bonds.

$$\Delta H = (436 + 611) - (350 + 410 + 410) = -123 \text{ kJ mol}^{-1}$$

For the second reaction, you are breaking two H–H bonds and two C=C bonds, and we are forming two C–C bonds and four C–H bonds.

$$\Delta H = (436 + 436 + 611 + 611) - (350 + 350 + 410 + 410 + 410 + 410) = -246 \text{ kJ mol}^{-1}$$

Could you have predicted that the second number would be twice the first?

Now here are the actual enthalpy changes for these reactions.[25]

+ H_2 ⟶ $\Delta H = -128$ kJ mol^{-1}

+ 2 H_2 ⟶ $\Delta H = -238$ kJ mol^{-1}

First of all, these numbers agree pretty closely with the calculated data above. But there is a discrepancy!

Using the experimental data for hydrogenation of 1-butene, we might expect ΔH for the hydrogenation of 1,3-butadiene to be −256 kJ mol^{-1}. From this, we conclude that 1,3-butadiene is 18 kJ mol^{-1} more stable than we would have expected. This is the stabilization due to conjugation.

[25] These reactions require a catalyst, which we aren't going to worry about. For now, all we are doing is balancing the numbers of atoms.

YOUR NEXT TASK

Draw an energy diagram with butane at the bottom, 1-butene + H_2 in the middle, and 1,3-butadiene + 2 H_2 at the top. Calculate all the energy differences based on the above data.

I'm not going to give you the answer to this. Discuss it with your friends. Make sure you understand these numbers, and 'own it'. Don't be a passive partner in learning this material.

ENTHALPY OF HYDROGENATION OF BENZENE

We can do the same for benzene. Use the data in the table above to calculate the enthalpy change for the following two reactions.

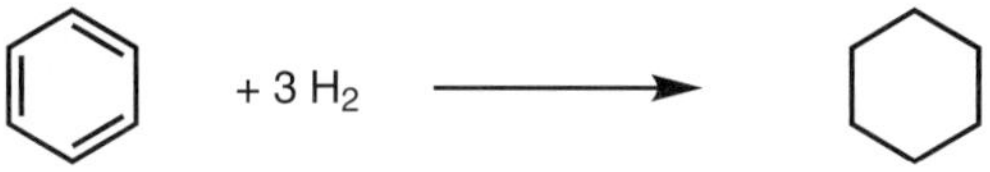

+ 3 H_2 ⟶

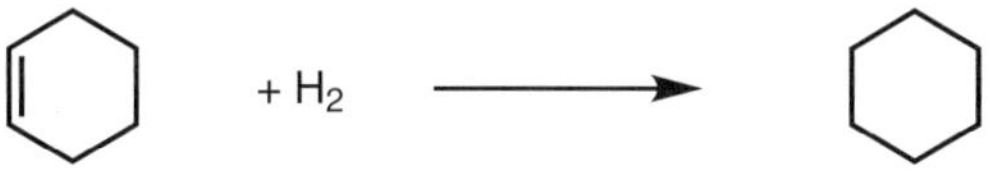

+ H_2 ⟶

Make sure you understand the assumptions that you are making with this type of calculation. The bond dissociation energies do not take conjugation into account.

Could you predict the numbers from the calculation for 1-butene?

Here are the experimental values.

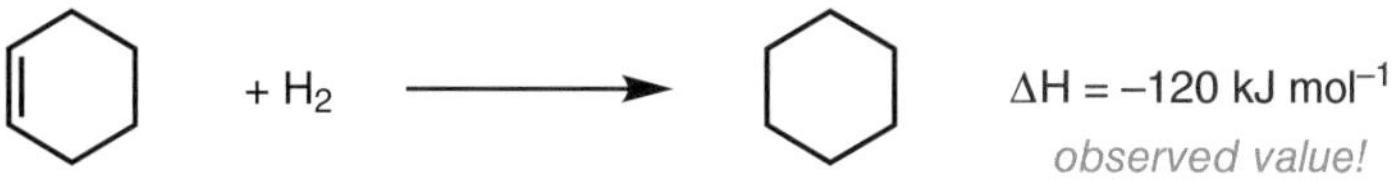

+ H_2 ⟶ $\Delta H = -120\ \text{kJ mol}^{-1}$
observed value!

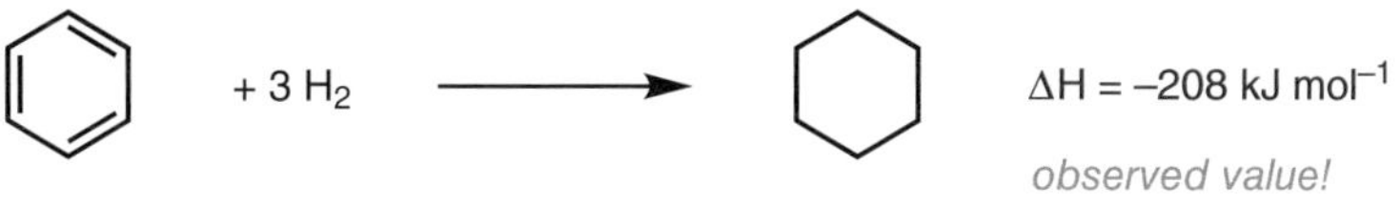

+ 3 H_2 ⟶ $\Delta H = -208\ \text{kJ mol}^{-1}$
observed value!

Of course, if there wasn't this mysterious 'aromatic stabilization', we would have expected the following, which is close (I hope!) to your calculated value.

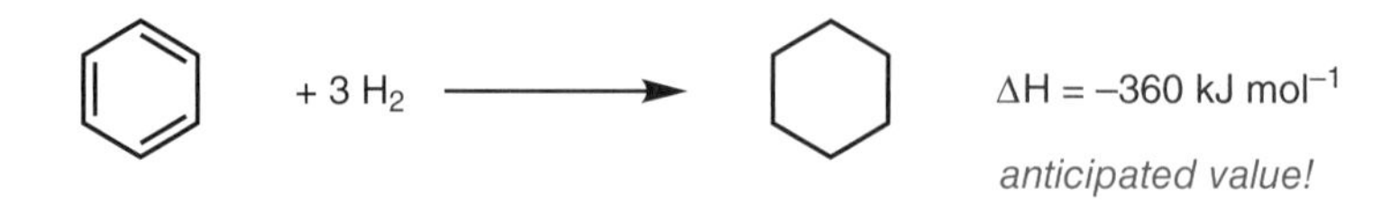

There is a rather large discrepancy of 152 kJ mol^{-1} between these two numbers. To put this into context, it is about half of a single bond dissociation energy. This is very significant.

We could do a similar calculation for hexa-1,3,5-triene, and we don't get anywhere near as much stabilization. The stabilization for benzene is a combination of two factors:

1. The number of double bonds (or π electrons).
2. The fact that we are dealing with a fully conjugated cyclic π system.

Fully conjugated is an important term, and worth a moment to consider. Think of conjugated double bonds as an electrical circuit carrying current. All you would need to stop this is one insulator—a CH_2 group. For a fully conjugated cyclic system, we need alternating single and double bonds all the way round, and back to the start.

WHERE ARE WE GOING WITH THIS?

In many reactions of aromatic systems, the aromatic stabilization is lost. This requires a large amount of energy, so that such reactions are slow, and require heat to proceed. There is also a strong driving force for regaining aromaticity.

When you learn about the reactions of alkenes and aromatic compounds, you will see that this extra stability accounts for all the differences in reactivity.

*That's why we are doing it this way! You apply a few basic principles consistently to **all** reactions!*

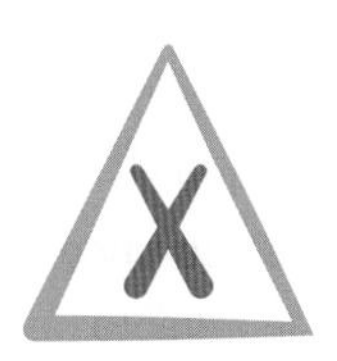

COMMON ERROR 5

Carbocations and Carbanions

THE MOST COMMON MISTAKE

As we have seen, and will continue to see, most problems arise from a fundamental disconnect between the structures and the words we use to describe them.

Early on you saw that *tertiary* carbocations are more stable than *secondary* or *primary* carbocations. There is a risk that you will learn the pattern '*tertiary* = stable'. Then, when you are given a series of carb*anions*, you state that the most substituted one is most stable.

What's the solution to this problem?

The solution is remarkably simple—slow down! Take your time and whenever you look at a carbocation or carbanion, consider the effect of each substituent in turn. Ask yourself whether it is stabilizing or destabilizing the charge.

Make sure you understand the mechanism by which it is stabilizing or destabilizing the charge—it will either be inductive or mesomeric.

I consider hyperconjugation to be a mesomeric effect. Well, it certainly isn't inductive, as we saw in **Common Error 3**.

Am I right or am I wrong? I don't think it matters, as long as I know why I view things in this way.

ONE SPECIFIC MISTAKE

Carbanion **1** is stabilized by the adjacent carbonyl group. Students sometimes claim that carbanion **2** is also stabilized—in fact it is significantly destabilized.

O, H_3C, ⊖ — **1**

O, H_3C, ⊖ — **2**

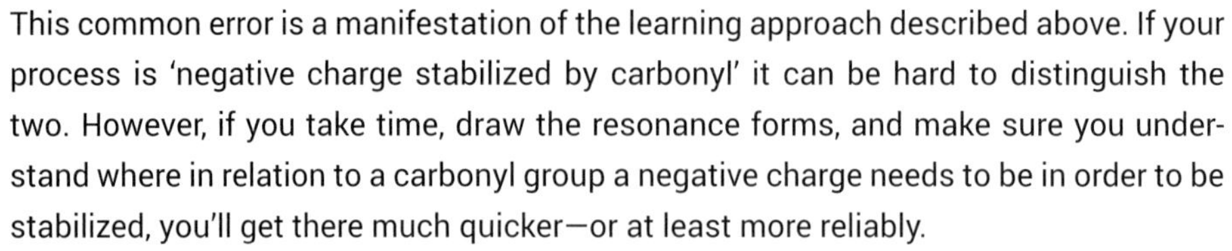

This common error is a manifestation of the learning approach described above. If your process is 'negative charge stabilized by carbonyl' it can be hard to distinguish the two. However, if you take time, draw the resonance forms, and make sure you understand where in relation to a carbonyl group a negative charge needs to be in order to be stabilized, you'll get there much quicker—or at least more reliably.

2.12

5

Keep pushing on with this one. Every reaction you see will involve a level of charge redistribution. If you don't know how to work out whether a given substituent will stabilize or destabilize this charge, you won't be in a position to tell whether you are looking at a fast reaction or a slow reaction.

There were many things I had to ask my lecturers to explain repeatedly. It's perfectly normal for this to take time to learn and internalize.

BASICS 21

Reactivity of Conjugated Systems

Let's join up some dots. We have seen that conjugated systems are more stable than non-conjugated systems. We have seen the evidence for this. We have also seen that conjugation leads to stabilization of carbocations and carbanions that are intermediates in a lot of organic reactions.

ARE CONJUGATED SYSTEMS LESS REACTIVE?

This is a fair question. After all, more stable should mean less reactive, right? Well, this would depend on what it is reacting to form. Let's take a look at this. We haven't covered much reactivity yet, but we have covered just enough to look at this problem.

If we add a proton to conjugated diene **1**, we will get the resonance-stabilized allylic carbocation **2**. If we do the same with the non-conjugated diene **3**, we will get a *secondary* carbocation **4**.

$H^{\oplus}$

1 → **2**

$H^{\oplus}$

3 → **4**

Have another look at the table of data in **Perspective 2**. It's all about finding the right data to compare. We do not have the stability of carbocation **2** or **4**. We do, however, have the data for the ethyl and allyl cations. That is, we have two *primary* carbocations, one of which is stabilized by a double bond.

Here, we have a secondary alkyl and secondary allylic carbocation. Get used to the way we refer to these.

The allyl cation is about 70 kJ mol^{-1} more stable than the ethyl cation. This represents an estimate of the difference in stabilization provided by a methyl group and by a carbon–carbon double bond.

Draw the reaction profiles for carbocation formation from diene **1** and diene **3**. We saw in **Basics 10** that conjugated systems are about 20 kJ mol^{-1} more stable than non-conjugated systems.

Did you draw the profiles? I'm not going to draw them for you!

2.13

You should have deduced, based on the numbers given in this chapter, that formation of carbocation **2** from conjugated diene **1** is more favourable by 50 kJ mol^{-1}.

*While resonance stabilization means that diene **1** is more stable than diene **3**, it can still be more reactive.*

Of course, these numbers do not refer to the rate of reaction, as we do not know the activation energy. But we can 'guess'. We saw the **Hammond postulate** in Basics 19. We know that for an *endothermic* reaction, such as formation of a carbocation, the transition state will resemble the carbocation, and will be stabilized by the same factors.

BASICS 22

Acid Catalysis in Organic Reactions Part 1

INTRODUCTION

We are going to have to do this one in two stages. We don't know enough (yet) to give a full treatment of the problem, but it's too important to not make a start!

Acid catalysis is widely used to increase the rates of chemical reactions. This happens because the activation barrier is lowered. In **Basics 15**, we saw the effect that changing the activation barrier has on the rate of a reaction.

Knowing that acid catalysis can be used does not tell you which particular acid will be most effective. This is very often dictated by factors such as solvent (polarity, solubility of the acid as well as of the reagents) and other functional groups present in a compound. With experience, you will learn to recognize which acid catalysts are most commonly used in particular reactions.

First of all, you need to understand the basic reasons why acid catalysts work. Fortunately, they are not too complicated.

In **Fundamental Reaction Type 1** we introduced the S_N1 mechanism for nucleophilic substitution. In **Basics 14** we saw reaction profiles.

In this chapter, we are going to look at how catalysis changes the energetics for carbocation formation.

If we want to analyse the catalytic effect, we must compare the catalysed process to the uncatalysed process. We cannot see the effect of the catalyst unless we have something to compare it to.

We will build this up in layers, starting by defining the problem and looking at it at quite a basic (but still useful) level. Later, in **Worked Problem 4**, we will ask (and answer) additional questions, and in doing so we will draw quite a complicated reaction profile.

The key point is that we are applying basic knowledge to a meaningful problem.

*The **real** key point is that with practice, you can take this approach to **all** organic chemistry problems. Throughout this chapter, I am going to be asking, then answering, questions. Pay attention to the questions. Learning to ask the right questions is the key skill.*

Let's introduce the problem.

ACID CATALYSIS IN THE FORMATION OF CARBOCATIONS FROM ALCOHOLS

Here is a simple equilibrium process.

$$R{-}OH \rightleftharpoons R^{\oplus} + {}^{\ominus}OH$$

This is the formation of a carbocation by heterolytic dissociation of a C–O bond. We generally think of a carbocation as an intermediate in a reaction. For now, we will consider it to be the product.

This is a simple process involving one bond breaking. There will be a transition state, but no intermediate in the process.

Now let's think about the energetics. Hydroxide is a poor leaving group, because it is not very stable. We know that the pK_a of water is 15.7, so that K_a for the following equilibrium is $10^{-15.7}$.

$$H_2O \rightleftharpoons H^{\oplus} + {}^{\ominus}OH$$

Once again, we see the value of a reference point from which we can extrapolate.

You see a lot more protons than you see carbocations. You can bet that **any** carbocation we encounter as organic chemists will be less stable than a proton. Therefore, K_a for the carbocation above will be **a lot** less than $10^{-15.7}$. With a K_a value such as this, the reaction **must** be *endothermic*.[26] We can now draw a simple reaction profile.

[26] Strictly speaking, it must be *endergonic*, but you will encounter the term '*endothermic*' far more often.

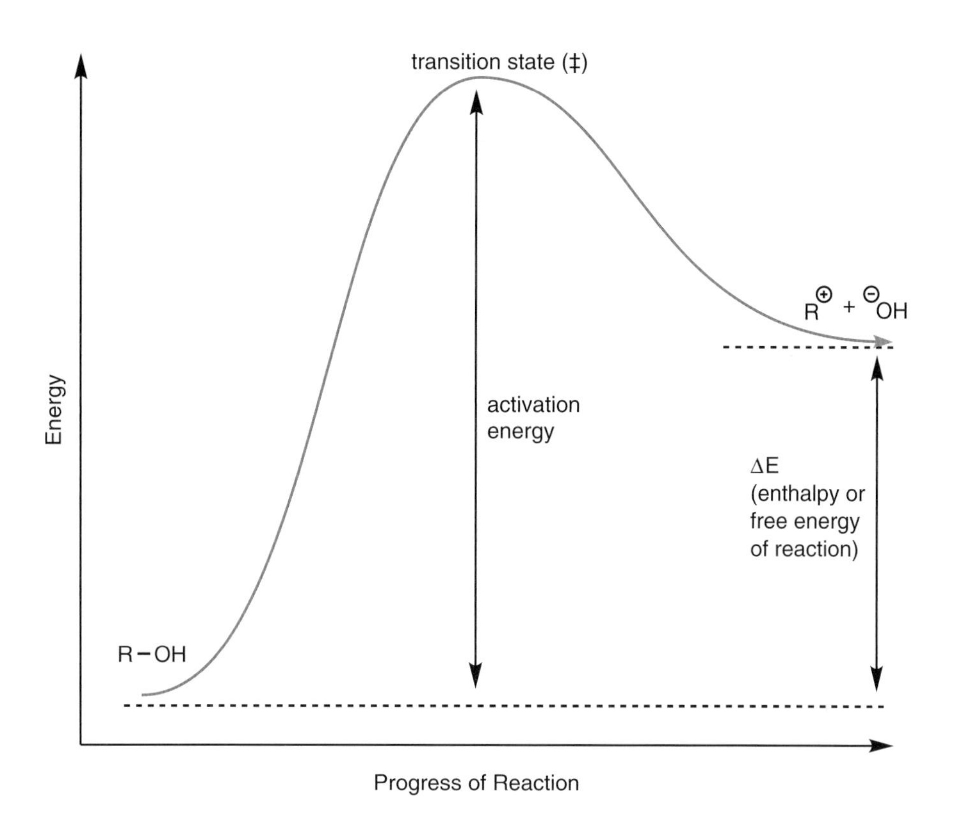

Now compare it with the following process.

$$R-\overset{\oplus}{O}H_2 \rightleftharpoons R^{\oplus} + H_2O$$

Here, we have protonated the alcohol first. Now we have neutral water as the leaving group, which is **much** better. The position of the equilibrium will be shifted to the right, even though we are forming the same carbocation.

Let's look at this a bit more carefully. The species on the left is protonated. It is less stable. The same carbocation is formed in both reactions, but the second reaction has water as the second product (more stable) while the first reaction has hydroxide as the second product (less stable).

Therefore, the acid-catalysed process will have a less stable reactant and a more stable product. It will still be *endothermic*. Just not quite as *endothermic*! We can draw a new reaction profile. I have left the 'old' reaction profile on there for reference.

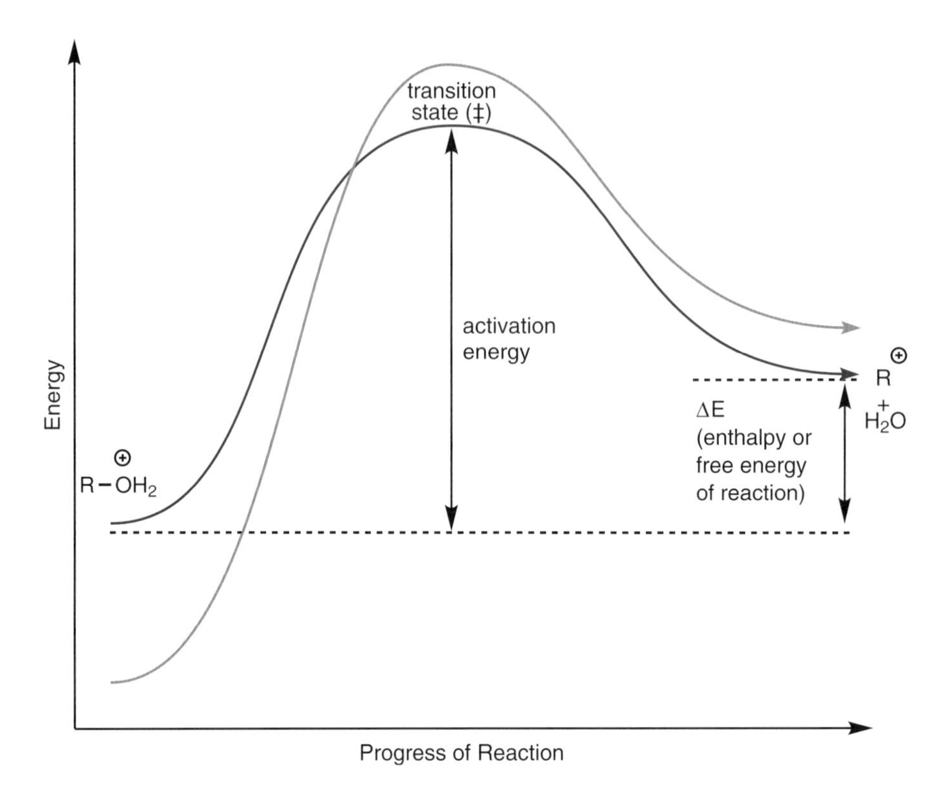

Since both reactions are *endothermic*, the transition states will resemble the products.

*Is it possible that the second process could be **slower** even though it is less endothermic?*

The Hammond postulate (**Basics 19**) tells us that the transition state will resemble the carbocation + leaving group more closely than it will resemble the starting materials. If we are stabilizing the product in the second process, we will also be stabilizing the transition state. We could not easily stabilize the product and then have a higher activation energy.

I think this is quite complicated to understand at first. Take your time with it. Carry on reading, and then come back to it. Persevere and you will get it.

I think this is a really good example for self-reflection. You need to be able to gauge your progress. If you really understand this, you are doing well.

Now, to get back to the point!

If we can compare activation energies, we can compare rates.

The second, catalysed, reaction will be faster!

SO WHY DO WE NEED TO COME BACK TO THIS LATER?

That's a good question. If we have fully addressed the problem, what more do we need to do? Let me answer this with another question.

How do we know that the protonated alcohol is less stable than the unprotonated alcohol?

If we use a really strong acid, we might expect to fully protonate the alcohol, and this would imply that the protonated alcohol **could** be more stable than the unprotonated alcohol (under these conditions).

Does this matter?

If we use a weak acid, don't we need to consider the extent of protonation of the alcohol **and** the rate of formation of the carbocation? With two activation barriers, could it be slower?

It can't, but we need quite a lot of detail in order to prove that this is the case.

You need to 'internalize' the **process** (not just the information) in this chapter before we come back to finish the job. We will do this in **Worked Problem 4**.

REACTION DETAIL 1
Nucleophilic Substitution at Saturated Carbon

INTRODUCTION

Now, we need to look at more detailed aspects of reactions. Inevitably, these chapters will be longer and include more factual information. However, the style stays the same. You will be guided to apply the basic principles to show you that you already know enough to predict much of the reactivity. Being able to apply basics to predict reactivity trends is far better than learning the trends by rote.

WHAT DO WE NEED TO DO?

In **Fundamental Reaction Type 1**, we saw two possible mechanisms for substitution reactions at saturated (sp^3 hybridized) carbon. Now we need to look at the reactions in more detail and to consider the following questions:

1. Which substrates will react by which mechanisms?
2. Which leaving groups will be best?
3. What are the stereochemical consequences of each mechanism?
4. What reaction conditions favour each mechanism?
5. What effect does changing the nucleophile have?

S_N1 OR S_N2?

It isn't easy at first to decide whether a given substitution reaction will proceed via an S_N1 or an S_N2 mechanism. At first, we will look at each individual mechanism and decide which structural parameters favour the mechanism.

We are going to consider the S_N1 mechanism first.

THE S_N1 MECHANISM AGAIN

Here is the reaction again.

$$R{-}X \xrightarrow{\text{slow loss of } X^{\ominus}} R^{\oplus} + Nu^{\ominus} \xrightarrow{\text{fast}} R{-}Nu$$

$X^{\ominus}$

For the following series of compounds, rank them according to the rate of substitution by an S_N1 mechanism.

Consider steric and electronic effects. Since the reaction has an intermediate, make sure to use the Hammond postulate. Sketch reaction profiles.

$(H_3C)_3C{-}Br$ $\quad$ $H_3C{-}Br$ $\quad$ $H_3C{-}CH_2Br$ $\quad$ $(H_3C)_2CH{-}Br$

You have to do quite a lot here. All of it needs to become habit. In a while, you will start to spot all of the relevant factors at a glance.

STRUCTURAL FACTORS IN S_N1 SUBSTITUTION

The rate-determining (slow) step in S_N1 substitution is formation of the carbocation intermediate. The property that determines the rate of reaction is the activation energy.

Here is the reaction profile again.

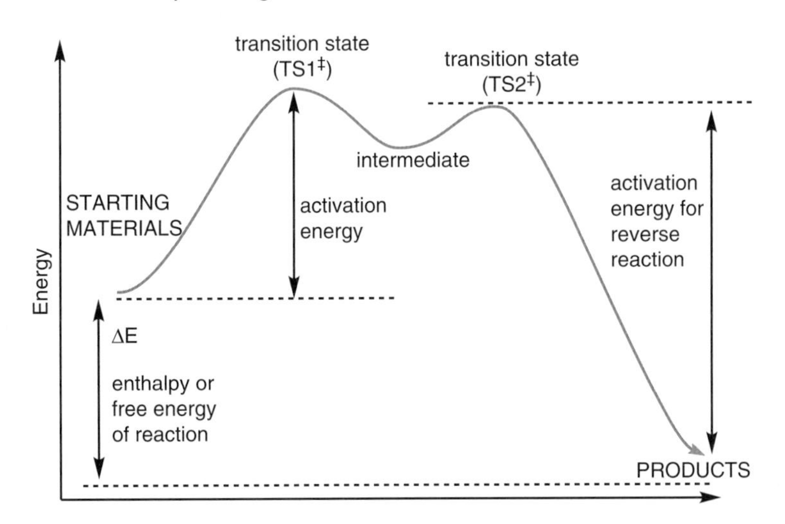

We know what stabilizes the carbocation. If the carbocation is more stable, then the activation energy for its formation will also be lower. Don't lose sight of the Hammond postulate (**Basics 19**).

So, factors present in the substrate essentially come down to what is attached to the carbon bearing the positive charge. We have already seen the following trend in carbocation stability.

$$(CH_3)_3\overset{\oplus}{C} > (CH_3)_2\overset{\oplus}{C}H > CH_3\overset{\oplus}{C}H_2 > \overset{\oplus}{C}H_3$$

1

Therefore, $(CH_3)_3C$–Br is much more likely to undergo S_N1 substitution than CH_3–Br.

This is an electronic effect. We should also ask if there is a steric effect. Here is a space-filling model of *t*-Bu-Br. It is pretty crowded.

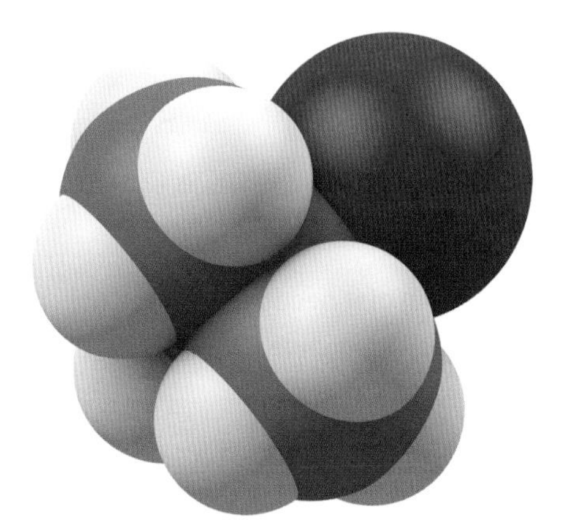

As we lose the bromide anion, the carbocation flattens out, so that the methyl groups move further apart. This will relieve steric crowding.

It isn't easy to separate the steric from the electronic effect in this case. Steric crowding destabilizes the *t*-Bu-Br,[27] raising its energy. This has the effect of lowering the activation energy as well. It would be difficult to determine how much of the stability of the *t*-butyl cation is steric and how much is electronic.

In reality, we need a substrate that is capable of forming a tertiary carbocation (or one of comparable stability) in order to have predominantly S_N1 substitution. Methyl, primary, and secondary substrates tend to prefer S_N2 substitution.

Don't get blinkered though. In **Basics 17** we saw that the following carbocation is very unstable, because it cannot become planar. Therefore, it won't be formed (the barrier to formation is high) and the corresponding bromide does not undergo S_N1 substitution.

[27] Again, we need to get used to seeing the same 'structure' represented in different ways.

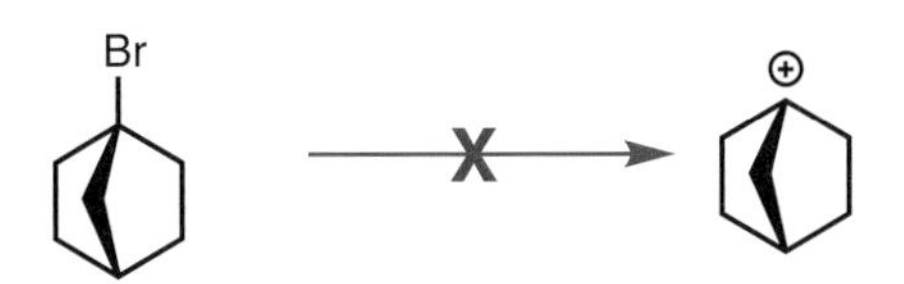

It's always useful, when planning a synthesis, to know which reactions don't work. In this case, the simple reasoning (it's a tertiary carbocation) is not sufficient. You have to also recognize that only planar tertiary carbocations are stable.

1

Even a fantastically stable carbocation doesn't guarantee a predominantly S_N1 pathway. Methoxymethyl chloride, CH_3OCH_2Cl, forms a pretty stable carbocation (we've seen it before, but draw it again for practice), and as such is a good substrate for S_N1 substitution reactions. The rate of S_N1 substitution from this compound is high. It turns out that this compound also reacts quickly in S_N2 reactions.

While an S_N1 mechanism does generally predominate, the relative proportions of product formed by each mechanism will depend on the precise reaction (e.g. nucleophile) and conditions (e.g. solvent).

As a general case, it is possible for a single compound to undergo some substitution by an S_N1 mechanism and some by an S_N2 mechanism. The amount of substitution by each mechanism will depend on the rates of the two processes. In fact, it gets more complicated than this. The precise extent of bond forming/breaking in a transition state means that some S_N2 substitutions have significant S_N1 character, and vice versa.

*This can be a bit subtle. We will come back to it in **Perspective 4**.*

LEAVING GROUPS IN S_N1 SUBSTITUTION

A better leaving group tends to be more stable. Therefore, the trends of leaving group ability will be the same in S_N1 and S_N2 substitution reactions. We will consider them in the section on S_N2 substitution.

There is just one added point that is specific to the sort of conditions we might use for S_N1 substitution reactions—acid catalysis. Let's look at the equation for formation of the carbocation. When we look at the 'intermediate' in the middle of the reaction profile, we shouldn't simply be thinking about the carbocation. We should be thinking about the carbocation, and the negatively charged (in this case) leaving group, X. So, if X^- is more stable, the 'intermediate' will be more stable, and hence more rapidly formed (Hammond postulate!).

$$R{-}X \longrightarrow R^{\oplus} + X^{\ominus}$$

We can compare the following two equations.

$$R{-}OH \longrightarrow R^{\oplus} + \overset{\ominus}{O}H$$

$$R{-}\overset{\oplus}{O}H_2 \longrightarrow R^{\oplus} + H_2O$$

Hydroxide is a terrible leaving group. It's a reasonably strong base. Water is a much better leaving group. We saw this in **Basics 22**. By protonating the alcohol (acid catalysis) we favour loss of the leaving group, so we can accelerate the S_N1 substitution.

SOLVENT EFFECTS IN S_N1 SUBSTITUTION

The S_N1 reaction has a polar intermediate, and therefore needs a polar solvent to aid in the formation and stabilization of this intermediate.[28] The solvent needs to be non-nucleophilic, otherwise it would attack the carbocation and give an alternative substitution product. Formic acid (HCO_2H) is often used to favour the S_N1 mechanism. Aqueous ethanol is much better than pure ethanol.

Why is formic acid not very nucleophilic?

I'm not going to give you the answer to this question now, but you can read ahead to the section on nucleophilicity if you need guidance.

The **dielectric constant** is a measure of solvent polarity.

In many cases, the solvent **is** the nucleophile. We refer to a reaction with solvent as a **solvolysis**.

STEREOCHEMICAL EFFECTS IN S_N1 SUBSTITUTION

We haven't started talking about stereochemistry in a formal sense yet, so we aren't ready to discuss the stereochemical implications of a mechanism. When we do this, I want to be confident that you associate the stereochemical outcomes of a reaction with shape, rather than with the words used to describe the shape. This is a subtle distinction, but an important one. Stereochemical aspects of substitution reactions get their own chapter, Reaction Detail 2.

[28] Strictly speaking, to accelerate the reaction we need to stabilize the transition state for carbocation formation. However, we have already established that the transition state resembles the carbocation, so we shouldn't be surprised to find that solvents that stabilize carbocations also stabilize the transition states for their formation.

EFFECT OF THE NUCLEOPHILE IN S_N1 REACTIONS

The nucleophile in an S_N1 reaction doesn't react until after the rate-determining step, and so a good nucleophile will react at the same rate as a poor one.

However, it is possible that a reaction which is S_N1 with a poor nucleophile will become partially or predominantly S_N2 with a good nucleophile. You cannot rule out changes of mechanism.

1

We could see this situation on a reaction profile, where the S_N1 profile is unaffected, but the energy of the S_N2 transition state is lowered such that it becomes competitive.

WORKED PRACTICE PROBLEM

Draw a reaction profile for an S_N1 substitution. On the same graph, draw two S_N2 substitution reaction profiles, with different nucleophiles. Draw the activation energies of the three processes such that with the better nucleophile the S_N2 reaction will predominate but with the poorer nucleophile the S_N1 reaction will predominate.

Here's the diagram. I'm not going to do all the work for you. Check that you understand why the shapes and energies of the profiles are as they are.

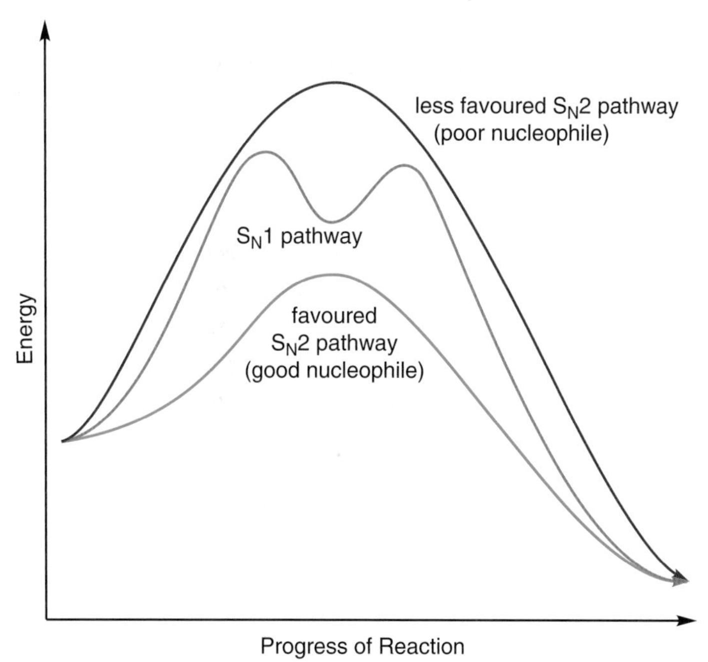

THE S_N2 MECHANISM AGAIN

Let's look at the S_N2 reaction again.

Nu⊖ → C–X → [Nu- -C- - -X]‡⊖ → Nu–C + X⊖

transition state

We need to consider the influence of various reaction parameters on the rate of reaction. Ultimately, this will help us to learn which S_N2 substitution reactions are good, and which are not so good.

STRUCTURAL FACTORS IN S_N2 SUBSTITUTION

Have yet another look at the mechanism!

Nu⊖ → C–X → [Nu- -C- - -X]‡⊖ → Nu–C + X⊖

transition state

We are developing an overall negative charge, and it is building on the carbon atom that is being substituted. If any of the groups attached to this carbon can stabilize a negative charge, they will lower the activation barrier to S_N2 substitution, accelerating the reaction. If any of these groups destabilize a negative charge, they will destabilize the transition state, so they will slow the reaction. I deliberately worded those two sentences differently, but effectively they say the same thing.

Here are some relative rates of substitution reactions by a purely S_N2 pathway. These relate to the use of chloride as nucleophile, and acetone as solvent.

Me–Br	Et–Br	Br (n-propyl)	Br (isopropyl)	Br (tert-butyl)
78	1.5	1	3×10^{-2}	7×10^{-3}

Let's focus initially on the following two compounds.

Me–Br   Br (isopropyl)

The one on the left undergoes substitution 2600 times more rapidly than the one on the right. When you see a big difference in reaction rates, you know there must be a good reason. Neither of these compounds will give a stable carbocation by loss of bromide, and the reaction is done under conditions that do not favour carbocation formation anyway. It has to be S_N2 in both cases.

The S_N2 transition state features a build-up of negative charge. The two methyl groups in isopropyl bromide will destabilize the negative charge, just as if we were dealing with a full negative charge.

*Re-read **Basics 16**, particularly the last part which refers to partial charges.*

Of course, those two extra methyl groups will also hinder the approach of the nucleophile in a purely steric effect. Once again, we are considering steric and electronic effects, and it isn't easy to separate the two.

Here is another compound that undergoes **very** fast S_N2 substitution.

How does the ketone stabilize the S_N2 transition state? Since we are dealing with a build-up of negative charge, let's consider the more extreme scenario—a full negative charge.

We know that a carbonyl group will stabilize an adjacent negative charge (**Basics 18**).

 Draw the resonance forms for stabilization of the negative charge.[29]

We might draw the transition state for substitution of the above chloride as:

Since the negative charge is building on the α-carbon atom, the same factors that stabilize the full carbanion (above) will stabilize the partial negative charge. Sure,

[29] If you put a negative charge anywhere on the benzene ring, you have made a mistake. Check!

2.14

you can't draw a resonance form to show this, but the resonance form is just one way to represent the stabilization.

STEREOCHEMICAL EFFECTS IN S_N2 SUBSTITUTION

1

Once again, this is covered in detail in Reaction Detail 2.

THE LEAVING GROUP

In order to understand leaving group effects, we should think about the overall reaction again.

$$R{-}X + Nu^{\ominus} \longrightarrow R{-}Nu + X^{\ominus}$$

If X^- is more stable, then the forward reaction will be more favourable. We can think in terms of an acid–base equilibrium:

$$HX \rightleftharpoons H^{\oplus} + X^{\ominus}$$

If HX is a strong acid, then X^- is more stable. So, a good leaving group is the conjugate base of a strong acid. The aqueous pK_a values for hydrogen halides are shown in the table below. HI is clearly the most acidic, while HF is the least acidic.

Hydrogen Halide	pK_a
HF	3.2
HCl	–7
HBr	–9
HI	–10

Therefore, for halide anions, the trend in leaving group ability is as follows:

$$I^{\ominus} > Br^{\ominus} > Cl^{\ominus} >>> F^{\ominus}$$

We have already seen the effect of protonation on alcohols as leaving groups in the section on S_N1 substitution above. I won't repeat the same argument here, as acidic conditions tend not to be useful for S_N2 substitution. In many cases, the nucleophile could remove the proton that you have just added.

Remember, nucleophiles can be bases, bases can be nucleophiles.

However, even if we cannot use this approach to making oxygen a better leaving group, we can certainly use the same principle.

4-Toluenesulfonic acid has a pK_a of –2.8. Water has a pK_a of 15.7.

$$H_3C\text{–}C_6H_4\text{–}SO_2\text{–}OH \underset{}{\overset{pK_a\ -2.8}{\rightleftharpoons}} H^{\oplus} + {}^{\ominus}O\text{–}SO_2\text{–}C_6H_4\text{–}CH_3$$

4-toluenesulfonic acid

$$H_2O \overset{pK_a\ 15.7}{\rightleftharpoons} H^{\oplus} + {}^{\ominus}OH$$

Therefore, we can expect the toluenesulfonate[30] anion to be a **much** better leaving group than hydroxide.

Draw resonance forms to explain why the toluenesulfonate anion is stable (and therefore a good leaving group).

It turns out that it is really easy to convert an alcohol into the corresponding 4-toluenesulfonate ester. We aren't going to worry right now about how we do this.

$$R\text{–}OH \longrightarrow R\text{–}O\text{–}SO_2\text{–}C_6H_4\text{–}CH_3$$

Conceptually this is the same as protonating the alcohol, so that when we carry out a nucleophilic substitution reaction, the leaving group has a lower pK_a.[31]

$$Nu^{\ominus} + R\text{–}O\text{–}SO_2\text{–}C_6H_4\text{–}CH_3 \longrightarrow Nu\text{–}R + {}^{\ominus}O\text{–}SO_2\text{–}C_6H_4\text{–}CH_3$$

We have already seen one advantage of this method. Protonating the alcohol will not be effective, since the nucleophile can simply remove the proton. A second advantage is that many organic functional groups are sensitive to acid, so that this method is much more tolerant of such functional groups.

[30] We usually call it 'tosylate'.

[31] Strictly speaking, the conjugate acid of the leaving group has a lower pK_a, but we tend to use these terms interchangeably.

The reasoning behind this approach needs to become second nature. Such 'activation' of functional groups is extremely common, and you need to see it for what it is, and understand the principles. Otherwise, every new example will be another disconnected fact to remember.

SOLVENT EFFECTS IN S_N2 SUBSTITUTION

The S_N2 reaction has a relatively polar negatively charged transition state. However, from a practical point of view it turns out that one of the most important factors for a good solvent is that it needs to be able to dissolve both starting materials well. Dimethylsulfoxide (DMSO) and dimethylformamide (DMF) both work well.

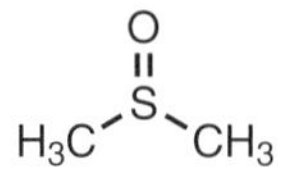

$H_3C-S(=O)-CH_3$ (DMSO) $(H_3C)_2N-C(=O)-H$ (DMF)

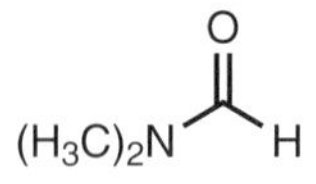

These solvents are classified as 'polar aprotic' solvents. They do not have hydrogen bond donors. Methanol is a polar protic solvent—the hydrogen of the OH group can form hydrogen bonds.

If a solvent is 'too good' it will solvate the nucleophile and not let it get to the substrate. For example, methanol and DMF both have the same dielectric constant, yet DMF is much better for substitution reactions using sodium azide as nucleophile, as the azide anion is much less solvated.

Some of these points may seem like complicated details to remember. If you focus on the principles, you will find that these additional points just slot into place. Remember, I'm not cleverer than you, and I did it!

EFFECT OF THE NUCLEOPHILE IN S_N2 REACTIONS

It's complicated! Nucleophilicity often depends on the type of solvent used. For now, we will keep it simple and look at the trends only in polar aprotic solvents. In polar aprotic solvents, where the nucleophile is not solvated to the point that its nucleophilicity is compromised, the nucleophilicity correlates well with the pK_a of the conjugate acid.

$$^{\ominus}NH_2 > {}^{\ominus}OH > F^{\ominus}$$

$$F^{\ominus} > Cl^{\ominus} > Br^{\ominus} > I^{\ominus}$$

Effectively, the less stable anion (higher pK_a conjugate acid) is more reactive. This is what you would have expected. This trend holds true for nucleophiles in which the attacking element is the same.

$$^{\ominus}OH > PhO^{\ominus} > AcO^{\ominus}$$

The negative charge in acetate[32] is more stabilized (draw the resonance forms!) than phenoxide (guess what!), which is turn is more stable than hydroxide.

All other things being equal, negatively charged nucleophiles are more reactive than neutral nucleophiles.

$$PhS^{\ominus} > PhSH \qquad ^{\ominus}OH > H_2O$$

Here's an interesting series of amine nucleophiles.

$$Et_3N > Et_2NH > EtNH_2 > NH_3$$

*We have seen that alkyl groups are electron-releasing. Although we saw this in the context of carbocation and carbanion stability (**Basics 16**) we can apply these same principles to the basicity of amines. The nitrogen lone pair (think, carbanion) is destabilized by the addition of each subsequent ethyl group. If the amine is less stable, it will be more basic/nucleophilic.*

Look at the following reaction. Would you ever be able to make ethylamine using this reaction? Discuss this with your friends, and see if you can convince them, one way or the other.

See how many principles you can apply, but don't get too bogged down. The point of this exercise is to try to decide what is relevant and what is not.

$$Et{-}I + NH_3 \longrightarrow EtNH_2 + HI$$

SOLVATION OF THE NUCLEOPHILE

If we use a solvent that is 'too good', it will solvate the nucleophile, and then it will be harder for the nucleophile to get to the substrate. This tends to apply to polar protic solvents (ethanol, for example). In that case, the nucleophilicity trends can be reversed. Look at it this way—if the nucleophile is so good that it will react well with the substrate, it will also 'react' with solvent (be solvated) and there is more solvent around.

[32] You can't draw the resonance forms if you don't know what acetate is. I'm not going to make it easy for you by putting the structure here.

TO FINISH (FOR NOW)

1

You can apply most of the fundamental principles to substitution reactions. If you cannot understand how these principles apply to substitution reactions, you probably won't understand how to apply them to other reactions.

Don't let this put you off though. Maybe you will see a principle applied to a different reaction, and then you'll understand how it applies to substitution. Keep coming back to this chapter and checking how well you understand the basics, especially after we have covered the stereochemical implications in more detail.

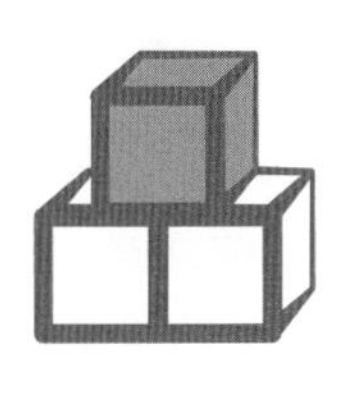

BASICS 23

What Defines a Transition State?

A COMMENT, BEFORE WE START

I find that most students are happy to accept that a transition state is the 'thing' at the highest energy point along the reaction profile. If you think about an S_N2 substitution reaction, such as the one shown below, then we are happy to draw something in the middle to represent the transition state.

$$HO^{\ominus} + H_3C{-}I \longrightarrow \left[HO\cdots CH_3\cdots I\right]^{\ddagger\ominus} \longrightarrow HO{-}CH_3 + I^{\ominus}$$

We say that the C–O bond is starting to form, and the C–I bond is starting to break. At no point do we have an intermediate, so that the reaction profile is as shown in the first example in **Basics 14**.

I would like to provide a little more detail, by drawing from infrared spectroscopy and computational chemistry, to try to provide a more detailed definition. Don't worry too much if you don't understand everything in this chapter on the first reading. Just come back to it later. Similarly, if you read this and think 'is that all?', then don't worry—the simplest ideas are often the most profound. I find that forcing myself to think about transition states allows me to make better predictions about rates of reaction. I want to ensure that you have the tools to do this.

WHAT HAPPENS WHEN A BOND VIBRATES?

Bear with me on this. Molecular vibrations are fundamental to transition states. Here is a 3D representation of a ketone, propanone. On this structure, I have added a double-headed arrow to indicate the C=O bond. It turns out that this compound gives a very significant peak in the infrared spectrum, at about 1710 cm^{-1}, corresponding to the vibration of the C=O bond.

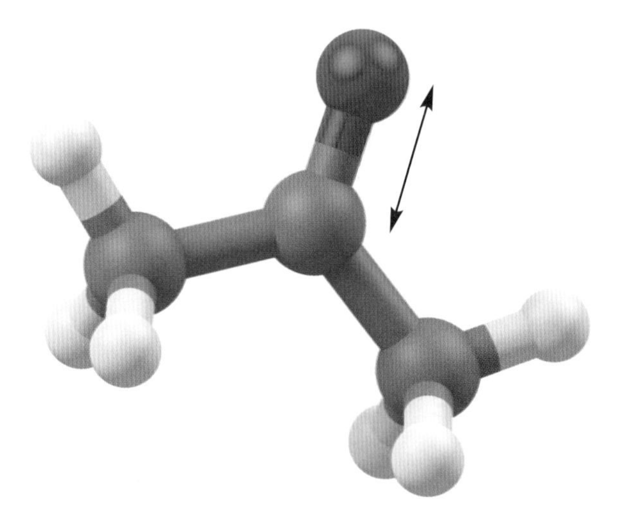

Consider what happens to the energy of the compound as the bond is stretched from its 'relaxed' position. The energy must **increase**. Similarly, if we compress the bond from its 'relaxed' position, this also **increases** the energy of the system. This is a 'normal' molecular vibration—a distortion from the equilibrium.

BOND VIBRATIONS IN TRANSITION STATES

A transition state is a little different. Let's have another look at the reaction of iodomethane with hydroxide.

$$HO^{\ominus} + H_3C{-}I \longrightarrow \left[HO{\cdots}\underset{H\ \ H}{\overset{H}{C}}{\cdots}I \right]^{\ddagger\ominus} \longrightarrow HO{-}CH_3 + I^{\ominus}$$

It is possible to calculate the structures and energies of transition states. Here is a computer-generated transition state for this process. In terms of the reaction coordinate, we are now right at the top of the curve. I have highlighted two distances, 'a' and 'b'.

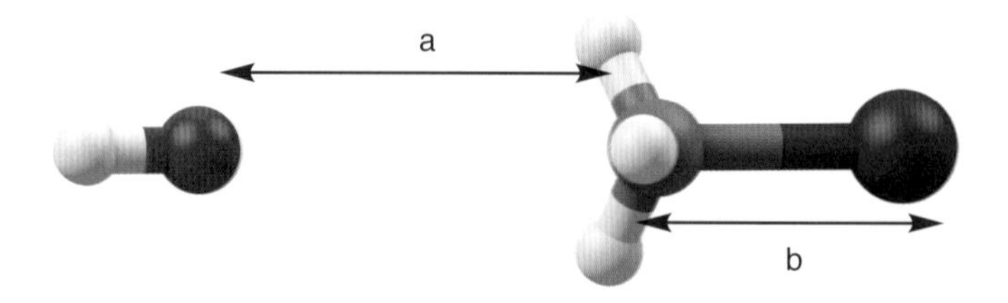

2.15

If we reduce the distance 'a', we will increase the distance 'b'. This is proceeding forwards to the product, and as such the energy in the system will decrease. If we increase the distance 'a', we will reduce the distance 'b', and we will be returning to starting materials, so once again the energy of the system will decrease.

So, a 'normal' vibration of a ground state molecule is one in which any distortion from the 'normal' geometry results in an increase in energy. A 'vibration' of a transition state along the reaction coordinate in either direction results in a decrease in energy.

That's about it really. I just want to add some terminology from computational chemistry. I have a really simple reason for including this here. It is normally found in computational chemistry textbooks, alongside complicated mathematics. I've usually given up reading them before I get to this point!

There is a very specific term for this type of vibration. It is referred to as being **imaginary**. We would know that a calculated structure is a transition state if it has only one **imaginary frequency**.[33]

IS A TRANSITION STATE THE HIGHEST POSSIBLE ENERGY OF THE SYSTEM?

If we look at the example above, what will happen if we take one of the C–H bonds of the iodomethane and stretch it? It will increase the energy of the system. Similarly, any other distortion of the transition state will increase its energy. The transition state is only the maximum energy along the reaction coordinate. It is a minimum energy in every other respect. This is quite a complicated idea and it may take you a little time to get used to.

Let's think of it another way. If you want to get from one side of a mountain range to the other, you probably won't go over the top of the tallest peak. If you want to make your life as simple as possible, you will go through the lowest mountain pass.

[33] Imaginary numbers are multiples of the square root of −1. This relates to the mathematics of the quantum chemical calculation. In my opinion, it is more important to have an intuitive feeling for why this one vibration is 'different'.

PERSPECTIVE 3
Bonding Beyond Hybridization

When we think of electrons and orbitals, we use quantum mechanics. It is impossible to accurately solve the Schrödinger equation for anything other than the simplest systems. For real organic molecules, you cannot do it at all. Therefore, everything you will ever be told about bonding is a **model**. Every model makes approximations. Some models make more approximations than others. So far, we have used hybridization as our model. We will continue to do so!

Before we progress, let me be absolutely clear about one thing. You need to understand hybridization, and to be fluent in your use of the language used to describe hybrid orbitals. Hybridization allows us to consider a single bond to consist of two electrons in one orbital that is the entire contribution to the bond between two atoms. It is a beautiful and elegant model.

There is a distinct advantage to thinking about two-centre two-electron bonds. This was stated in **Basics 6**, but it is worth reiterating. In describing chemical reactions, we will often be talking about forming or breaking one bond in a structure. To be able to consider this bond as a discrete entity, with two electrons in one orbital, is a real convenience.

As with all models, it has its limitations. Consider the structure of diborane, B_2H_6. You have hydrogen atoms bonded simultaneously to two boron atoms.

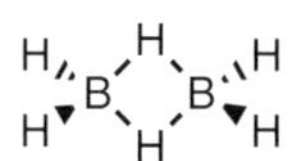

2.16

So far, we have stated that hydrogen can only form one bond, and this is definitely true for stable organic molecules. For diborane, we talk about three-centre two-electron bonds, in which each B–H–B bridge has two electrons in a bonding orbital. Many people find this confusing—I was one of them.

To be honest, I also found hybridization to be confusing! It felt rather random to take three p orbitals, all at 90° to one another, plus an s orbital which lacks directionality, and 'mix them up' to get new molecular orbitals at 109.5° to one another.

First things first, if you did find hybridization confusing, try not to think too deeply about it. Accept it and don't question it too much. Use the language of hybridization. Recognize the convenience of considering a bond to be two electrons in a single orbital. But also recognize that it is only a ***model*** *for the bonding.*

It is actually valid to 'hybridize' the orbitals, as the original orbitals and the hybrid orbitals are both valid solutions to the Schrödinger equation. We are okay to use this 'more convenient' solution. However, there are alternative explanations for the bonding that do not require the use of hybridized orbitals. What I want to do in this chapter is describe the bonding in organic molecules in a way that uses the 'original' atomic orbitals rather than hybridized orbitals. We can compare and contrast hybridization with this (possibly more rigorous) molecular orbital approach. We can see why hybridization is equivalent to this model in many respects.

In giving you this alternative view, I don't want to stop you using hybridization. Actually, it's quite the opposite. I want you to continue to talk about hybridization ***without worrying about it****. In most cases, hybridization is perfectly fine. But there will be times when you need to consider alternative models of bonding. I want you to be prepared for this.*

A MOLECULAR ORBITAL REPRESENTATION OF METHANE

As with hybridization, we are going to completely ignore the 1s shell. So, in methane, CH_4, the carbon has a share in 8 outer electrons (the '2' shell). Four of these electrons were its own, and each hydrogen atom provides one electron. We can put two electrons into each molecular orbital, so that we will fill a total of 4 orbitals.

The carbon had one 2s orbital and three orthogonal (at 90° to one another) p orbitals from which to form the molecular orbitals. Here is a molecular orbital diagram for methane.

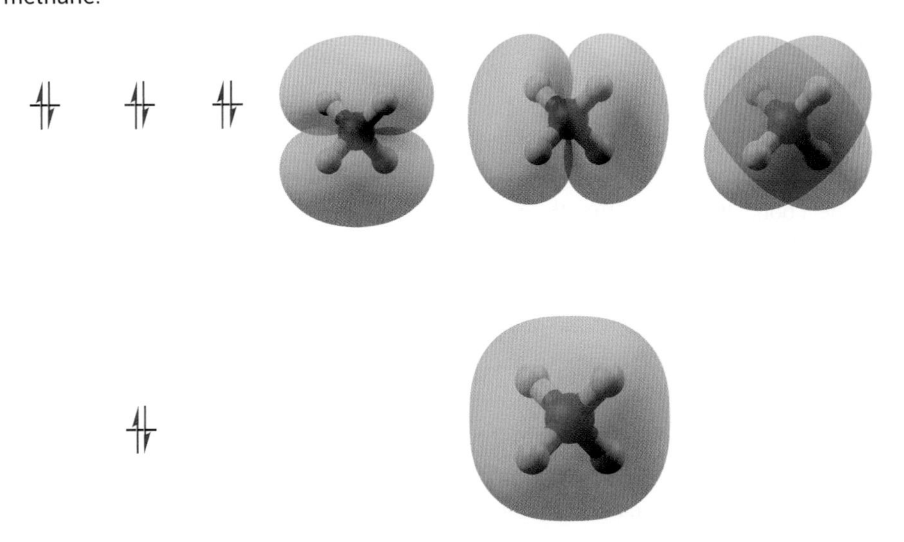

The lowest energy molecular orbital in methane is shown at the bottom. It is derived from the s orbital of carbon overlapping with the s orbitals on **all** of the hydrogen atoms, and it is therefore completely symmetrical and covers the entire molecule. It provides a contribution to all four of the C–H bonds. This means that we need to dispense with the idea that we have one orbital for each bond.

Above this are the other three orbitals. Effectively they are the three p orbitals on carbon, each lobe overlapping with the s orbitals on two of the hydrogen atoms.

Overall, the most energetically favourable spatial arrangement for the four hydrogen atoms around the carbon is tetrahedral. We don't actually need to use hybridization to explain this. The change of directionality of the orbitals that we use in hybridization arises naturally from these molecular orbitals that are derived from the atomic orbitals on carbon.

*Pick a hydrogen atom at random and look at these four orbitals again. Hopefully I won't have to work too hard to convince you that the C–H bonding for this atom (indeed all of them) has a contribution from **all** these orbitals.*

I can't quite decide whether this explanation is simpler than hybridization, or more complex. It feels simpler, because you can use the atomic orbitals without changing their symmetry. But it is more complex because you need to dispense with the idea of a two-centre two-electron bond. Whichever way you look at it, I hope you can see the advantage of hybridization. With the 'real' molecular orbitals, you can no longer think about 'the bond' as being a clearly defined entity with two electrons. Sure, there are eight electrons in total, and four bonds, so it averages out.

A MOLECULAR ORBITAL REPRESENTATION OF ETHENE

Let's now look at the molecular orbitals for ethene. When we looked at this from the perspective of hybridization, we 'reserved' two p orbitals, one from each carbon atom, to form the π bond. It turns out that the same happens here, but not as a result of a choice.

Let's look at the σ bonding. We aren't going to draw a molecular orbital diagram as we did for methane, but we will still consider the orbitals. We are going to use three electrons from each carbon atom, and one electron from each of the four hydrogen atoms, giving us a total of ten electrons to accommodate. We are going to fill five molecular orbitals.

Two of the orbitals are derived from the s orbitals on carbon. The first one spans the entire molecule, as we saw for methane. The second one has a node between the two carbon atoms, so it doesn't contribute to C–C bonding, but it does contribute to the C–H bonding as shown below.

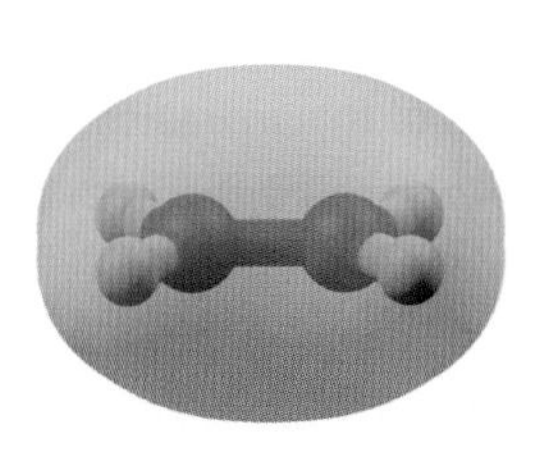

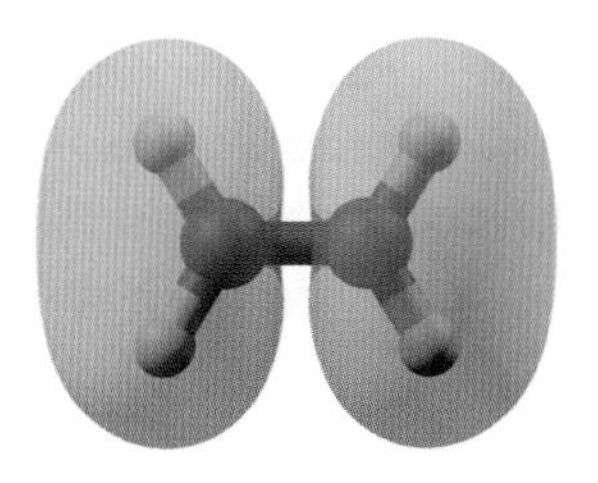

Then we have three orbitals derived from the remaining p orbitals. These are shown below, with the lowest energy orbital on the left.

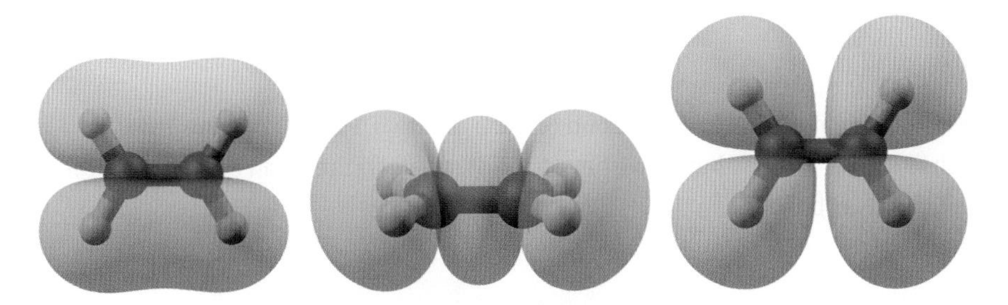

The first orbital is considered to be derived from the p_y atomic orbitals on the two carbon atoms, and clearly contributes to the C–C bonding, but not in a way that we would conventionally think of as a σ bond. The second orbital is derived from the p_x atomic orbitals, and it is spherically symmetrical about the C–C bond axis. The third orbital is also derived from the p_y atomic orbitals, and it does not contribute to C–C bonding. We will come back to this one after we have looked at the π orbitals. All the orbitals contribute to the C–H bonding to various extents.

We explained the increasing σ bond strength in going from sp^3 to sp^2 to sp hybridization by referring to the increasing contribution of the s orbitals. We would draw the same conclusions from this molecular orbital approach. With ethene, we have taken away one of the p orbitals from each carbon to form the π bond, so that we have one s orbital and two p orbitals (per carbon) to form three bonds. Effectively, these bonds are still made up of 33 per cent s and 67 per cent p (which is the same as sp^2). The outcome is the same, whether we use hybridization or not.

Finally, we can consider the π bond. We haven't used the p_z orbitals yet. We can overlap these to give a new bonding orbital, shown below. The π bond can be considered as a two-electron two-centre bond.

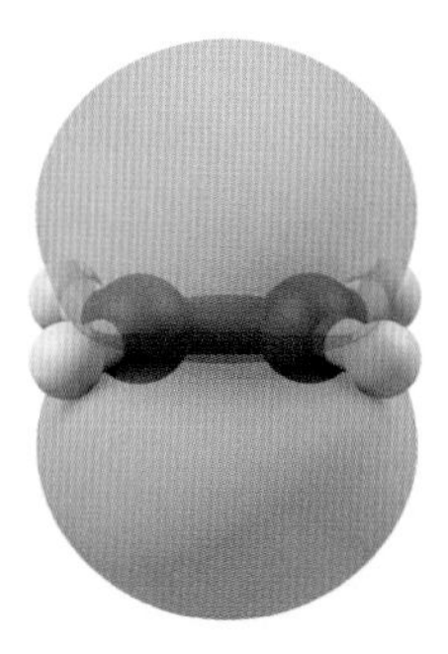

Whenever we have occupied, bonding molecular orbitals, we will have unoccupied, anti-bonding molecular orbitals. We saw the implications of this in **Basics 9**. The π^* orbital of ethene is shown below. It is exactly the same as we would see from a hybridization model, because we consider these orbitals to be un-hybridized.

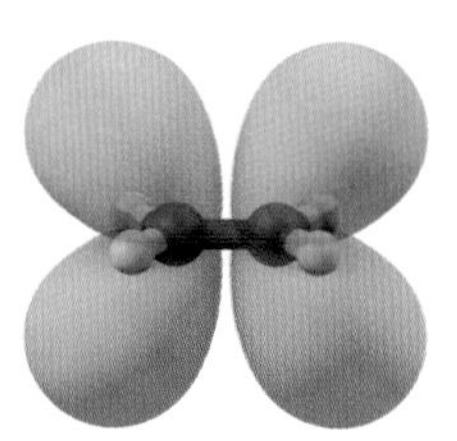

Now we can come back to the second of the p_y orbitals. Here it is again. This looks very much like the π^* orbital, but as we have seen, it is filled rather than empty. It contributes to all four C–H bonds.

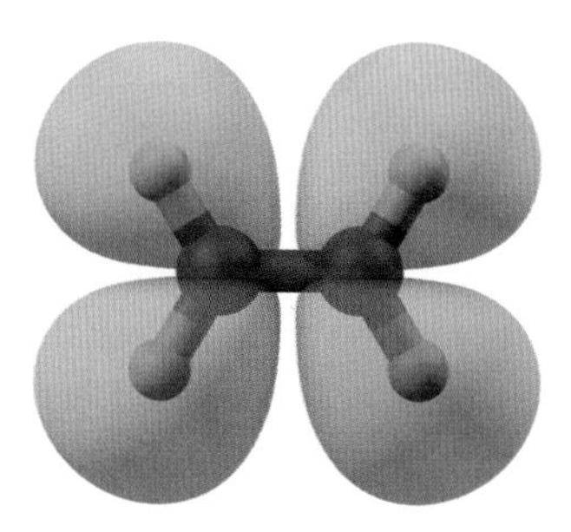

A MOLECULAR ORBITAL REPRESENTATION OF WATER

Water is an important molecule, and we can use it to illustrate key points about the bonding in the vast array of organic compounds that contain an oxygen atom. Molecular orbital theory can also be used to explain why water is bent rather than linear (hybridization can also be used to explain this!).

First of all, we have eight electrons to accommodate in four orbitals. The 'conventional' explanation, which is supported by hybridization, is that you have two O–H bonds, and two lone pairs of electrons.

You will doubtless have been told that oxygen lone pairs resemble 'rabbit ears'.

Here are the two lowest energy molecular orbitals of water. The one on the left is derived from the s orbital on oxygen overlapping with the s orbitals on both hydrogen atoms. The one on the right, which is a bit higher in energy, is derived from overlap of one of the p orbitals on oxygen with the s orbitals on the hydrogen atoms. So far, so good. Overall these represent two single bonds, although it would not be clear why water is bent. Surely the second orbital would give better overlap if water was linear?

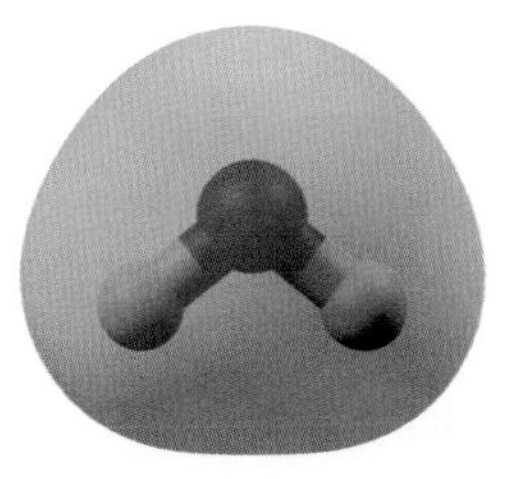

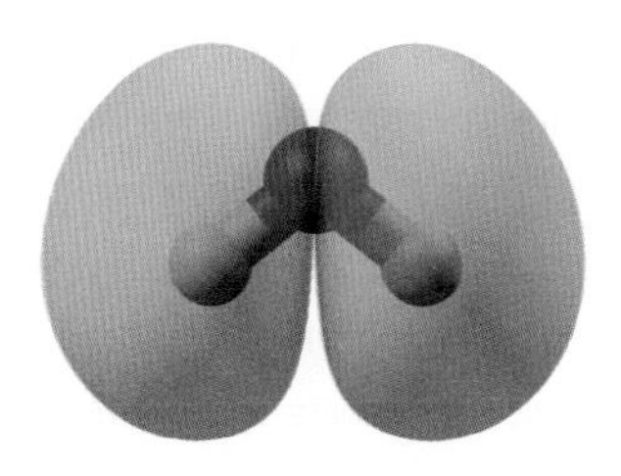

The next orbital explains this rather nicely. It is derived from a p orbital on oxygen, and one lobe of this orbital overlaps with the s orbitals on both hydrogen atoms. In the hybridization model we would expect this to be a lone pair, whereas the molecular orbital explanation shows that this orbital does contribute to O–H bonding. We couldn't retain this interaction (as well as the one above right) if water was linear.

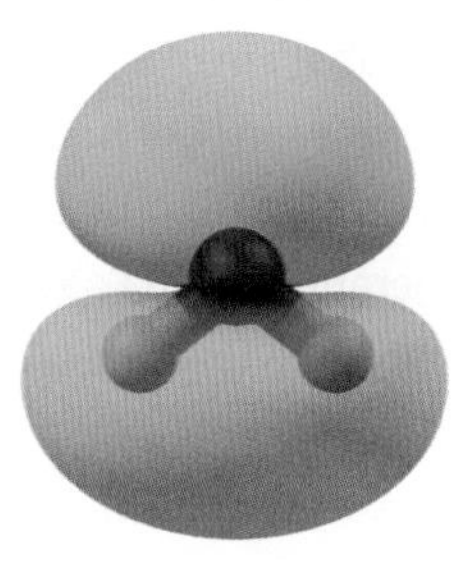

The final orbital is shown below. I have turned the structure sideways to show it more clearly. In essence, this orbital is a p orbital on oxygen that does not contribute to bonding. This is definitely a lone pair.

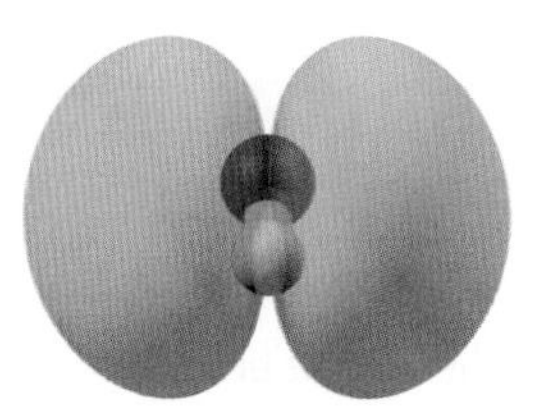

Why does this matter? Let's have a look at a very simple process, protonation of water to give the hydronium ion.

The hydronium ion is trigonal pyramidal. You can think of it as being tetrahedral with one site being occupied by a lone pair (although naturally the molecular orbitals will provide an alternative interpretation). So, we could think of water as being tetrahedral with two sites being occupied by lone pairs, one of which we are protonating.

In using this interpretation, we are implying a geometry to the protonation. We are tending to assume that the proton is being delivered along such an axis.

In fact, the favoured geometry of protonation of water is such that all three hydrogen atoms and the oxygen are in a plane. It is only when protonation is almost complete that the geometry distorts to become trigonal pyramidal.

I ask again, why does this matter? We get to the same end point, and it is much simpler to think of oxygen having two entirely non-bonded pairs of electrons. Well, there are many reactions that rely on protonation of an oxygen atom. Carbohydrates are a major part of your diet, and your metabolism of carbohydrate relies on the hydrolysis of glycosidic (sugar) linkages. The enzymes in your body that carry out this task protonate the sugars prior to hydrolysis, and they need to be optimized to deliver a proton in the correct orientation. I am not exaggerating when I say that the geometry of oxygen lone pairs and their protonation is a matter of life and death!

Having said that, don't worry about it too much! Your body knows how to break down carbohydrate. If you find it more convenient to think of water as having two equivalent lone pairs, go for it. Most of the time, when you are drawing mechanisms, it won't matter. Just be aware that it ***can*** *matter, and you have an alternative explanation that you can use.*

A MOLECULAR ORBITAL REPRESENTATION OF CHLOROMETHANE

In Basics 9 we first encountered the idea of breaking a bond by putting electrons into its antibonding orbital. We have seen this idea with the S_N2 substitution reaction, focusing on hybridized orbitals. What I would like to do now is show you that we can do exactly the same thing using 'real' orbitals.

When we looked at the bonding in methane, we found that none of the C–H bonds were aligned on the same axis as a p orbital. This meant that there were no σ bonds that were derived mainly from one of the p orbitals.

When we look at chloromethane, CH_3Cl, we find that the situation is slightly different. We are going to ignore the s orbital on carbon, as it is similar to the situation in methane.

Now let's look at the p orbitals. The p orbitals on carbon can overlap with the p orbitals on chlorine (green). This adds a little complexity, because there is a bit of π bonding between C and Cl. You wouldn't have expected this from hybridization. However, we can compare this to hyperconjugation. We can see that there is overlap of the p orbitals on carbon with the s orbitals of the hydrogen atoms. This is in addition to the overlap from the s orbital on carbon that I haven't shown you.

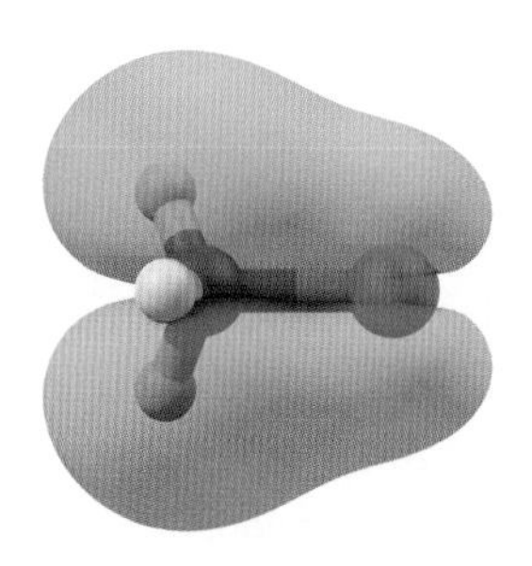

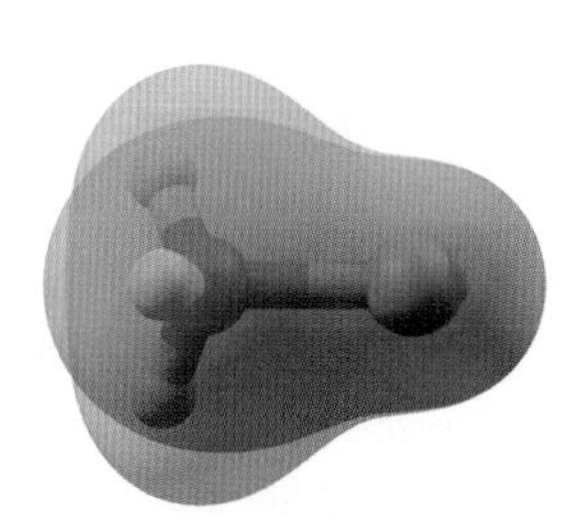

The remaining p orbital on carbon is more interesting. It is aligned along the C–Cl bond, and overlaps with a p orbital on Cl. This bonding orbital contributes mainly to the C–Cl bond, although there is also some contribution to C–H bonding.

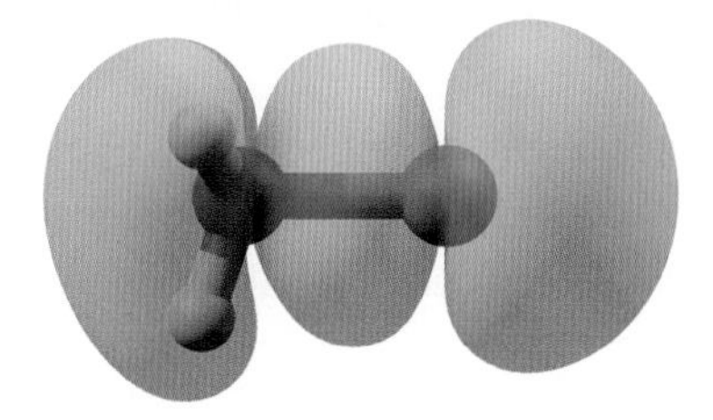

However, for an S_N2 substitution, it isn't the bonding orbital that we are interested in. It's the antibonding orbital. The antibonding orbital that corresponds to the bonding orbital we have just looked at is shown below. There is a node between the C and the Cl. If we want to break the C–Cl bond, we simply need to put some electrons into the antibonding orbital. We get these from the nucleophile, which could in principle attack at carbon on the left, or on Cl at the right. Energetically it is much better to attack at carbon and have Cl^- as the leaving group.

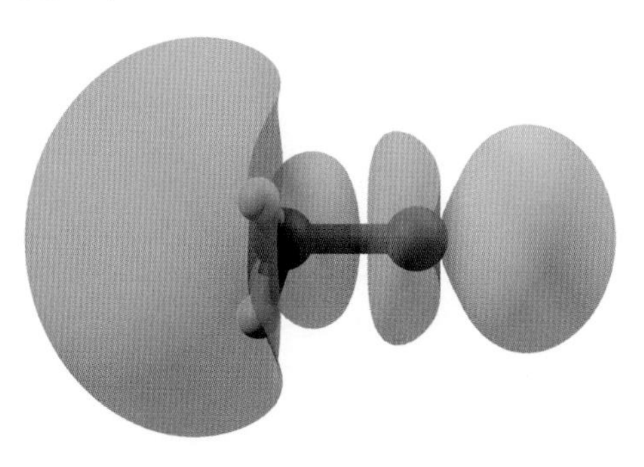

It is reassuring that we would expect the same outcome using hybridized orbitals and 'real' orbitals. It should give you more confidence in using hybridization as a model for bonding!

WHAT SHOULD YOU DO WITH THIS INFORMATION?

The short answer to this question is 'very little'.

If, like me, you didn't like the idea of hybridization, I want you to become more comfortable with it by understanding that it is only a model for the actual bonding. There may come a point where you need to move beyond hybridization to a more complete molecular orbital theory, and I'd like the information to be there for you.

For me, there is an elegance to molecular orbital theory. We don't need to pick and choose when we use three-centre two-electron bonds. **Every** bond is multi-centre, multi-electron, even in a simple molecule such as methane. However, this comes at a cost. When you describe a reaction in which one bond is broken or formed, it is convenient to think of this bond as a discrete entity. With molecular orbital theory, you lose this convenience.

Hopefully, with the example of chloromethane above, you will see that actually you don't lose **all** of the convenience, and the similarity between the sp^3 hybridized σ^* orbital of chloromethane and the 'real' molecular orbital will convince you that it is perfectly acceptable to use the former.

In 99.9 per cent of situations, it is absolutely fine to continue to use (and talk about) hybridization so that it becomes a natural part of your language.

FUNDAMENTAL KNOWLEDGE RECAP 1

Bond Lengths and Strengths

Much of the focus of this book is building patterns and habits. I have tried to keep the factual information to a minimum, but of course it is still there.

Here is the problem. A chemical bond is between 1.0 and 1.6 Å in length. I don't know about you, but I cannot imagine what that means.

It's such a small number, 10^{-10} m.

The only way to deal with this is to learn the numbers. The same applies to bond strengths, about 300–450 kJ mol^{-1} for single bonds.

If you don't know what a typical bond length/strength is, you won't know whether a bond in a given structure is stronger or weaker than the average (or shorter or longer).

Here are our representative bond lengths again.

1.57 Å
1.09 Å
$H_3C{-}CH_3$
1

1.35 Å
1.07 Å
$H_2C{=}CH_2$
2

1.21 Å
1.06 Å
H−C≡C−H
3

And the table of bond dissociation energies.

Average Bond Dissociation Energy/kJ mol^{-1}					
H−H	436	**C−C**	350	**C=C**	611
H−Cl	432	**C−H**	410	**C≡C**	835
H−Br	366	**C−O**	350	**C=O**	732
H−I	298	**C−Cl**	330	**C≡N**	898
H−O	460	**C−Br**	270		
H−S	340	**C−I**	240		

You need to get to grips with these, so that when you see some of the weaker interactions (such as hydrogen bonding, about 25 kJ mol^{-1}), you will be able to put them into context.

There are quite a few other non-covalent interactions, but we don't want to get bogged down with detail.

FUNDAMENTAL KNOWLEDGE RECAP 2

pK_a

Arguably, this is even more important than the previous one. Remember that we use the pK_a scale to quantify the stability of carbanions.

If it isn't stable (high pK_a) then it won't be easy to form.

I'm not going to define pK_a again. Whenever I think about the acidity of the hydrogen atom in an organic molecule (which is the same as thinking about the stability of the resulting anion) I would not think through the definition and the equations.

I would think quite naturally about what is a high number and what is a low number.

The meaningful process is whether a particular base is 'strong enough' to remove a particular proton.

Fortunately, after a while you will get used to seeing which bases are used in particular reactions. The first step is to get used to a selection of the numbers, so that you will find it easier to 'slot in' new examples.

Here is the table of pK_a values that we first encountered in **Basics 18**.

Compound	pK_a
CH_4	48
$CH_2=CH-CH_2-CH=CH_2$ (H H)	32
$H_3C-C\equiv C-H$	25
$H_3C-C(=O)-OCH_2CH_3$	25
$H_3C-C(=O)-CH_3$	19

Compound	pK_a
H_2O	15.7
cyclopentadiene (H H)	15
$H_3C-C(=O)-CH_2-C(=O)-CH_3$	13
H_3C-NO_2	10

You **do** need to know all of these numbers!

SECTION 3

A FOCUS ON SHAPE

INTRODUCTION

We are making good progress. We have looked at some aspects of shape. We have seen that the steric crowding in a *tertiary* alkyl halide makes it easier to form a carbocation—this steric effect is in addition to the electronic reasons.

Now it's time to add more information about **stereochemistry**—how the 3D properties of molecules in general relate to their reactivity.

*Stereochemistry covers **all** aspects of the shape of molecules.*

Most aspects we will consider at this level concern chirality, in which a molecule is not identical to its mirror image. We will see that this has implications for the S_N1 and S_N2 substitution mechanisms.

More fundamentally, you need to get good at drawing and recognizing the three-dimensional representations of molecules, and determining whether two structures represent the same, or different, **stereoisomers**.

The rules that govern this are easy to summarize and easy to understand. But you need to practise applying them until you can do it without thinking about it.

Whatever else happens in this section, don't ever tell yourself that you can't do it. You can. It may take more effort for you than for someone else, but that doesn't change the fact that you absolutely **must** 'get' stereochemistry.

This is a topic that you cannot leave until 'revision'.[1] You need to identify the key skills and to understand how (and when!) you will build them.

[1] Much like the rest of organic chemistry. I know I'm nagging, but I'm trying to make your life easier.

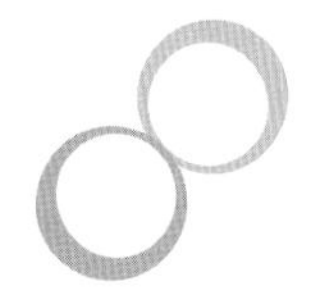

HABIT 4

Representing Stereochemistry—Flying Wedge and Newman Projections

We need conventions that allow us to see the shape of molecules. Flying wedge projections, in which atoms not in the plane of the page are drawn with either a wedged (coming out of the plane of the page) or a dashed (going into the plane of the page) bond, are the standard.

In addition, Newman projections are very good when we want to look directly down bonds. Let's define what these are.

Knowing what they are is the easy bit. The difficult bit is to use them often enough so that you can do it without thinking. Then that becomes easy as well!

FLYING WEDGE PROJECTIONS

We already mentioned the use of wedge and dash bonds in **Habit 1**, but it's time to build on this notation. Here is a structure, this time an amino acid such as the ones found in proteins in your body. There are three tetrahedral carbon atoms, but one of them is a CH_3 group, so we don't need to worry about that.

The other two have, in the orientation shown, the OH/NH_2 groups sticking up towards us and the H atoms pointing down into the page.

There really isn't a lot more to say about this. Here is a 3D representation of the same molecule.

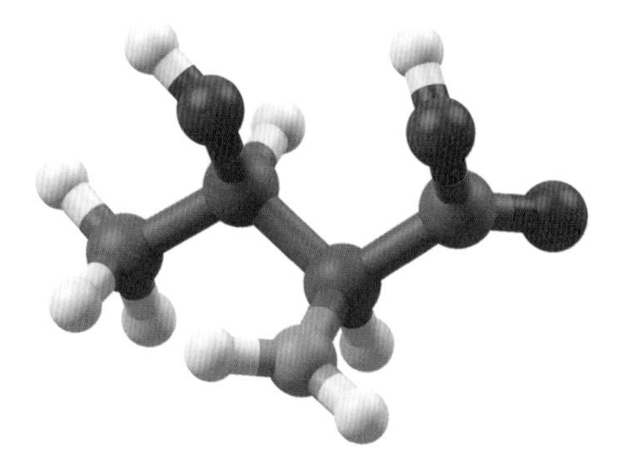

We could simplify the drawing as follows.

Now we have omitted the hydrogen atoms, as we have done before.

Of course, the two carbon atoms that we removed them from are still tetrahedral, and the hydrogen atoms are still there on the structure.

If we look at the carbon with the –OH group attached with a wedged bond, we can see two 'straight' bonds that are in the plane of the page, so that the 'missing' substituent **must** be in the position shown in the first structure.

Again, there isn't much more to understand. The key point is that you must be able to look at a structure and picture how it would look in three dimensions. This gets easier with practice, and at first you will find that making a molecular model helps you connect the structure you have drawn with the three-dimensional 'picture' in your mind. After a while you will find you don't need to make a model to see the shape of a molecule.

Once we have established all of the key principles, we will practise drawing flying wedge structures until you get the hang of it.

NEWMAN PROJECTIONS

A Newman projection is a stereochemical representation designed for looking directly down a bond. The Newman projection for ethane looks like this.

There is an important convention with this representation. The three lines that join in the centre represent bonds to the **front** carbon atom. So, the three 'H^a' are on the front carbon. The three shorter lines that join the circle are on the rear carbon atom. It is important that you know this convention. As with all conventions, if you use it lots, you will get used to it and you will get better at interpreting/drawing it.

Is this an intuitive representation with realistic perspective? I don't think so! Does it confuse me? Not anymore, because I am used to it!

Make a model of ethane, look down the bond and convince yourself that the Newman projection looks about right.

We will come back to Newman projections in Section 5.

BASICS 24

Isomerism in Organic Chemistry—Configurational Isomers

INTRODUCTION

The study of the three-dimensional shape of compounds is really important. The three-dimensional shape of a compound can affect many properties, most significantly the biological properties. In the most extreme case, one stereoisomer of a compound could cure a person of a disease. Another stereoisomer of the same structure could kill them. It is impossible to overstate the importance of stereochemistry. However, that doesn't necessarily make it complicated.

We have seen how to use wedged and dashed bonds to indicate the shapes of molecules. We now need to build on these ideas and introduce some terminology.

We will start with some definitions.

DEFINITION—STEREOISOMER

Stereoisomers are isomers with the same connectivity, but having a different arrangement of atoms in space. There are several types of stereoisomer, but for now we will only worry about one type—configurational isomers. We will define the term in a minute! Let's look at the two different types of configurational isomer first.

DEFINITION—ENANTIOMER

An enantiomer is a type of stereoisomer that is a mirror image of the original structure while having a different arrangement of atoms in space. We also add the qualifier 'non-superimposable' to the definition—this simply allows us to state that the mirror image is not the same as the original compound.

Here are the enantiomers of the amino acid alanine.

It isn't easy to explain why these are mirror images.

Make a model of each structure and convince yourself that they are!

DEFINITION—DIASTEREOISOMER

This is a type of stereoisomer of a compound that is **not** a mirror image. It would also have to not be the same compound. We use the concise wording 'non-superimposable non-mirror image'. I find that this wording confuses some people. Don't over-think it.

Here is an example. We will discuss this compound in detail in Habit 5. For now, try to convince yourself that the two structures do not represent the same compound, and they are not mirror images.

KEEPING IT SIMPLE!

Let's make a very clear point about stereochemistry. If you are given two structures that possess the same connectivity of all atoms, you have only three possibilities.

1. They are exactly the same compound.
2. They are not the same, but they **are** mirror images.
3. They are not the same and they **are not** mirror images.

If they are not the same, but they are mirror images, they are ***enantiomers****. If they are not the same and they are not mirror images, they are* ***diastereoisomers****.*

If you understand (and internalize) this, you will have a strong basis for your understanding of stereochemistry.

DEFINITION—CHIRAL

A compound is **chiral** if it is not superimposable with its mirror image. It can be said to possess **chirality**.

If a compound has an enantiomer, it is chiral. If a compound is chiral, it has an enantiomer!

Now we can get to the heart of the matter. What structural feature do we need in order to have configurational isomers?

DEFINITION—STEREOGENIC CENTRE

A stereogenic centre is a tetrahedral atom, most commonly carbon (sp^3 hybridized), with four **different** substituents attached. Having a stereogenic centre is the most common way for a compound to be chiral.

DEFINITION—CONFIGURATIONAL ISOMER

Configurational isomers are stereoisomers that can only be interconverted by a process that involves the breaking and forming of bonds. For example, two enantiomers of a chiral compound are configurational isomers. The 'configuration' is the sense of stereochemistry. We will classify the configuration of stereoisomers in Habit 6.

Look at the pairs of stereoisomers above. Make models of them, and try to interconvert the isomers. You will find that the only way to do this is to break two bonds and swap the substituents.

If you have to break bonds to interconvert the stereoisomers on a model, you would have to do the same chemically. Configurational isomers really do live entirely separate existences!

WHERE ARE WE GOING WITH THIS?

You'll build your understanding of stereochemistry better if you appreciate what you might need to do with stereochemical representations.

- You will need to determine if two structures represent the same stereoisomer, or different stereoisomers.
- You will need to determine whether a particular compound (structure) is chiral.
- You need to know how many different stereoisomers could exist for a given compound.
- You will need to determine the stereochemical outcome (which stereoisomer is formed) of reactions.

The last two are linked. How can we know **what** outcomes are possible if we don't know **how many** outcomes (stereoisomers) are possible?

It can sometimes appear complicated because the same structure can be drawn in several different ways. After a while you will simply get used to it. It is one of those skills that you should develop through consistent practice. If you do a little each day (or week), you'll build your skills and confidence.

You don't want to get to your exam and have to work too hard on stereochemistry. Stereochemistry should be as much a part of your language as the words on the page.

There is just one more term relating to types of stereoisomers. I deliberately left it to the end, as we don't encounter it too often. I don't want you to be confused by this.

DEFINITION—EPIMER

'Epimer' is a specific stereochemical term for a diastereoisomer of a compound in which all stereogenic centres but one are the same. Here are the structures of two sugars, glucose and galactose. They both have five stereogenic centres. Four of them are the same and one is different (shown in purple on the galactose structure). Therefore, glucose and galactose can be described as **epimers**. They are also diastereoisomers, but **epimers** is the specific term for these types of diastereoisomer.

You would not describe two diastereoisomers with only two stereogenic centres as epimers. That would just be pointless and confusing.

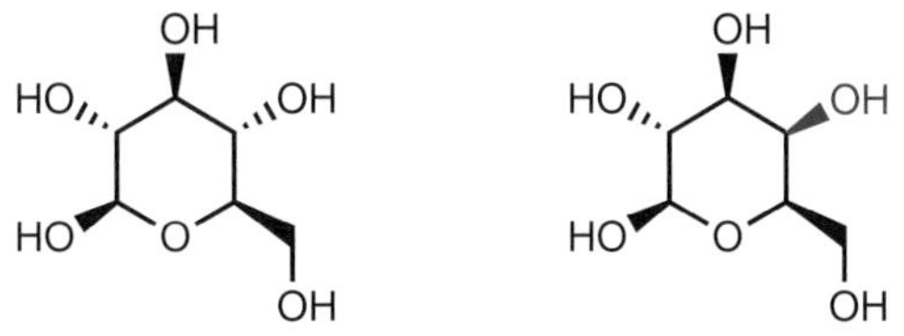

glucose

galactose

You probably won't encounter this term very often, and **diastereoisomers** still includes epimers, so it isn't an 'essential' term.

We will come back to this example in Basics 25 when we discuss the properties of stereoisomers.

HABIT 5

Getting Used to Drawing Stereoisomers

THE SAME STRUCTURE DRAWN IN DIFFERENT WAYS

We are going to use amino acids as examples in the following discussion. The proteins and enzymes in your body are polymers of amino acids, and with only one exception, the amino acids are chiral. Only one enantiomer of each amino acid is found in your body.

Here are two representations of the amino acid 'alanine'. These are actually both the same structure. It doesn't actually matter whether I have the NH_2 or the H to the left or right. We really want the NH_2 directly in front of the H, but this would clutter the representation, so you need one to be slightly to the left and the other slightly to the right. It's just a convenience.

3.1

H_3C CO_2H / H_2N H H_3C CO_2H / H NH_2

Here are two more representations of the same structure.

HO_2C CH_3 / H NH_2 HO_2C CH_3 / H_2N H

All I have done is rotate these structures by 180° so that the CO_2H moves to the CH_3 and vice versa. Don't over-think this. The point is that you learn to recognize the wedged and dashed bonds as indicating shape, and that you simply get used to them. The words on this page are simply 'abstract' shapes that you have trained your brain to recognize as words and letters. You **can** do the same with stereochemistry. Don't take short cuts either.

Make this structure with your molecular model kit and twist it around in your hands so that you convince yourself that all of the structures above represent the same stereoisomer.

Here are four more representations of the same structure (with the same stereochemistry).

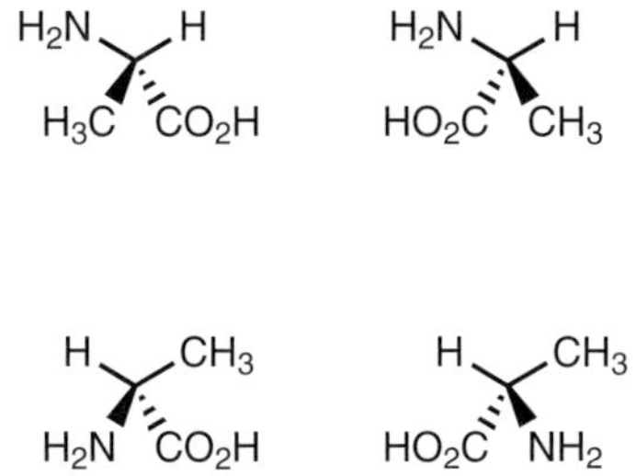

We didn't need to use an example of a chiral compound to make this point. Any structure, chiral or not, can be drawn in different orientations. But alanine is chiral, so we will now look at its mirror image.

ENANTIOMERS

We defined enantiomers above in Basics 24. In order for a compound to have an enantiomer, it needs to be chiral. The simplest way for a compound to be chiral is if it has one stereogenic centre.[2] So, all of these definitions are interlinked.

It's probably easier to just jump in and have a look at some structures. Here are two mirror image structures of the amino acid 'alanine'. We are going to see what the '*R*' and '*S*' in the compound names mean in Habit 6. For now, we will simply say that they are labels that are used to describe the stereochemistry.

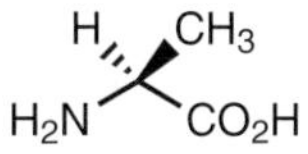

(*S*)-alanine

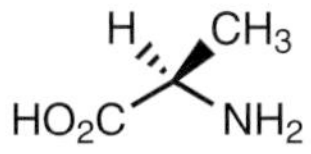

(*R*)-alanine

Here is a really important point. These are mirror images. I've drawn this with a mirror plane vertically between the two structures. However, it doesn't matter which mirror plane we choose. This means that two enantiomers of any given structure can be drawn in a number of ways. We saw this above.

Now we come to two really important points.

[2] A structure with more than one stereogenic centre will also be chiral, subject to some additional criteria that we will come to in due course.

HOW DO WE CHECK IF TWO STRUCTURES ARE MIRROR IMAGES?

We will assume for the purpose of this discussion that we already know we are dealing with two structures with the same connectivity. That is, we are either dealing with the **same structure**, **enantiomers**, or **diastereoisomers**.

In the example above, you can hopefully see that the two structures are related by a vertical mirror plane between them. But what if I draw them in a different orientation that is not so easy to see? I'm going to leave the names and labels on there.

(*S*)-alanine (*R*)-alanine

*Here's an important point. If I have a compound with **one** stereogenic centre, and I swap two substituents (any two!), I will make the mirror image of the structure.*

 Prove this for yourself with your molecular models.

In the right-hand structure, I really want the NH_2 to be where the H is. So, I simply swap them around. This makes the mirror image of the structure on the right, so I need to lose the name. It won't be '*R*' any more.

(*S*)-alanine

Now we have the mirror image structure on the right, but we have the CO_2H and NH_2 substituents where we want them. I can now swap another two substituents (in this case H and CH_3) to make the mirror image again. The mirror image of the mirror image is the original structure!

(*S*)-alanine (*R*)-alanine

We are back to our original structures, which you can hopefully see are mirror images.

Of course, the really quick way to check is to make a model of each structure and turn them around in your hands until you can see whether they are mirror images or not. Don't take the approach that you will make a model of the structure if you can't see it on paper. In the first instance, making the model should be your 'go to' strategy.

HOW DO WE PROVE THAT TWO 'STEREOISOMERS' ARE NOT SUPERIMPOSABLE?

This is not a silly question. If you have two structures and you find that they are identical, then this is clear. If you are having trouble superimposing then, does this mean that you have missed something, or you need to try harder? With practice, you will get hang of when a compound is chiral and when it is not. I know this looks like I am putting the emphasis on you, the reader, to do all the work. But if I just give you a set of 'rules' to learn, you will either learn them correctly or incorrectly. Your success in organic chemistry will then depend on how well you can learn a set of rules. However, if you can develop a deep understanding of principles, you will have to spend less time memorizing facts!

Let's start with one fact. A compound with one ***stereogenic centre***,[3] *and no other stereochemical features (such as a double bond) will exist in two stereochemical forms. These forms will be* ***enantiomers*** *of one another. This is an absolute, and anything that you can be absolutely certain of is good!*

After this chapter, I'm going to give you a lot of structures to draw. There is no point giving you a load more examples without engaging you directly in the learning process.

DIASTEREOISOMERS

These are non-superimposable non-mirror images. Sounds complicated? It doesn't need to be. Here is the key point again!

If two compounds have the same connectivity, but a different orientation of atoms in space (that is, they are not the same) then they are either enantiomers or diastereoisomers. If they aren't enantiomers, they are diastereoisomers.

[3] A stereogenic centre is often referred to as a **chiral centre**. I prefer the term **stereogenic centre**. There are many compounds with stereogenic centres that are not chiral. It seems strange to talk about **achiral** compounds having chiral centres. Similarly, the absence of a stereogenic centre doesn't necessarily mean that the compound will not be chiral. We will see an example of this in **Worked Problem 6**.

The most common way for a compound to have a diastereoisomer is if it has two or more stereogenic centres. However, this isn't the only way.

The structure below is the naturally occurring stereoisomer of the amino acid 'threonine'. We saw this before (Habit 4), but it was drawn in a different orientation.

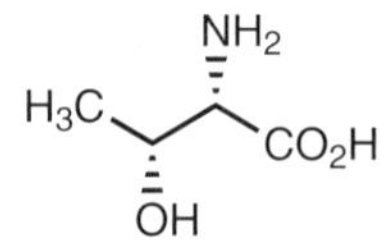

You may notice that I haven't explicitly drawn the hydrogen atoms on the stereogenic centres on this structure. There is a plan here! By mentioning hydrogen atoms and stereogenic centres, I want you to be looking for the stereogenic centres in the structure. A more 'complete' structure is shown below, but this can look a bit cluttered, and doesn't add any more information. You **need** to be thinking about shape and recognizing that there are hydrogen atoms present and where they must be based on the tetrahedral (sp^3) nature of the carbon atoms.

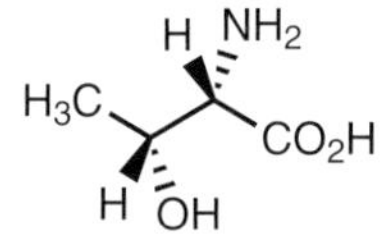

The following structure is one representation of the **enantiomer** of this structure. You can hopefully see that both stereogenic centres have been **inverted**. Make a model of both structures and compare them!

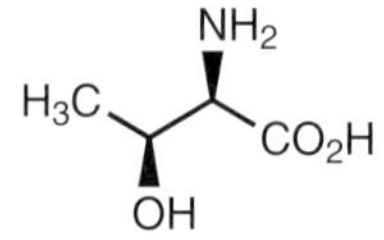

In the next structure, only one stereogenic centre is inverted compared to either of these structures. Therefore, the following structure is not a mirror image of either of the above structures. However, it possesses the same connectivity, so it is clearly a stereoisomer. If it isn't an enantiomer, it has to be a **diastereoisomer** (of both of the above structures).

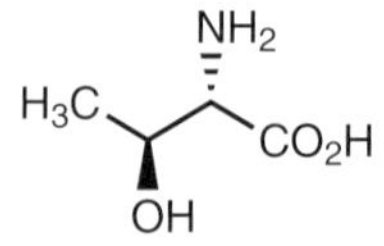

Again, we will look at lots of structures of stereoisomers in Practice 9.

NEWMAN PROJECTIONS OF STEREOISOMERS

In Habit 4 we looked at the flying wedge representation of threonine, and we also defined Newman projections. Let's link the two. Here is the threonine structure again.

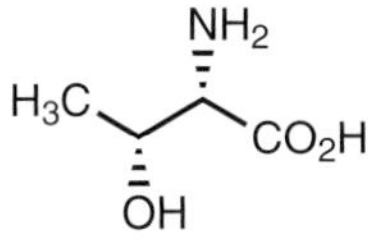

I have not explicitly drawn the H atoms on carbons 2 and 3. We could add them as follows. I've added some numbers to atoms so you will know which I am talking about.

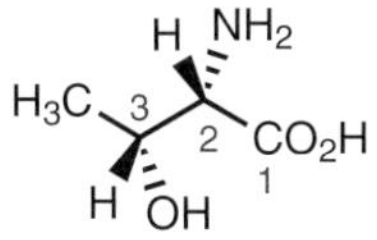

Adding the H atoms clutters the diagram, and it doesn't add anything. Since we have a straight bond from 2 to 1 and 2 to 3, and we have a dash to the NH_2, it **must** be a wedge to the H.

We can draw a Newman projection of this. Mine looks like this, but yours may be different (and still correct!) depending on which way you look at it.

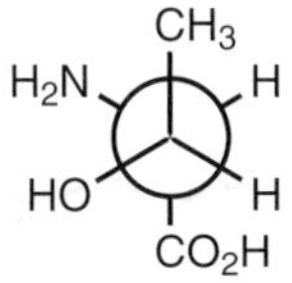

I chose to look at the structure above from left to right along the central bond, and I drew everything in the same orientation above. Once I've done this, I can rotate around the central bond to give a conformational isomer.

We are going to look at bond rotation and conformational isomerism in Section 5. For now, use your molecular models and convince yourself that you can rotate around single bonds.

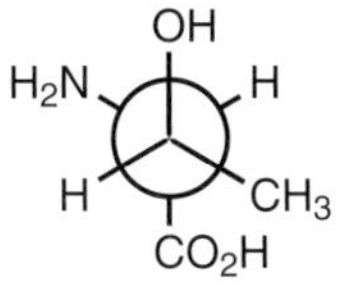

Now you could convert this back into a flying wedge projection. Here it is.

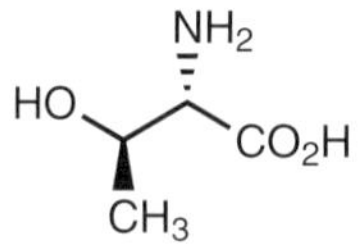

With practice (and I will give you lots of examples, and you can make up more for yourself) you will be able to convert the first flying wedge projection into the second without using a Newman projection between. Until you are confident with the procedure, do it in a way that you can be absolutely confident of getting the right answer.

The trouble with stereochemistry is that you are either right or you are wrong.

AVOIDING AMBIGUITY

When you draw a flying wedge projection, it should represent one structure and one structure alone. This isn't always easy, but there are some things you can avoid. Have a look at the following structure.

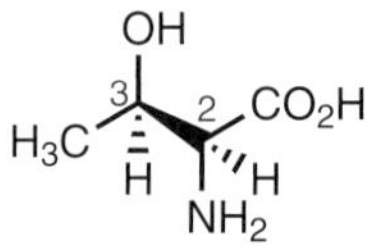

Here, we have a wedge bond between carbon 2 and carbon 3. But, does this mean that carbon 2 is in front of carbon 3? Based on the perspective, you would be tempted to assume this, but it is not entirely clear. If it is, then carbon 3 and the H on carbon 2 are both behind carbon 2. It can be confusing. The best way to avoid ambiguity is to ensure that whenever two stereogenic centres are linked, we link them with a straight bond, not a wedge or a dash. Sometimes this isn't possible, but most of the structures you see will allow this.

There is one further ambiguity that you may encounter. Look at the following two structures. The one on the right actually has the dashed bond with better perspective, since the CH_3 is moving into the page and the bond gets smaller as you get closer to it. Some chemists insist on this representation. Others (including me) are less bothered by this. If we were to apply 'perspective' to the dashed bond on the left, then we would assume that the dashed bond meant that the 'rest of the molecule' is going into the page relative to the methyl group. In that case, we could just as easily use a wedged bond. We know what we mean by the representation on the left. It's just a matter of seeing it enough times that you get used to what it means.

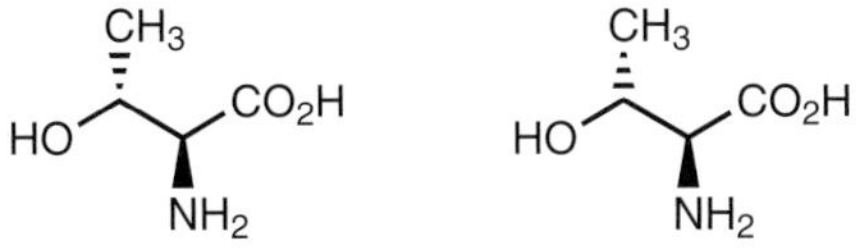

SUMMARY AND RECAP

Compounds have shape. You must know how to represent this correctly. You don't know what representation you will encounter, so you must be confident with all of them. So far, we have only encountered the flying wedge and Newman projections. There are others!

Here's a reiteration of something we said in the last chapter. If you have two structures with the same connectivity of atoms, then there are three possibilities.

1. They are the same structure. This might seem like a trivial case, but it is not. You could draw the same structure in several different ways, so we have to check if they are the same.
2. They are not the same compound, but they are mirror images.
3. They are not the same compound, and they are not mirror images.

This might seem as though it is simplifying the situation, but in fact it is reducing it to the minimum number of possibilities.

If they are the same, then they are the same! If they are not the same, and they are mirror images, then they are enantiomers. If they are not the same and they are not mirror images, they are diastereoisomers! And that's it!

It is quite common in the early stages to remember the word 'enantiomer' but not the word 'diastereoisomer'. I think this is because you tend to encounter enantiomers more. There are 'enantiomers' and 'that other possibility'!

Now, if you've got this far, you know what I'm going to say!

You need to keep working on the stereochemical definitions until you are just as confident with diastereoisomers as you are with enantiomers. A couple of structures each week will reinforce the point and reduce the effort of revising it later.

PRACTICE 9

Getting Used to Stereoisomers

THE PURPOSE OF THE EXERCISE

Up to now, some of the exercises have been a little bit artificial, in the sense that I am asking you to do 'something' in order to get you to practise the basic skills. Now, it is getting a bit more real. Organic chemists manipulate stereochemical structures of compounds all the time, and change drawings from one orientation to another. They need to be able to do this with 100 per cent confidence.

This really does sum up the problem with learning organic chemistry. If you can successfully manipulate structures 70 per cent of the time, that's good, but when you get one wrong, you will then draw incorrect conclusions and make further mistakes. The key skills need to be 100 per cent reliable.

QUESTION 1

For the following structures, draw three more stereochemical representations depending on which substituent is in the position marked as '**a**'.

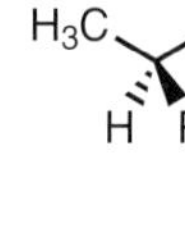

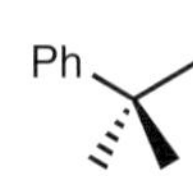

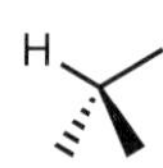

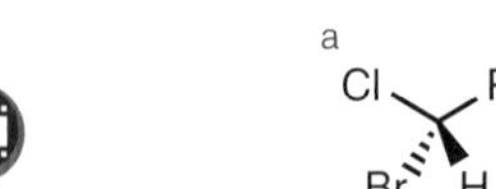

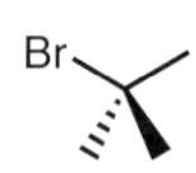

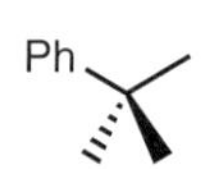

a
Ph *i*-Pr
Cl H

Cl

i-Pr

H

a
H_3C CHO
Br H

H

OHC

Br

a
H Me
Ph $CONH_2$

Ph

Me

H_2NOC

Initially that was quite difficult. However, I suspect you got into a bit of a routine, and you could see where it was going.

Have another look at this. The positioning of the atoms is ***not*** *the same for each compound. For example, in the first compound, the substituent on the left in the second structure is the ethyl group, which was in plane and to the right in the first compound. For the third structure, the H is on the left in the second structure, and this was on the wedged bond in the first structure. Go back and check that you didn't get into so much of a routine that you got them wrong without thinking!*

In fact, for each structure, there are three possible correct answers.

*What is your strategy for solving this problem? Do you make a model of the compound and rotate it? This is a good approach. Do you just look at the structure on the page and visualize where the substituents are? This is where you need to get to, but I would say it isn't a good strategy for learning. I think the 'substituent swapping' strategy outlined in the previous chapter is a good one. Assigning the stereochemistry as R or S (**Habit 6**) is also a good way to check if two structures are identical. There are lots of good approaches. Try a few and see which works best for you!*

Check that you understand what functional groups are present as well.[4]

QUESTION 2

For the following structures, work out how many **different** structures there are. Group all of the same ones together.

[4] It would be rude not to!

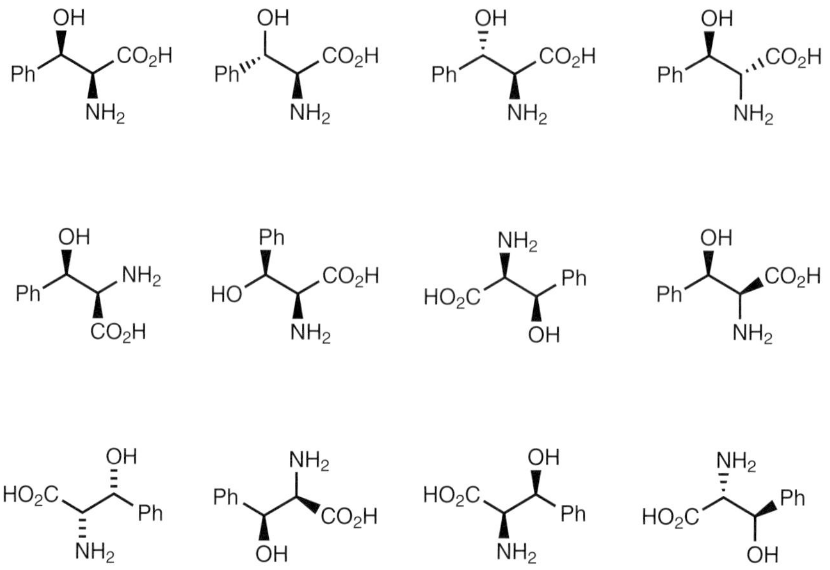

In this question, you have to rotate the entire structure through various axes, or to rotate individual bonds within the structure.[5] *Read ahead to the next chapter—the section on total number of possible stereoisomers. If you know how many stereoisomers you are looking for in total, it may help you identify which of them each structure is. Personally, I think Newman projections are a great method for solving this problem. It is easier to rotate bonds in a Newman projection (or at least it is easier to not make a mistake) than with a flying wedge projection.*

3.3

[5] Again, we won't formally discuss bond rotation until Section 5, but don't let that stop you using your molecular models and rotating bonds!

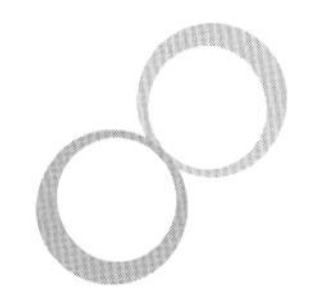

HABIT 6

Assignment of Stereochemistry—The Cahn–Ingold–Prelog Rules

It takes a little time to get used to assigning stereogenic centres. After a while, you will be able to do it without thinking. As with so many other aspects of organic chemistry, the route to competence is practice. Don't learn this once, think it is okay, and then wait until two days before an exam to 'revise' it. At that point, you will just be re-learning what you have forgotten. It's much easier to invest a little more time to work through one or two examples every week and keep it topped up.

If you find that working through one or two examples is a lot of effort, then you definitely need to do it. It ***will*** *get easier!*

INTRODUCTION

We have discussed stereochemistry and chirality in Basics 24 and Habit 5. In **Basics 3** we also discussed nomenclature of organic compounds, and we have recognized that it is important to have a name for a compound that uniquely identifies that compound. Different stereoisomers of a compound are different compounds, so we need to have names that also include the stereochemistry. That means we must have labels for the stereochemistry.

Over the years there have been a few types of nomenclature applied to stereogenic centres. Some of the older ones are still in common usage for certain types of compounds. We will mention one of these in Basics 25 but I don't want to 'dilute' the message by introducing too much terminology. Let's focus on the most important, universally accepted, nomenclature.

The **configuration** of a stereogenic centre in an organic compound can be assigned using the labels '*R*' and '*S*'. We can assign any stereogenic centre in any compound. All we need are the rules.

At first, you learn the rules. Then you practise applying the rules. After a while, you will apply the rules with little effort or conscious thought. If you don't practise, it will always be difficult.

THE CAHN–INGOLD–PRELOG RULES

There are three relatively straightforward steps.

1. Assign each substituent on the stereogenic carbon atom a priority **a–d**, where '**a**' is highest and '**d**' is lowest.
2. Align the molecule so that you are looking down the bond from the carbon atom to the '**d**' substituent (so that '**d**' is pointing away from you).
3. Look at the other three substituents and decide whether **a** → **b** → **c** is going clockwise or anticlockwise. If it is clockwise, the stereogenic centre is '*R*'. If anticlockwise, it is '*S*'.

STEP 1—ASSIGN THE PRIORITIES

We will do this entirely by example. Look at the following structure. We are starting with a compound with one stereogenic centre, but we can apply the same approach to any number of stereogenic centres, no matter how complex the molecule.

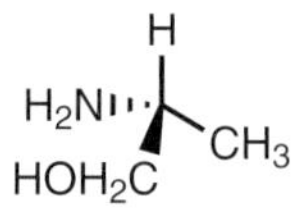

We definitely have a carbon atom with four different groups attached. It's always worth checking! We assign priorities **a–d** based initially on atomic number of the atoms attached. Nitrogen is highest and hydrogen is lowest. Carbon is in between. So, we have **a** and **d** really quickly. Label these on the structure.

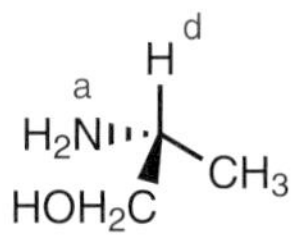

The remaining two substituents both have carbon atoms attached, so we then need to look to the next atoms attached. A carbon attached to oxygen takes priority over a carbon that is only attached to hydrogen. So, we have:

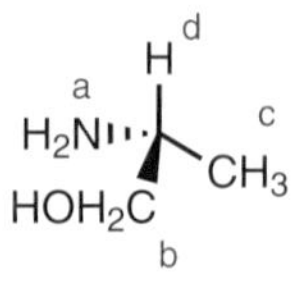

It is a common misconception that the 'total atomic weight' of the substituent determines priority. This is not true. Just look at one atom at a time until you find a difference.

Now we have determined the priorities, we are ready for step 2.

STEP 2—ALIGN THE MOLECULE

We now need to look along the bond from the stereogenic centre to the H (substituent **d**). Here is a 3D representation of the molecule in the above orientation.

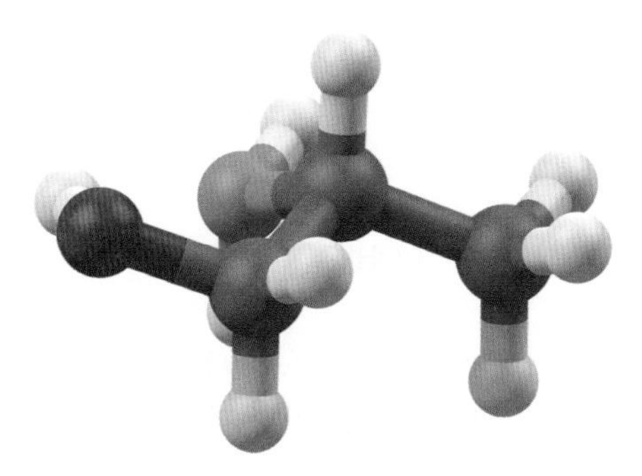

Using molecular modelling software, I can rotate this molecule so that I am looking directly down the C–H bond. You can do the same with your molecular models. This is what we see. I have angled it slightly so that you can still (just) see the hydrogen atom.

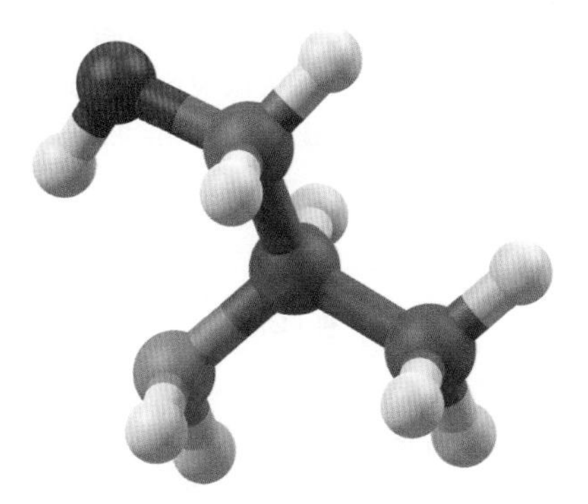

STEP 3—ASSIGN THE STEREOCHEMISTRY

From this, we can see that the **a** → **b** → **c** direction, NH_2 → CH_2OH → CH_3, is clockwise. Therefore, this is '*R*'. Had it been anticlockwise, it would have been '*S*'.

The key expression in the above paragraph is 'we can see that'.

Making a model of the compound and looking at it is the best approach. You will 'just see' it! But this isn't always practical, and we are looking at the journey you need to take. Ultimately, you need to be able to simply look at the original structure, and 'see' the direction of the substituents. Until you can confidently do this, there are some strategies that you can use.

SWAPPING SUBSTITUENTS TO CHANGE THE ORIENTATION

Let's have another look at the structure again.

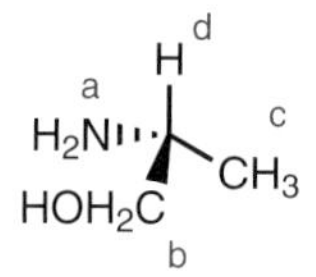

We need to look along the C–H bond from directly below the structure. In this orientation, it isn't always easy to visualize whether the substituents going from **a** → **b** → **c** are clockwise or anticlockwise.

If we want to look at the molecule above in the 'correct' orientation, we need to have the H on the end of the dashed bond. We can get it there, but we need to understand what happens when we swap two substituents on a stereogenic centre. If the centre was '*R*', swapping two substituents will make it '*S*', and vice versa. Make two 'copies' of a model of a simple chiral molecule. Swap two substituents (any two!) on one of them and convince yourself that the two models are now non-superimposable mirror images.

Now you've done that, let's swap **a** and **d** so that the 'H' is going away from us.

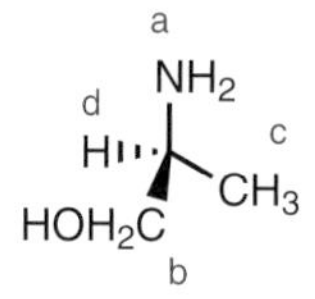

This has changed the stereochemistry of the compound. Fortunately, swapping two more substituents (any two!) will change it back to the original stereochemistry. We still want the 'H' to be going away from us, so we will leave that where it is. Let's swap **a** and **c**.

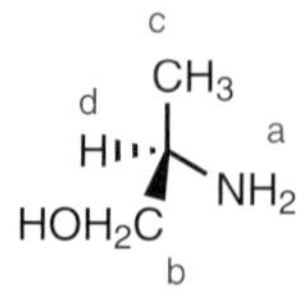

Because of what we have done, this structure has the same 'absolute stereochemistry' (*R* or *S*) as the one we started with. It is the same compound with the same stereochemistry. However, the orientation makes it easier to see whether it is *R* or *S*. In this case, going from **a** → **b** → **c** is clockwise.

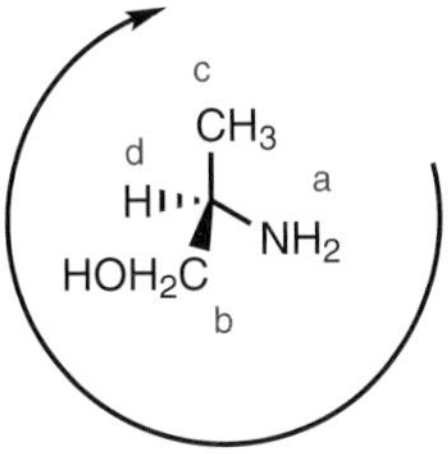

c
CH_3
d
a
H
NH_2
HOH_2C
b

Once again, this is '*R*', exactly as we saw from the 3D representation.

In terms of assigning the priorities, there isn't much more to say about it. If two substituents are identical at the first atom, look to the second. If they are identical at the second, look to the third, and so on.

There are just a few key pointers.

1. We are looking for the highest atomic number of the highest priority substituent. Therefore, CH_2OH takes priority over $C(CH_3)_3$ even though the latter has more substituents attached. One oxygen takes priority over three carbons.

Again, it isn't the 'total weight' of the substituent that matters.

2. Different is different! If the only difference is 100 atoms away, the carbon atom would still be stereogenic and you can still assign '*R*' or '*S*'.

There is one significant added complication. It is only a complication when you are first learning. Once you understand and accept it, you can practise it and internalize it. Then you'll just use it with confidence. This complication relates to how we treat double bonds.

THE CAHN–INGOLD–PRELOG RULES AND DOUBLE BONDS

We have established how we assign the priority to substituents on a stereogenic centre. There are a couple more rules you need to be aware of. The most important one focuses on double bonds.

Let's look at a quite tricky example.

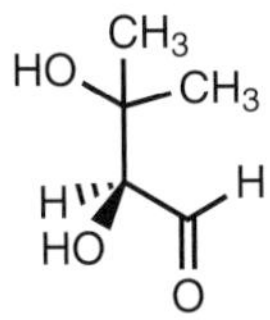

CH_3
HO
CH_3
H
H
HO
O

We have a carbon atom with four different substituents. One of them is H. I have oriented this so that it is pointing away from you, to make it a little easier. The H will be the lowest priority, **d**.

We have an 'OH' group, which is higher in priority than either of the carbon groups attached. Therefore the 'OH' group is priority **a**. So far, nothing new!

Now the new bit.

The top group has a carbon atom as the point of attachment. This is then joined to two more carbon atoms and one oxygen. The lower-right group is a carbon atom that is attached to oxygen and hydrogen.

So, you might imagine that the top group has the higher priority.

You'd be wrong. There is something we need to do to account for the double bond.

We have two bonds to oxygen. **We treat it as if** there were two oxygen atoms attached, as in the following structure.

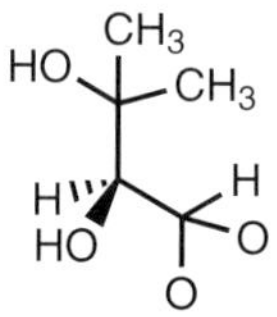

CH_3 HO CH_3 H\ HO H O O

Note that I have deliberately not added hydrogen atoms to these oxygen atoms. We are not talking about a real structure. We are talking about formalized rules.

So, for the purpose of these rules, the lower substituent has two oxygens and one hydrogen attached to the carbon. The upper substituent has one oxygen and two carbons attached to the first carbon.

Once again, we do **not** add up all of the atomic weights.[6] We take the simple view that two oxygen atoms take priority over one oxygen atom. Therefore, the priorities are as follows.

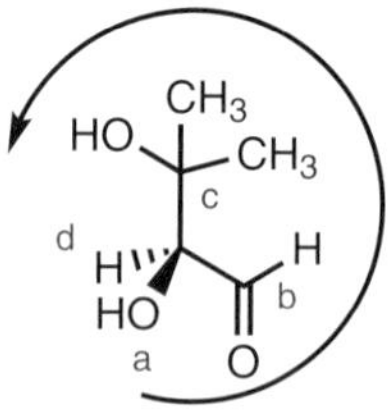

CH_3 HO CH_3 c d H\ H HO b a O

The **d** substituent is pointing away from us. Going from **a** to **b** to **c** is anticlockwise. Therefore, this is '*S*'.

[6] Anyone reading this book is capable of understanding this point. The only reason to get this wrong is if you have not practised it sufficiently to make it part of your toolkit.

So, to summarize, when you have multiply bonded substituents, you treat it as if the same substituent was attached by a single bond twice (double bonds) or three times (triple bonds).

STEREOISOMERS WITH MULTIPLE STEREOGENIC CENTRES

If a compound has more than one stereogenic centre, you simply assign them one at a time.

Here is an example. The compound below has two stereogenic centres. The priorities are shown below for the one on the left. You'll see I have taken a short cut. There is an H attached to this carbon. It is the **d** substituent. It might look like I have ignored it.

I definitely haven't. I know it is there, but I don't need to draw it to keep track of it.

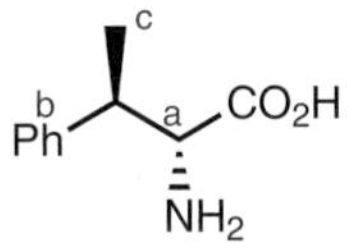

c
b a CO_2H
Ph
NH_2

The substituents **a**, **b**, **c** are arranged clockwise. The H is down. Therefore, in this orientation we are looking from the stereogenic carbon towards the H. This is the correct orientation so this stereogenic centre is '*R*'.

Remember what Ph is![7]

Now for the second stereogenic centre. I've taken the same short cut. However, this time the H is up. Therefore, we are looking from H towards the stereogenic carbon. We really need to be looking in the opposite direction. In this orientation **a**, **b**, **c** are anticlockwise. Looking from the other side they would be clockwise, so this is also '*R*'.

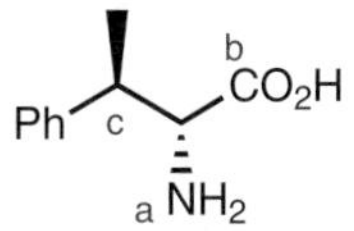

b
CO_2H
Ph c
a NH_2

As always, make a model of it if you can't see it yet. After a while this will become second nature, but only if you 'train' it.

[7] Hint—it isn't phenol!!!

STEREOGENIC CENTRES IN RINGS

This often causes confusion. What if the stereogenic centre is in a ring? Well, don't over-complicate it. If two of the 'substituents' on the stereogenic centre form the ring, then go clockwise around the ring for one and anticlockwise for the other. When you find a difference, assign the priority.

Here is an example.

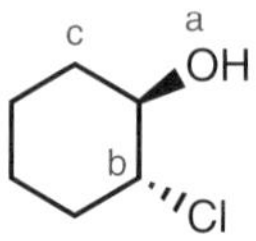

c a OH b Cl

This one also has two stereogenic centres. The priority for the upper one is shown. You simply view **b** and **c** as substituents. The carbon atom labelled **b** has a Cl attached, so it takes priority over **c**. There is an H on the stereogenic centre and this is **d**. Since it is down (it must be!), this stereogenic centre is '*R*'. Because the H is pointing down, I didn't need to draw it to be sure I had the direction of rotation correct.

Now for the second stereogenic centre. The assignment of priorities is shown below. The carbon atom labelled **b** has an OH attached, so it takes priority over **c**. There is an H on the stereogenic centre and this is **d**. Again, we have not drawn it. However, it is pointing up, so it is clear that we are looking at the compound in the wrong orientation.

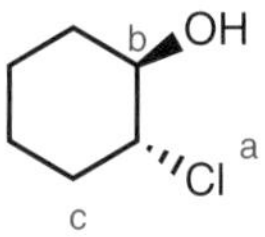

b OH Cl a c

When we started using the Cahn–Ingold–Prelog rules, we identified a method of swapping pairs of substituents to ensure that the lowest priority substituent is pointing away from us. In a ring, two substituents are linked, so this isn't easy to do. We need alternative strategies.

You could always go back to your failsafe strategy of making a model.

Alternatively, you could say that since we are looking from the **d** substituent **TO** the stereogenic centre (*i.e.* exactly the opposite direction to that needed) and we see the substituents **a**, **b**, **c** anticlockwise, then if we were looking in the correct direction they would appear clockwise so this stereogenic centre is '*R*'.

As a further alternative, you could rotate the molecule 180° about a horizontal axis. This is shown below. Now it is easier to see that the substituents are clockwise.

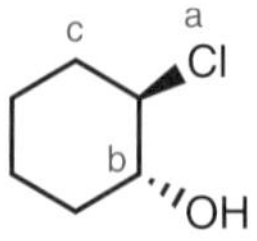

c a Cl b OH

USING *R* AND *S* TO FORMALIZE THE RELATIONSHIP BETWEEN STEREOISOMERS

If we only have one stereogenic centre in a compound, it can either be *R* or *S*. If it is *R*, the mirror image, or enantiomer will be *S*, and vice versa.

Now let's consider the case of a compound with two stereogenic centres that are not identical (you'll understand why I restrict it this way soon). Each stereogenic centre could independently be '*R*' or '*S*'.

Here is a table of the relationships between them.

	RR	*RS*	*SR*	*SS*
RR	**same**	Diastereoisomer	Diastereoisomer	**Enantiomer**
RS	Diastereoisomer	**same**	**Enantiomer**	Diastereoisomer
SR	Diastereoisomer	**Enantiomer**	**same**	Diastereoisomer
SS	**Enantiomer**	Diastereoisomer	Diastereoisomer	**same**

If they aren't the same and they aren't enantiomers, they must be diastereoisomers! That's it!

TOTAL NUMBER OF STEREOISOMERS

For any given structure, you need to know how many stereoisomers could possibly exist. If you have one stereogenic centre, it can be either *R* or *S*. Therefore, there are two possibilities, or two stereoisomers in total.

If we now add a second stereogenic centre **independently** of the first, then it can be *R* or *S* while the first is *R* or *S*. We then have a total of four possibilities, which we could write as *RR, RS, SR*, or *SS*. Note that *RS* and *SR* are different as the first letter relates to the first stereogenic centre and the second letter to the second.[8] In adding the second stereogenic centre, we double the number of total stereoisomers that are possible. If we now add a third stereogenic centre, we will double the number of possible stereoisomers again, to eight. This is because for each of the four stereoisomers we listed above, we can add either *R* or *S* on the end.

In general, with 'n' stereogenic centres, the total number of possible stereoisomers is '2^n' (2 to the power n).

[8] Think about what would happen if the stereogenic centres both contain the same substituents. Make up an example and draw it. Analyse it carefully and see if you can spot anything! We will come back to this point later!

'How many total stereoisomers exist for a compound with "n" stereogenic centres' is a common question on year 1 exams, usually with a given structure so you have to work out the number of stereogenic centres first.

It is very common to incorrectly report the number of stereoisomers as 2n or n^2.

You might make this mistake if you memorize the formula (incorrectly) rather than understanding it—you would know there is a '2' and a 'n' in the formula. Once you fully understand and internalize the logic, you won't get this wrong. You'll be relying on your understanding rather than on your memory. Then you'll be prepared whenever you encounter any added complications.

In fact, 2^n is the **maximum** number of possible stereoisomers. It is very common to not have quite as many stereoisomers. We are going to encounter this in Habit 7. There is a hint to this in footnote 8, and it relates to the fundamental definitions of enantiomers and diastereoisomers. Whatever else you observe, the fundamental definitions always have priority.

GEOMETRICAL ISOMERS OF ALKENES

Because of the π bonding in alkenes, the bonds have a high barrier to rotation. If we were to rotate the bond, we would lose the overlap between p orbitals on the two carbon atoms. We already know how strong this overlap is (it is the π bond energy, which we discussed in **Basics 6**). Therefore, alkene isomers are separable, and could be described as **geometrical isomers**. These are shown below for but-2-ene.

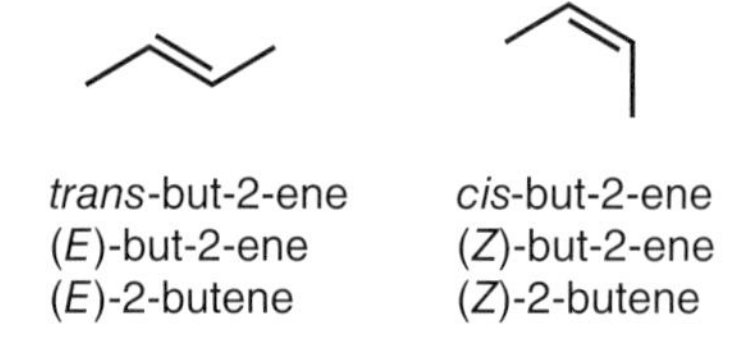

Regarding the names, we have two types of stereochemical descriptor in common use. Strictly, we use '*E*' and '*Z*'. '*E*' comes from the German word 'entgegen', meaning 'against', and '*Z*' from 'zusammen', meaning 'together'.[9] As we can see in the names above, sometimes you can put the number indicating the placement of the double bond at the start of the name, but you can be clearer if you put this number immediately in front of 'ene'.

[9] I don't think about the German words, despite having lived in Germany. I just know what *E* and *Z* mean for alkene geometry.

There is a 'trivial' nomenclature for alkene geometry, '*trans*' and '*cis*'. These refer to '*E*' and '*Z*' respectively.

In order to determine whether a double bond is '*E*' (*trans*) or '*Z*' (*cis*), look at the substituents on the two ends. Assign priorities 'a' and 'b' at each end, exactly as in the Cahn–Ingold–Prelog rules. If the two groups you assigned 'a' are on opposite sides, it is '*E*' (*trans*). If they are on the same side, it is '*Z*' (*cis*). And that's it.

Although these are commonly referred to as **geometrical isomers**, they are simply stereoisomers. Since they possess the same connectivity but are not mirror images, they **must** be **diastereoisomers**. Yes, that's right. They are not chiral. Neither of them has a stereogenic centre, but they are diastereoisomers. Remember the fundamental definition. Not the same, not mirror images.

Here's a cautionary note on how you might get 'blinkered' by questions. On occasion, I have set exam questions on the stereochemical outcome of a reaction involving formation of double bonds. What I am actually wanting in this case is the double bond geometry, *E* or *Z*. It is relatively common to equate 'stereochemistry' with 'chirality', and therefore look for a reaction that forms a stereogenic centre.

*Remember that 'stereochemistry' refers to **all** aspects of the shape of molecules.*

PRACTICE 10

Assigning Stereochemistry

THE POINT OF THE EXERCISE

During various chemical transformations, the stereochemistry of a given stereogenic centre may change. If it does, you need to know about it. The labels are very useful for keeping track.

Let's not have any illusions about this. If you give a cat a choice of two cushions, one with 'R' and one with 'S', it will sit on one of the cushions. The cat will get the stereochemistry right 50 per cent of the time! You need to do better! With experience you will get it right 99 per cent of the time.[10]

QUESTION 1

Assign the stereochemistry of the following compounds as *R* or *S*. These aren't too bad, as the lowest priority substituent is in the right place!

3.4

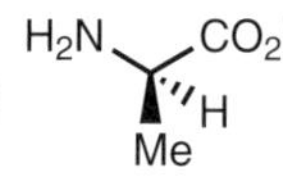

QUESTION 2

More of the same, but now the lowest priority substituent isn't always in the right place, and there are two stereogenic centres. It's all about identifying effective strategies.

[10] Everyone is entitled to make a mistake once in a while!

For each of the above structures, how many possible stereoisomers are there? That includes the one drawn! Are you sure? I want you to think about the fundamental definition of stereoisomers, not just the formula (2^n)! Have another look at this after you've read Habit 7. Did you miss anything?

Make sure you have identified and assigned all stereogenic centres! It is very common for students to miss one or more stereogenic centres in a complex molecule.

Speaking of which ....

QUESTION 3

Here is the structure of erythronolide B. Assign all of the stereogenic centres as *R* or *S*. How many stereoisomers exist in total for this structure?

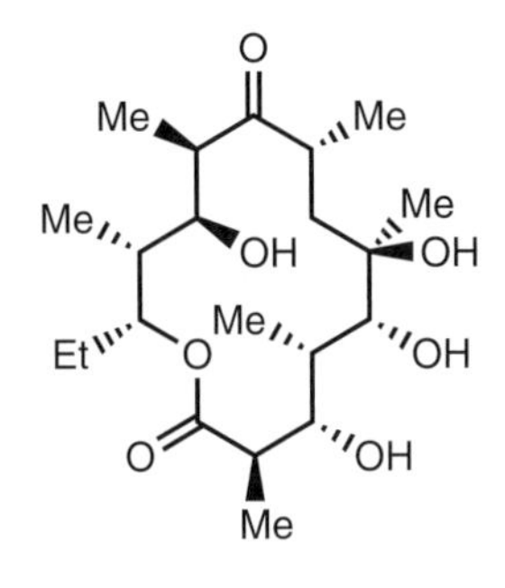

3.5

QUESTION 4

There are eight natural product structures in **Practice 3**, with (I think!) 59 stereogenic centres between them. Assign all of those as *R* or *S* as well. Assign any double bonds as *E* or *Z*.

You don't have to do all of these at once. I would recommend doing a few every week, and then keep coming back to them again. You ***can*** *find the answers online, although it isn't easy.*

HABIT 7
Stereoisomers with Symmetry

This one is another habit. It is relatively easy to look at a structure, count the number of stereogenic centres, and state the number of possible stereoisomers as 2^n (2 to the power n). Most of the time you will be correct—but not always. Sometimes, stereoisomers have internal symmetry that reduces the number of possible stereoisomers. Fortunately, you are very good at spotting symmetry.

People with symmetrical faces are considered to be more attractive. Symmetry pervades the natural world. Buildings are designed with strong elements of symmetry. Quite simply, we are used to it, and spotting symmetry in organic structures can become natural.

7

INTRODUCTION

Chirality is all about symmetry, or the lack of it, in molecular structures. Don't lose sight of the fundamental definitions though. A chiral compound is one with a non-superimposable mirror image.

This usually arises due to the presence of a stereogenic centre in a molecule, but there are other ways for a compound to be chiral. Similarly, the presence of stereogenic centres does not automatically mean that a molecule is chiral.

YOU DON'T ALWAYS GET 2^n STEREOISOMERS

In Habit 6, we reached the conclusion that a compound with 'n' stereogenic centres can have **up to** 2^n stereoisomers. It turns out that you don't always get all 2^n stereoisomers. Let's see why by looking at four structures, **1**, **2**, **3**, and **4**. All of these structures have 2 stereogenic centres, which we will assign in a moment.

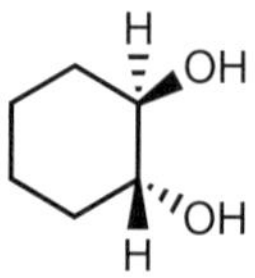

H OH OH H   H OH OH H   H OH OH H   H OH OH H

1 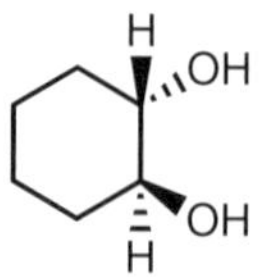   **2** 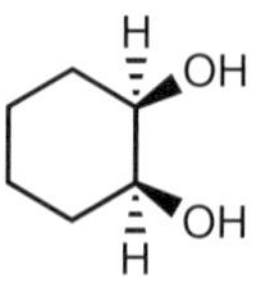   **3** 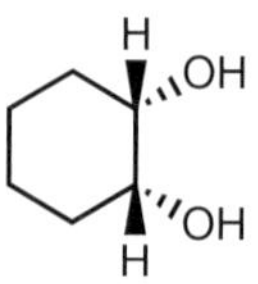   **4**

We will deal with **3** and **4** first. If we put a mirror plane horizontally through the centre of either of these structures, hopefully you can see that the top part reflects into the bottom part. They both have a plane of symmetry.

If they have a plane of symmetry, then they cannot have a non-superimposable mirror image. These structures do not represent chiral compounds!

If you were to put a pin horizontally through the centre of structure **4** and rotate the molecule 180° around it, you would bring the OH groups to the front. This would give you structure **3**.

So, **3** and **4** represent the same structure, which we have established is not chiral. We can get rid of structure **4**.

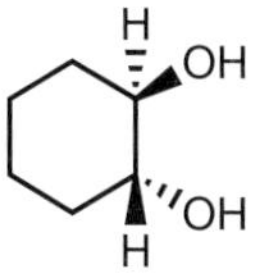

H OH OH H **1** H OH OH H **2** H OH OH H **3**

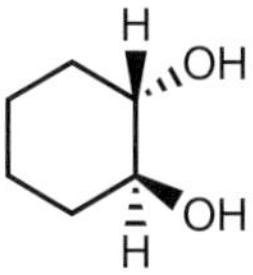

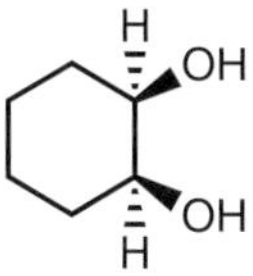

It can be productive to think about the sort of simple statements we could make about these compounds. Always keep it simple.

*In **1** and **2**, the two hydroxyl groups are on opposite sides of the ring.*

*In **3**, the two hydroxyl groups are on the same side of the ring.*

Straight away, we have determined that **3** is not the same as **1** or **2**. As long as we cannot rotate bonds so that they become the same (why do you think I selected an example with a ring?) then we have established that not all of these molecules are the same.

Are **1** and **2** the same as each other? Well, let's put a mirror plane below the structure **1**. Reflection in this plane gives the following:

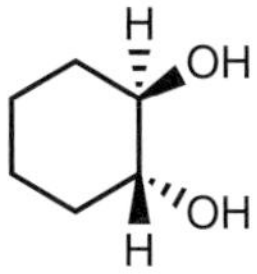

H OH OH H **1**

H OH OH H

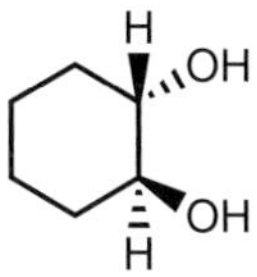

The structure below the line is **2**, so that **1** and **2** are mirror images.

But are they enantiomers? It is harder to explain that they are not superimposable, as we are having to prove a negative. Try to rotate the structures or move them in different directions. Whatever you do, you cannot superimpose them. They are non-superimposable mirror images. They are enantiomers.

What about **1** and **3**? Well, we established that they are not the same. They are not mirror images. Therefore, they are diastereoisomers.

We can do this another way. We can assign the stereochemistry *(R/S)* of each stereogenic centre. This is what we get. Check it!

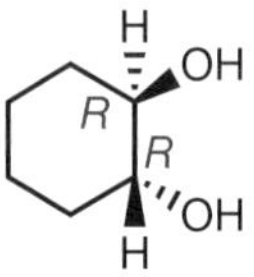

1 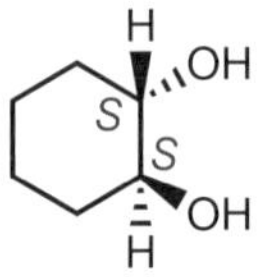 **2** 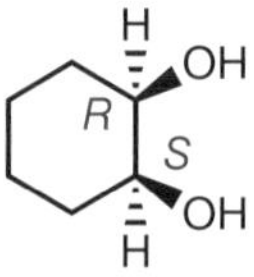 **3** 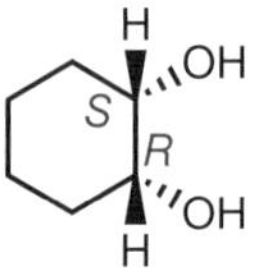 **4**

Now we can relate this back to the table we saw in Habit 6. Here it is again.

	RR	*RS*	*SR*	*SS*
RR	same	Diastereoisomer	Diastereoisomer	Enantiomer
RS	Diastereoisomer	same	Enantiomer	Diastereoisomer
SR	Diastereoisomer	Enantiomer	same	Diastereoisomer
SS	Enantiomer	Diastereoisomer	Diastereoisomer	same

The only 'complication' is that since the stereogenic centres are identical, *RS* and *SR* are now identical. The reason for this is quite simple. We cannot state which stereogenic centre is *R* and which is *S*.

Here is the revised table for this special case.

	RR	*RS*	*SR*	*SS*
RR	same	Diastereoisomer	Diastereoisomer	Enantiomer
RS	Diastereoisomer	same	**same**	Diastereoisomer
SR	Diastereoisomer	**same**	same	Diastereoisomer
SS	Enantiomer	Diastereoisomer	Diastereoisomer	same

Stereoisomer **3** is known as a ***meso*** isomer, and as a result, there are only three possible stereoisomers, rather than four. So, 2^n is the **maximum** possible number of stereoisomers.

Don't lose sight of the important definitions. **3** and **4** are achiral (not chiral) because they are superimposable with their mirror images.

The ideas in this chapter are simple, but they are often found to be confusing. As always, if you have to keep re-learning facts because you didn't internalize the principles, it will be hard work. Don't stop practising this when you think you've got the hang of it. If you really have got the hang of it, working a few more examples for 10 minutes, a couple of times a week, will keep things ticking over. This will be far more efficient than leaving it until just before the exam and having to work twice as hard.

3.6

BASICS 25
Properties of Stereoisomers

In Basics 24 we saw some definitions that relate to stereochemistry. These primarily related to the structures of compounds. As we now start to talk about reactivity, we need to introduce definitions that relate to reactions.

We will also take this opportunity to look at some of the properties of the stereoisomers that we encountered previously.

PROPERTIES—ENANTIOMERS

Enantiomers are exact mirror images. There is no reason why one enantiomer of a compound would be more stable (or less stable) than the other, so that two enantiomers will be of **equal stability**. They will have equal energy.

Two enantiomers of a compound will also have identical physical properties, such as melting point, boiling point, or solubility in the same solvents. There is only one physical property that differs between enantiomers, the **optical rotation**, which we will discuss below.

PROPERTIES—DIASTEREOISOMERS

Diastereoisomers are, at the most basic level, different compounds. They will have different stability. They can have different physical properties, such as melting point, boiling point, solubility. If you draw a reaction profile of a process involving diastereoisomers, you would normally assume that the diastereoisomers would be of different energy. If a reaction can produce two different diastereoisomers of a product, there is the distinct possibility that they will be formed in unequal amounts. We will look at the criteria for this in Basics 28.

PROPERTIES—OPTICAL ROTATION

Chiral compounds can rotate the plane of plane-polarized light. This is known as optical rotation, and the amount of optical rotation is a characteristic of a compound.

The optical rotation, or more precisely the **specific rotation**, is given the symbol $[\alpha]_D$. The amount of rotation is different at different wavelengths and temperatures, so we need to report the wavelength and temperature used for the measurement.

We don't need too much detail at this level, but you should be aware that the two enantiomers of a compound will rotate light in opposite directions, one clockwise and one anticlockwise. We refer to these as the (+) and (–) enantiomers.

Here are the two enantiomers of the amino acid alanine.

H_3C CO_2H H NH_2 HO_2C CH_3 H_2N H

The one on the left has a positive specific rotation. The one on the right has a negative specific rotation. We'll add the names. We put a (+) or (–) before the name. Instead of this, we could put an (*R*) or (*S*) before the name. Here are both types of name.

H_3C CO_2H H NH_2 HO_2C CH_3 H_2N H

(+)-alanine
(*S*)-alanine

(–)-alanine
(*R*)-alanine

*Students often confuse this 'rotation' with the 'rotation' that we use to determine configuration in the Cahn–Ingold–Prelog rules. What we are dealing with here is a physical property—interaction of matter with electromagnetic radiation. The Cahn–Ingold–Prelog rules are empirical rules to apply a descriptor to a stereogenic centre. There is absolutely **no connection** between the two.*

Measuring the optical rotation of a compound is very useful, as it will tell you how stereochemically pure a compound is. A 1:1 mixture of two enantiomers will have a zero optical rotation, since one enantiomer will 'cancel out' the other enantiomer.

A 75:25 mixture of two enantiomers will have 50 per cent of the specific rotation for a pure enantiomer of the compound. This is because the 25 per cent of the minor compound will cancel out 25 per cent of the major compound, so that in effect only 50 per cent of the major compound is contributing to the measurement.

REINFORCING THE POINT

We looked at the oxidation of glucose in **Basics 15**. We saw in **Basics 24** that glucose and galactose are diastereoisomers.

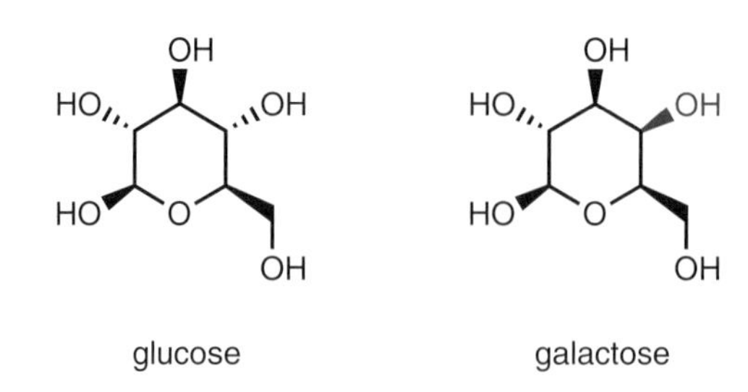

glucose galactose

 Calculate the enthalpy change for the oxidation of glucose (to $CO_2 + H_2O$) and the same for the oxidation of galactose.

I don't mean 'look up your previous answer for the oxidation of glucose'. Do it again. See if you get the same answer. See if you get there quicker!

What answer do you get in each case?

Hopefully you got the same answer in each case. You are breaking and forming the same bonds. **But** ...

 What can you say about the energy of glucose and that of galactose?

They will be different. Perhaps not by much, but they will not be identical.

 Draw a reaction profile for the oxidation of glucose and that of galactose. Don't worry about what happens in the middle. We only really need starting materials and products.

Hopefully you drew glucose and galactose at slightly different energy levels, so that the enthalpy change for oxidation in each case will be slightly different.

We don't need to worry (for now) about which is more stable. We don't have enough information to allow us to predict this with enough confidence. For now, it is enough that we know that diastereoisomers will have different energies, and they can react differently.

REACTION DETAIL 2

Stereochemical Aspects of Substitution Reactions

Right, now we are going to consider the stereochemical outcome of a reaction. There will be words that you associate with a given outcome, but we need to look at the fundamental principles, so we can see when the words apply and when they do not.

It is common in the early stages to connect the outcome of a reaction with the words used, and then apply those same words to another reaction. As a result, you might predict the wrong outcome. I will try to highlight examples of this so that you can learn to avoid the problem.

STEREOCHEMISTRY AND S_N1 SUBSTITUTION

The differing stereochemical outcomes in S_N1 and S_N2 substitution reactions is one clear piece of evidence that allows us to distinguish them. It is important that we consider the outcome.

If you follow the reaction pathway from start to finish, you will understand the outcome. If you do this, there is less to learn!

In the S_N1 substitution, there is a carbocation intermediate. We know that carbocations are planar. We can show this as follows with wedges and dashes.

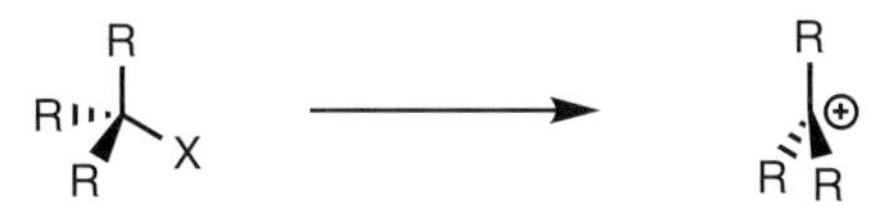

Now, here's the key question. If we have a nucleophile attacking a planar carbocation, is there any possible reason why attack from the left would be preferred over attack from the right (or vice versa)?

There is absolutely no reason. The two 'sides' of the carbocation are exactly equivalent.

Now, here's a key experiment. Let's suppose that the substrate is chiral. We need to go one step further. It needs to be a single enantiomer. Because we get a planar carbocation that is attacked equally from both sides, we will get a 1:1 mixture of enantiomers of the product. Here is a result from the pioneering work of Ingold.

Cl
Ph CH$_3$ → 60% aqueous acetone → OH Ph CH$_3$ + OH Ph CH$_3$

1:1 ratio

So, here are the words. S_N1 substitution reactions proceed with racemization.

The words are only the tip of the iceberg. The word 'racemization' is only applicable if you start with a single enantiomer of a chiral substrate. If the substrate is not chiral, or it is a racemic chiral compound (already a 1:1 mixture of enantiomers; make sure you understand the difference! We will define 'racemic' properly in Basics 29) the same principles apply, but you would not observe a particular outcome.

2

It is important, when testing mechanisms, to design experiments that will give a different outcome depending on the mechanism.

If we have a 1:1 mixture of enantiomers of the starting material, we will also get a 1:1 mixture of enantiomers of the product.

The key point at this stage is to understand, at a very fundamental level, why you could never predict which enantiomer of the product would be formed from a single molecule (which would have to be a single enantiomer!) of the starting material. Stereochemistry is information! The information is lost upon formation of the planar carbocation.

ADDED COMPLICATIONS—FORMATION OF DIASTEREOISOMERS

I really don't want to add to the complexity too much at this point. However, there is one more point worth making.

Let's consider a reaction of a substrate with two stereogenic centres. In the following example, the starting material has two stereogenic centres, although only one of them has its stereochemistry specified. We will lose the acetate group to form a carbocation, so the stereochemistry here doesn't matter.

This example is complicated. We have an allyl group as a nucleophile, which you haven't seen. We have acetate as the leaving group. We have a catalyst, bismuth triflate. We have a rather unusual solvent, nitromethane. None of this changes the simple fact that acetate leaves and allyl comes in.

In this case, you get an 87:13 mixture favouring the stereoisomer shown. This is a long way from 1:1.

When we lose acetate, we form a nice stable carbocation. The triple bond stabilizes it. If you draw the resonance forms, they will look a bit strange. If you consider the shape of the triple bond orbitals, and the stabilization they provide, you will better appreciate the resonance forms.

Draw out the carbocation, and the resonance forms. If you don't 'use' the methoxy group on the benzene ring, you have missed something.

I am not going to give you the answer to this part.

Now consider what happens if the nucleophile (formally an allyl anion) attacks from either side. What is the stereochemical relationship between the two products?

You should first of all draw the two possible product stereoisomers.

Once you have drawn them, consider the stereochemical relationship between the stereoisomeric products, and then link this to the properties that we saw in Basics 25. It might help to look ahead to Basics 28. It's all connected.

Once you've worked through this, you can decide if you will get a 1:1 mixture of the two possible stereoisomers, or if one might be favoured over the other.

Here are the two possible products.

The methoxy group isn't involved in the reaction, so the stereochemistry here doesn't change.

If the connection you make is 'S_N1 = racemization' then you could draw the second stereoisomer being the mirror image of the first. This would make no chemical sense.

The two structures above are not the same. They are not enantiomers. Therefore, they are diastereoisomers. Diastereoisomers are **not** the same compound. They have different energy. Therefore, they can be formed at different rates and in different proportions.

We will come back to this point in **Basics 28**. *Don't worry about it too much for now. The key point is that the term 'racemization' would not apply to the above example, even if we did get a 1:1 mixture of stereoisomers.*

2

STEREOCHEMISTRY AND S_N2 SUBSTITUTION

We have looked at a molecular orbital representation for bond breaking in **Basics 9**. When a nucleophile attacks, the electrons from the nucleophile have to be donated into a molecular orbital on the substrate. Since we are breaking a σ bond, the relevant orbital must be the antibonding orbital of the σ bond—the σ* orbital.

The antibonding orbital of the C–X bond is shown below. The largest lobe of this orbital is on the opposite side of the carbon to X, so this is where the nucleophile attacks.

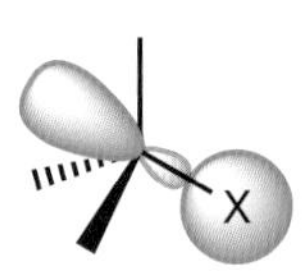

Very often, students will try to tell me that this is a steric effect—that there is more room for the nucleophile to attack at the back. We are looking at substitution at a tetrahedral carbon atom. All of the bond angles are approximately 109.5°.[11] It would be difficult to argue that any one angle of attack is better than another for **all possible substrates** purely on **steric** grounds.

So, if the correct reason is not steric, it must be electronic.

Now we can briefly consider the consequences of this.

Once again, here are the words. S_N2 substitution reactions proceed with inversion of configuration.

We need to look at this from a couple of angles. First of all, let's consider the simplest case, a single enantiomer of a chiral substrate.

[11] There will be slight variations depending on the sizes of the four different substituents.

The nucleophile attacks from the back of the leaving group, so that the net result is **inversion of stereochemistry** at the stereogenic centre at which substitution takes place.

*This is important! You should get into the habit of **always** drawing an S_N2 substitution reaction with the nucleophile attacking opposite to the leaving group. Once you've developed the habit, drawing it correctly is no more effort than drawing it incorrectly.*

As with S_N1 reactions, we have a number of possible scenarios to consider.

1. Achiral substrates.
2. Chiral substrates as single enantiomers.
3. Chiral substrates as a racemic mixture.

If the substrate is not chiral, the nucleophile still attacks from the back of the leaving group. The carbon atom still undergoes inversion.

There are just no observable consequences.

If the substrate is chiral, and a single enantiomer, then all of the molecules have the same stereochemistry. Upon substitution, they will all be inverted, so that the product will also be a single enantiomer, albeit inverted compared to the substrate.

If the substrate is a chiral compound, but a racemic mixture of the two enantiomers, then the product will also be a racemic mixture of two enantiomers. Each individual molecule will have undergone inversion of stereochemistry, but the bulk compound **must** still be racemic.

Remember—an individual molecule cannot be racemic!

If you have an achiral substrate, you will probably still talk about inversion of configuration.

This is okay, as long as you understand that you are talking about the direction of attack, and that there are no measurable consequences.

Similarly, if you have a racemic substrate, you will get a racemic product, even though the stereochemistry of each individual molecule will have been inverted.

WHAT IF WE CANNOT GET INVERSION?

Here is a compound that we looked at in the context of S_N1 substitution in **Reaction Detail 1**. Remember that we said that this compound doesn't undergo S_N1 substitution because it cannot form a stable planar carbocation.

Now we need to consider S_N2 substitution reactions of the same compound.

There is no way that the nucleophile can get 'inside' the molecule to attack from behind the Br. Even if it could, there is no way it could 'invert' this carbon atom. Think about the bond angles that would be needed!

We established that this compound does not undergo substitution by an S_N1 pathway. It also does not undergo substitution by an S_N2 pathway. Therefore, it *does not undergo substitution reactions at all.*

THERE'S ONE MORE IMPORTANT POINT ABOUT S_N2 SUBSTITUTION

Students sometimes see the words 'inversion of configuration' and automatically assume that this means that an (*S*) substrate will give an (*R*) product. This is true more often than not, but does not necessarily have to be the case.

Here's an example reaction. You only need to consider the bromine as the leaving group.

Draw the outcome of this reaction and assign the stereochemistry (*R* or *S*) of the starting material and the product.

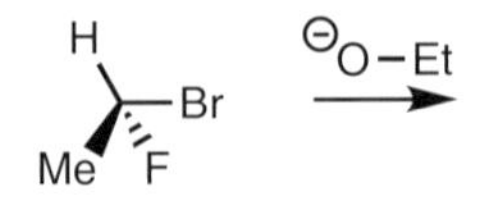

The starting material and the product are both (*R*) (or they should be!). But inversion of stereochemistry has definitely taken place (or should have!). All this tells us is that the priority according to the Cahn–Ingold–Prelog rules has changed during the reaction. In the real sense of the word, there **has** been inversion of stereochemistry.

Let's see how this might relate to an exam question. A student is given the above scheme and asked to draw and assign the stereochemistry of the starting material and the product. They will assign the stereochemistry of the starting material, in this case (R). Since they need to assign the stereochemistry of the product, and they know that inversion of stereochemistry takes place, they will state that the product will be (S) and then draw the (S) enantiomer of the product!

The problem here is associating the outcome with the words rather than with the structures. Make sure you understand why this is wrong. Learn to draw the outcome of a reaction based on shape!

SUMMING UP S_N2 REACTIONS AND STEREOCHEMISTRY

The S_N2 pathway is the most common pathway for substitution reactions at sp^3 carbon. If you have gone to all the effort of making a single enantiomer of a substrate, and you subject it to a substitution reaction, you are now able to determine whether it will react via an S_N2 mechanism. If it does, it will react with inversion of configuration, so you will get a single enantiomer product. You would not want to end up with a racemic product.

If you want a given stereoisomer of product, you would be able to decide which enantiomer of the substrate you need. Perhaps you can buy a starting material with this stereochemistry? Perhaps you need to use a stereoselective reaction to make it?

RETURNING BRIEFLY TO S_N1 SUBSTITUTION

Now you have got the basic point, I need to come back very briefly to S_N1 substitution. It turns out that even in 'simple' cases (single enantiomer substrate, only one stereogenic centre) it isn't **quite** as simple as I led you to believe.

Here is the reaction we looked at above. It turns out that the ratio of products isn't quite 1:1. We need to understand why this is the case.

Cl
Ph CH3
60% aqueous acetone
OH
Ph CH3
+
OH
Ph CH3
47.5:52.5 ratio

We get a little bit more inversion of stereochemistry. Why?

We could suggest that some of the reaction is proceeding via an S_N2 mechanism. Investigation of the kinetics tells us that this isn't the case.

It's all about timing!

Consider the following. The C–Cl bond starts to lengthen and break. The carbocation and the chloride anion begin to be solvated. If the nucleophile attacks before the chloride has *completely* moved away from the carbocation, there is more chance of it attacking the face of the carbocation opposite to the Cl. The longer the carbocation survives, the more chance it has to move away from the chloride, or to change its orientation.

*So, in practice you don't tend to get **perfect** racemization. That is a limiting case when the leaving group becomes fully separated. Most of the time you get a bit more of the product with inversion of stereochemistry.*

We will return to the idea that the fundamental mechanisms are more of a limiting case in **Perspective 4**.

COMMON ERROR 6
Substitution Reactions

THE BASICS

Good news! There isn't much here. Once you've learned the basics, you probably won't make too many mistakes with substitution reactions.

Of course, you have two major mechanisms. They have different stereochemical outcomes.

You need to be able to tell, at a glance, whether a particular substitution reaction is S_N1 or S_N2.

There are solvent effects that favour one or the other. There are steric effects that favour one or the other. But, above all of this, there is one much more important factor.

You will only ever get S_N1 if you form a stable carbocation!

WHICH MECHANISM IS WHICH?

The S_N1 mechanism has two steps. The S_N2 mechanism has one step. If you learn this by rote, it's easy to get them confused.

Make sure you really understand which is which!

When you answer an exam question, if you happen to use the wrong term, but you clearly demonstrate good understanding, I'm sure you'll do fine.

If your thought process is '*tertiary* = S_N1' and you then draw and explain why a one-step (*i.e.* S_N2) mechanism is 'correct', it won't end well.

As always, you need to make the shift from words to structures as your primary vehicle to demonstrate knowledge and understanding.

STEREOCHEMICAL CONSEQUENCES

Again, the problem here is that you use words to describe reactions. Of course, there is a reason for this. You are really good with words!

An S_N1 substitution reaction proceeds with racemization.

That's fine, but most of the S_N1 substitution reactions you will encounter are with achiral substrates, so the term 'racemization' does not apply. Try to visualize what is happening on a molecular level with any given substrate. We have seen (Reaction Detail 2) that even with chiral single enantiomer substrates, racemization is a limiting case.

An S_N2 substitution reaction proceeds with inversion of configuration.

This leads to the flawed logic that an (*R*) substrate will give an (*S*) product, and vice versa. Make sure you understand 'inversion' as something that is happening as a consequence of the direction of attack of the nucleophile, and you recognize that this happens because of the shape of the LUMO of the substrate (Reaction Detail 2).

6

That way, you are making all the right connections, and giving yourself something to build on before you deal with the 'special cases' we are going to look at.

When you look at them this way, they aren't special cases at all—just inevitable consequences of the basic mechanisms.

REACTION DETAIL 3

Substitution with Retention of Configuration

Since we considered all possible orders of bond formation for nucleophilic substitution in **Fundamental Reaction Type 1**, perhaps you were not expecting another new mechanism.

Fortunately, this one is only a variation on one of the previous mechanisms. It just happens that the timing is different, for a very good reason. It is important that you see how 'new' mechanisms relate to those you already know.

THE S_Ni MECHANISM

This is a slightly unusual variant on the S_N1 mechanism. Let's ask a question.

Suppose you had a substrate that would react in an S_N1 pathway, but the nucleophile was 'fixed' on the same side as the leaving group, what would happen?

This may seem like a strange and purely hypothetical question. However, we should look at the reaction scheme below. Reaction of alcohol **1** with thionyl chloride (**2**) gives a single enantiomer of alkyl chloride **6**. The sense of stereochemistry is the same as in alcohol **1**.[12]

[12] The configuration of the alcohol and the chloride are also the same, but of course this did not have to be the case. Work out the configuration (*R* or *S*) in both cases. Take every opportunity to practise!

There are only two ways that you can get retention of configuration in a substitution reaction. Either you really do get retention of configuration, or you could get inversion of configuration ***twice****. We will encounter this scenario in Applications 2.*

Here, it really is retention of configuration! Let's work through this multi-step mechanism. The reaction of alcohol **1** with thionyl chloride (**2**)[13] gives a compound (**4**) which is a lot like an ester—it's a sulfinate ester. Once you have encountered a few more reactions, you might find you can draw a more complete curly arrow mechanism for the formation of intermediate **3** from compounds **1** and **2**.

So far, so good. We have made the alcohol into a better leaving group, just as we did with the tosylate groups in **Reaction Detail 1**. This group then leaves to form a carbocation.

3

Here's the subtle point. Structure **5** exists as a 'tight ion pair'. The leaving group never makes it far enough away before the next step, attack of chloride, happens. This is because the chloride anion is 'delivered' by the leaving group.

As a result of this, chloride attacks from the same side—retention of configuration!

This reaction is known as **S_Ni**—internal nucleophilic substitution. Since it is a variant of S_N1 substitution, we would only expect this mechanism for those systems that form a very stable carbocation—it is pretty much limited to *secondary* benzylic systems. As such, it is not very useful in synthesis (unless this happens to be the exact reaction you want!), and I don't recall ever seeing it used.[14] Nevertheless, it is an interesting case that shows you that reaction mechanisms are a continuum in terms of the precise timing of bonds forming and breaking.

Incidentally, during the writing of this book, I found what appeared to be an authoritative web page which discusses the S_Ni mechanism and gave examples of this reaction that could never give stable carbocations, and (in my opinion, of course!) are more likely to react through mechanisms similar to those described in Applications 2. I don't think it is productive for me to cite the web page. I just want you to be aware that the internet is not always the best source of reliable information. If the proposed intermediates are not stable, a proposed mechanism is unlikely to be correct!

[13] Did you notice that twice now I have put the number **2** in brackets, but not the number **1**. This is deliberate. Thionyl chloride is the exact name for compound **2**. We could leave the number out of the sentence and it would still make sense (although it might be less clear). If we leave out the number **1**, you would not know which compound we were discussing.

[14] It will have been used, of course. Just not as frequently as S_N2 substitution.

COMMON ERROR 7

Stereochemical Errors

INTRODUCTION

Stereochemistry causes a lot of problems! The reasons for this are very simple, but they are not always easy to fix—at least not quickly!

A good grasp of stereochemistry requires you to be able to look at a structure and to visualize what it would look like in three dimensions. Once you've done this, you need to be able to rotate around a bond—all in your head—and visualize what it will look like now!

Once you can do this, you'll wonder why you found it difficult! Until you get there, you'll be convinced you will never do it.

HOW TO FIX THE PROBLEM

You already know what to do! Look at structures, draw structures, make models of structures. Keep going until you get it!

Believe, deep down, that you will get it!

Because this isn't about whether you are clever enough. It just takes time for your brain to adjust to this really important skill. I know I'm going on a bit, but maybe you'll be convinced!

Let's just recap the common issues.

THINKING THAT STEREOCHEMISTRY EQUALS CHIRALITY

When it comes to an exam question that mentions stereochemistry, students often look for an answer relating to chirality. Remember that stereochemistry relates to all aspects of the shape of molecules, including alkene geometry. Don't overlook this!

DRAWING THE WRONG STRUCTURE

We can consider this on several levels. If you draw the wrong stereoisomer, you have drawn the wrong structure. It's as simple as that. It doesn't matter which stereochemical representation you are using. You need to be equally good[15] with all of them.

In addition, when students are asked to draw stereoisomers of compounds, they do sometimes alter the connectivity so that they are drawing constitutional isomers rather than stereoisomers.

When you are drawing stereoisomers, you need to check, every single time, that what you have drawn makes sense. After a while, you'll find you no longer actually need to check. This is how you know you have got it!

DRAWING AN AMBIGUOUS STRUCTURE

We looked at this in Habit 5. You will do this lots at first. That's why you need to work with your colleagues and lecturers to ensure you are corrected. After a while, you will break the habit.

When you have drawn a structure, make a model of it. If you (or your friends) cannot be confident about the model, consider re-drawing the structure to be clearer.

Get the bond angles as close to realistic as you can.

DESCRIBING THE STRUCTURE INCORRECTLY

If you call two structures enantiomers when they are diastereoisomers, or vice versa, you're wrong. And that's it.

If you say that two structures are enantiomers, you are, in effect, saying that they have the same energy.

Always relate structures and properties.

If you say that two structures are diastereoisomers, you are saying that they have different energy and can react at different rates, or give different products.

[15] *i.e.* perfect!

IN CONCLUSION

Nag warning! Seriously, not 'getting' stereochemistry is the one thing that will hold you back. Everything in organic chemistry is related to some aspect of stereochemistry. You need to practise until you stop needing to think about it.

Work smarter, not harder!

You may think I am telling you to work harder. Nothing could be further from the truth. I am telling you what you need to do and when you need to do it.[16]

My way is easier. Trust me! It will save you time in the long run!

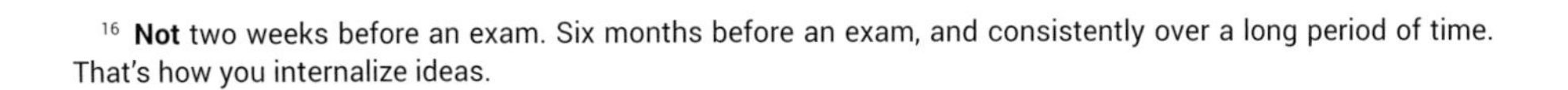
[16] **Not** two weeks before an exam. Six months before an exam, and consistently over a long period of time. That's how you internalize ideas.

SECTION 4

TYPES OF SELECTIVITY

INTRODUCTION

This is the shortest section in the book. That doesn't make it any less important. If anything, it is more important—so important that I dedicated an entire section to four short chapters focusing on one idea!

Selectivity is a fundamental concept in organic chemistry. Many reactions take place with various types of selectivity.

At the most basic level, if a reaction could give two (or more) different products, and it gives anything other than a statistical mixture of the possible products, we would say that the reaction is selective.

Of course, a reaction that gives 99 per cent of one product and 1 per cent of another product is much more useful than a reaction that gives 60 per cent of one product and 40 per cent of the other product.

The reason for this is simple—a reaction that gives a large amount of something you do not actually want is wasteful!

There are several different types of selectivity. In this section, we will first of all consider **chemoselectivity** and **regioselectivity**.

You absolutely **need** to recognise what type of selectivity is being exhibited in a given reaction, and to be able to correctly apply the terminology, so that if you are asked to give the outcome of a reaction, you discuss the correct thing.

Don't look for reasons for this to be complicated. Don't try to memorize the outcome of reactions either. If a given reaction can give two (or more) different products, and it gives more of one product than another, then there are two possible reasons.

1. One reaction has a lower activation barrier than another, so that the product formed by this reaction is formed faster.
2. The two products are in equilibrium, and one of them is more stable than the other.

Don't assume that the most stable product will always be formed. This is a common mistake.

We will consider another type of selectivity, **stereoselectivity**, in Basics 28. We will contrast stereoselectivity with **stereospecificity**, which we discuss in **Basics 36**.

BASICS 26

Selectivity in Organic Chemistry—Chemoselectivity

CHEMOSELECTIVITY

Chemoselectivity is observed where one outcome is preferred in a reaction.

There are actually two definitions that we need to consider.

*A **chemoselective** reaction is one in which one functional group reacts preferentially to another in a particular reaction.*

*A **chemoselective** reaction is one in which a functional group reacts preferentially to give one of two (or more) possible outcomes.*

Let's have a quick look at the first possibility. Consider the following reaction. We won't look at this reaction in this book, but don't let that worry you. The starting material has a ketone group and a carboxylic acid group. Lithium aluminium hydride is a very powerful reducing agent, and it reacts with both.

O O OH LiAlH_4 THF, 0 °C OH OH

What if we only want to reduce the ketone group? We can use a less reactive reducing agent, sodium borohydride. This is able to reduce the ketone, while leaving the carboxylic acid untouched.

O O OH NaBH_4 EtOH 2 h, 20 °C OH O OH

Less reactive = more selective!

When you are planning a synthesis, you need to be able to assess whether a reaction will exhibit the desired selectivity. If it does not, you might change the order of steps in a synthesis, or you might consider an entirely different approach.

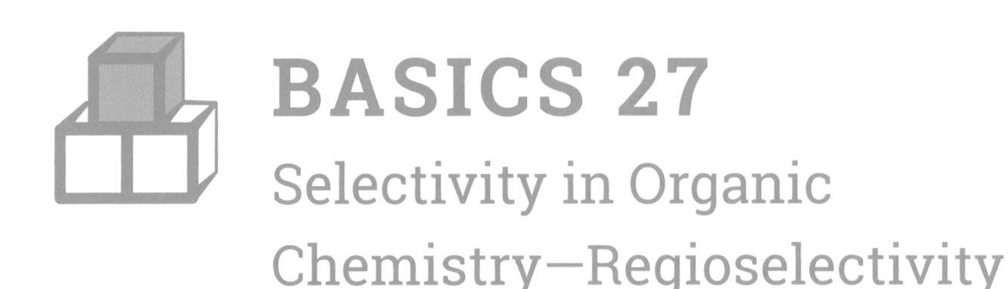

BASICS 27
Selectivity in Organic Chemistry—Regioselectivity

REGIOSELECTIVITY

A **regioselective** reaction produces more of one isomer than another. We haven't looked at the hydration (addition of water) of alkenes, but this doesn't matter. Each isomer is produced by the **same** reaction but at a different end of the double bond.

H_2O

Major Minor

The two products are **constitutional isomers** having the same functional groups. It is tempting to refer to these isomers as **regioisomers**, and many people do so. The term 'regioisomer' isn't easy to define, as it only applies when we look at a reaction.

If a reaction at a single functional group can produce two products as a result of reaction at (for example) the two ends of an alkene, this is regioselectivity.

The most common problem I see is that where a reaction has aspects relating to chemoselectivity, regioselectivity and stereoselectivity, students will answer a question on one type of selectivity by discussing a different type of selectivity.

Chances are, they would answer the question correctly if they knew which selectivity was which.

Every time we encounter selectivity, it is important that you articulate what type of selectivity it is. As with everything else you do, with constant reinforcement this will become natural. It sounds like a lot of effort, but it is actually more efficient to adopt a strategy that is certain to be successful. The things you found difficult (or tedious!) at first will become habits, and they will then become easy and natural.

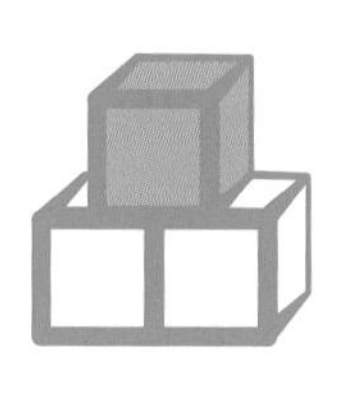

BASICS 28

Selectivity in Organic Chemistry—Stereoselectivity

INTRODUCTION

Before we start, let's establish one thing. There is **nothing** special about stereoselectivity. It follows the same 'rules' as the other types of selectivity.

We are going to do a lot more work in this chapter, because stereoselectivity often causes problems. In the early stages, it can be difficult to understand when a reaction can be stereoselective, and (often more importantly) when it cannot.

The key point we will address is the link between the properties of stereoisomers (Basics 25), reaction profiles (**Basics 14**), and rates of reaction (**Basics 15**).

Let's take a quick recap, and establish exactly what we are talking about.

WHAT ARE WE NOT TALKING ABOUT?

We have encountered stereo**specific** reactions. An S_N2 substitution reaction proceeds with inversion of configuration. The nucleophile attacks from the opposite side to the leaving group. Only one stereoisomer can be produced. We will reiterate this point in **Basics 36**.

This is not a stereoselective reaction, although it only produces one stereochemical outcome. There is no 'choice' here.

WHAT IS STEREOSELECTIVITY?

A stereoselective reaction is one that could potentially produce more than one stereoisomer, and it produces more of one stereoisomer than another.

There are only two possibilities within this. The two isomers could be diastereoisomers, in which case we are talking about **diastereoselectivity**. They could be enantiomers, in which case we are talking about **enantioselectivity**.

Most of the mistakes students make is with the types of isomers, rather than the type of selectivity.

Just because stereoselectivity isn't 'special' doesn't mean it isn't important. One stereoisomer of a compound can cure a disease. Another stereoisomer of the same compound could be highly toxic.

If only one stereoisomer of a compound has the properties you want, you need a synthetic strategy that allows you to make only that one stereoisomer.

DIASTEREOSELECTIVITY

Let's start with the simpler possibility. Diastereoselectivity is a type of stereoselectivity where the reaction concerned is producing two (or more) diastereoisomers in unequal amounts. In general, diastereoisomers have different energies.[1] Therefore, most reactions that produce diastereoisomers as products will exhibit some level of selectivity. Production of a 1:1 mixture of diastereoisomers is the exception rather than the rule.

Here is a reaction we first encountered in Reaction Detail 2. This reaction produces two products in unequal amounts.

Because the two possible products are diastereoisomers, and diastereoisomers have different energies (Basics 25), this reaction can be selective.

[1] Perhaps not much different though!

It actually goes slightly further than consideration of the products. The transition states leading to each diastereoisomer are themselves diastereoisomers. Therefore, the two diastereoisomers of product can be formed at different rates.

Yes, the definition (and consequences) of diastereoisomers doesn't just apply to stable compounds. If the transition states are not superimposable and not mirror images, then they will be diastereoisomers, and hence may have different energies.

I love this! It makes life easier. You don't have to think about when the stereochemical definitions apply. They apply **everywhere**, all along the reaction coordinate from starting materials to products and everything in between.

Draw a complete reaction profile for the reaction of compound **1** to give compounds **2** and **3**.

In order to do this, you might need to ask yourself (and answer) a few questions.

Is compound **1** (as drawn) a single compound?

No! One stereogenic centre is specified (the wedge bond) but the other is not. It could be either configuration. That is, compound **1** could also be a mixture of diastereoisomers. Think about what this means for the number of starting materials you need to draw, and their energies.

Does the reaction have an intermediate?

We established that this isn't an S_N2 substitution. Therefore, it does have a carbocation intermediate. The carbocation intermediate will have only one stereogenic centre, of fixed configuration. It will be a single species.

Having answered these two questions, we are in a position to draw a reaction profile.

We have two different starting material diastereoisomers, with different energies. These will form a common carbocation intermediate, possibly at different rates. The intermediate carbocation will then be attacked by allyltrimethylsilane to give the product diastereoisomers, again at different rates.[2]

[2] The rates must be different, since we established up front that this reaction is stereoselective.

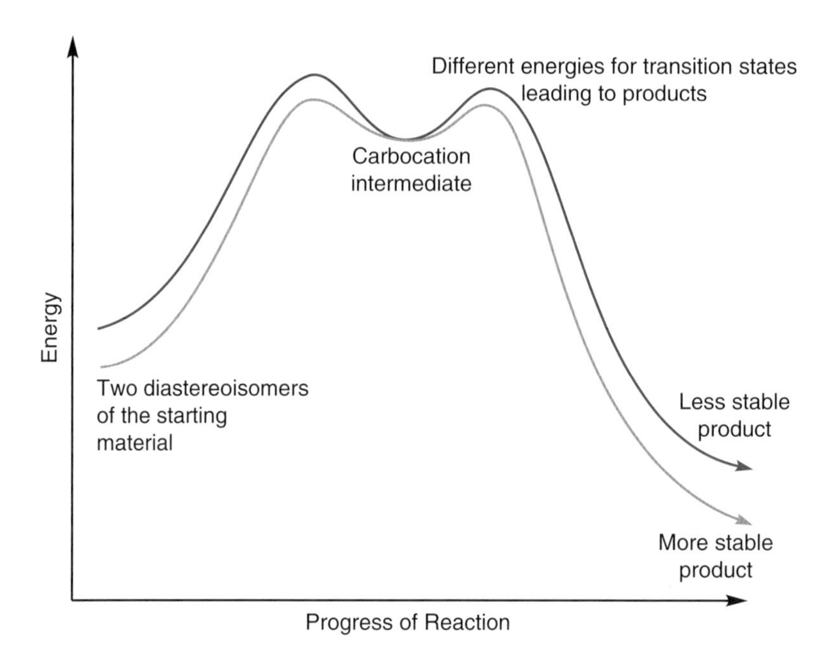

What is the take-home message?

Two diastereoisomers of starting material can react at different rates. Two diastereoisomers of product can be formed at different rates.

28

Let's see another, similar, example. Here is an S_N1 reaction of a cyclohexane.

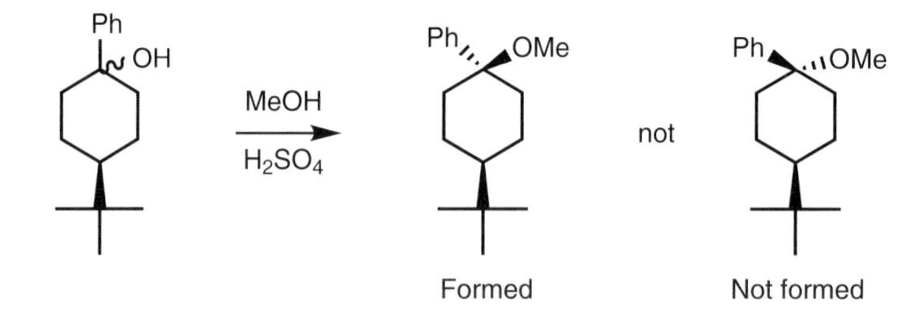

The starting material is represented by a single structure. The wiggly bond indicates that the compound is a mixture of isomers at this centre.[3] *I have indicated the stereochemistry of the t-butyl group with a wedge, so we have something for the alcohol stereochemistry to be relative to.*

4.1

We only get one of the two possible products, even though the starting material is a mixture of two stereoisomers.

[3] We could have done the same with compound **1**, but I wanted to show you two different ways in which stereochemistry can be incompletely specified.

The two possible starting materials are diastereoisomers. The two possible products are diastereoisomers. Since there is an intermediate in this reaction in which the stereochemical 'information' in the starting material is lost, it is entirely reasonable for both diastereoisomers to give the same stereochemical outcome. Because the products are diastereoisomers, with different energies (Basics 25), it is entirely reasonable for one stereoisomer to be favoured.

In this example, the compounds are cyclohexanes, and we will see (Basics 31) that cyclohexanes occupy a special place in organic chemistry. In Applications 3 we will come back to this example, and although we will draw it differently, all the same principles will still apply.

There is another point in all of this that is easy to miss. We are forming diastereoisomers that have no stereogenic centres. Stereoselectivity is very often about chirality, but it doesn't have to be.

ENANTIOSELECTIVITY

At first, enantioselective reactions seem more complicated. Let's try to reduce the problem to its simplest form. Here is epoxidation of an alkene.

Ph–CH=CH–CH_3 —(*m*-CPBA)→ Ph(epoxide, O)CH_3 + Ph(epoxide, O)CH_3

Again, we haven't looked at this reaction. That doesn't matter. We are focusing on principles.

The reaction starts with a flat, achiral molecule. The reagent, *m*-CPBA, can deliver an oxygen atom to an alkene. That's all you need to know for now, apart from the fact that it is achiral (not chiral). Epoxidation can take place on either the top face or the bottom face of the alkene, to give the two enantiomers of the product as shown.

Because these are enantiomers, they absolutely must have the same energy.

Because the reagent is achiral, the transition states will also be enantiomeric, and therefore of equal energy. The reaction profile looks like this.

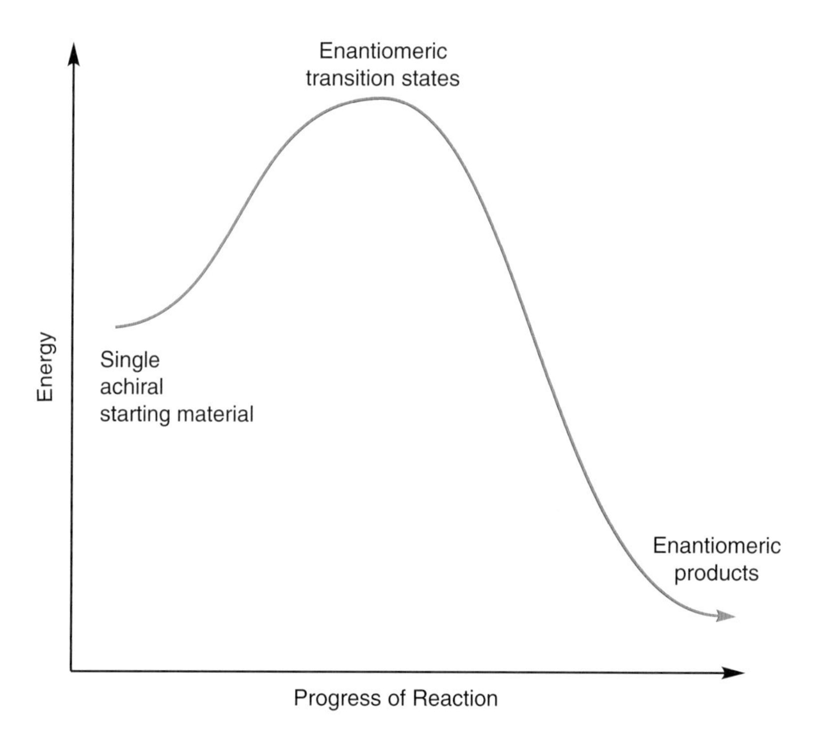

There **is** a red line and a blue line. They just happen to be at the same energy level at all points. There is absolutely no way that this reaction can be stereoselective at all.

But we might want a single enantiomer of an epoxide. There are many reactions that use such compounds, so they are valuable intermediates. The two enantiomers of a chiral compound may have profoundly different biological properties.

So, how can we modify this reaction to produce an excess (ideally 100 per cent excess!)[4] *of one enantiomer over the other? How can we make the reaction enantioselective?*

Let's consider what the reaction profile has to look like in order for the reaction to be enantioselective. We cannot change the relative energies of the enantiomeric products. We cannot change the fact that we have an achiral starting material. The only thing we might be able to change is the transition state.

If we can make the transition states diastereomeric, they can have different energies, and the activation barrier for the production of each enantiomer can be different. As long as we are only talking about stereogenic centres as the 'element' of stereochemistry, diastereoisomers have more than one stereogenic centre. If there isn't any stereochemistry in the starting material, where must the stereochemistry be?

[4] We measure the stereoselectivity of a reaction by the 'excess' of one stereoisomer over another. We will see this in Basics 29.

We will leave it here for now. We aren't going to answer this question in this book. That gives you lots of time to think about possible solutions. Focus on the definitions of enantiomers and diastereoisomers and you will understand the principles behind the solution.

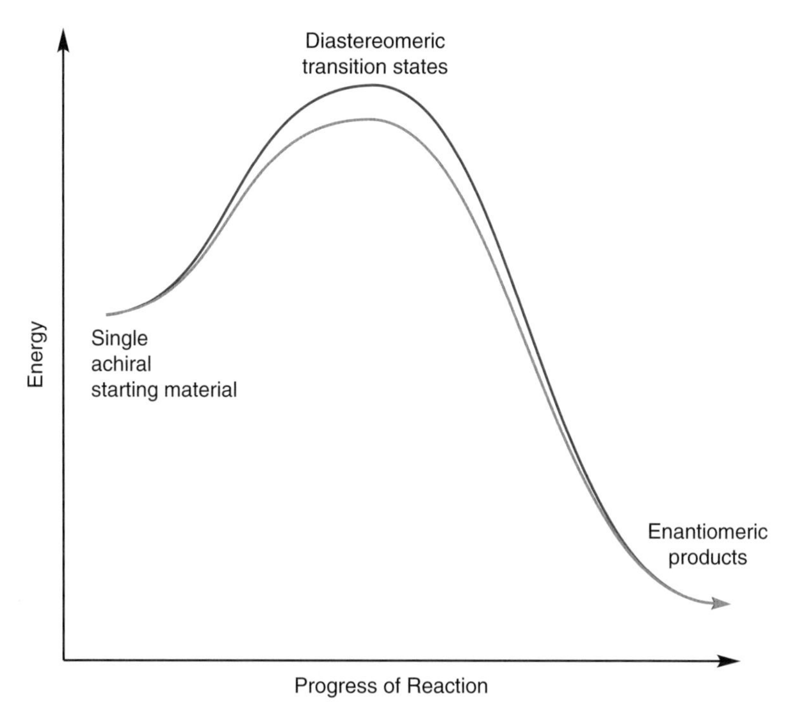

It has taken many people many years to develop reactions that give high levels of stereoselectivity. It would be almost impossible to predict which reagents and conditions will lead to high selectivity. But with experience, we will see the structures that tend to give high selectivity, and we will understand why they do this. We will improve our ability to predict based on empirical data.

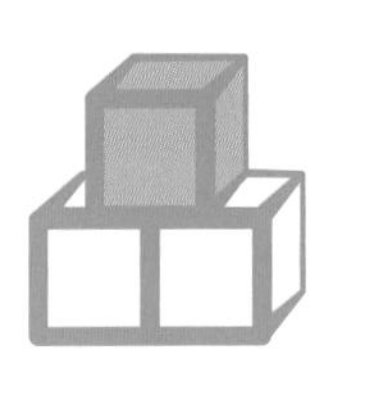

BASICS 29

Stereochemical Definitions Relating to Reactions

INTRODUCTION

There are a number of terms that are associated with the stereochemical outcome of reactions. If a reaction is chemoselective or regioselective, we can simply state which products are formed, and in what yield. For stereoselective reactions, we tend to talk about the excess of one stereoisomer over another.

We now need to take as read the stereochemical definitions related to individual compounds that we encountered in Basics 24 and Basics 25.

DEFINITION—ENANTIOMERIC EXCESS

If you have a mixture of two enantiomers, there can be any ratio, from 100:0 through 50:50 to 0:100. You would refer to this as an enantiomeric ratio, which can be abbreviated '**e.r.**'

A pure enantiomer will have the maximum possible **optical rotation**, or 100 per cent of the theoretical value (you'll see why I express it like this in a minute). A racemic mixture will have zero optical rotation, or 0 per cent of the theoretical maximum.

If you have a racemic mixture, the 'excess' of one enantiomer over the other is 50−50 = 0. This term is the **enantiomeric excess**, and is reported as a percentage.

A pure enantiomer has an enantiomeric excess of 100 per cent (100−0).

A 75:25 mixture of enantiomers will have an enantiomeric excess of 50 per cent (75−25). You can think of this 75:25 mixture as being 50 per cent of one enantiomer and 50 per cent of the racemate (25:25). The 'racemic bit' has zero optical rotation, so that the rotation of this overall mixture is 50 per cent of the theoretical maximum.

This is really useful. The optical rotation is directly proportional to the enantiomeric excess. If you measure the specific rotation of a sample, and you know the specific rotation of the pure enantiomer, you can calculate the enantiomeric excess. From this, you can calculate the ratio of the enantiomers.

Enantiomeric excess is generally abbreviated as '**e.e.**'

DEFINITION—DIASTEREOMERIC EXCESS

We have seen why the enantiomeric excess is a useful measure. If you have a mixture of two diastereoisomers, you can calculate the diastereomeric excess in exactly the same way.

You cannot relate the diastereomeric excess to a physical property in the same way you can with enantiomeric excess. However, this is still a useful measure.

Diastereomeric excess is generally abbreviated as '**d.e.**'. You can also talk about a diastereomeric ratio (**d.r.**)

DEFINITION—RACEMIC MIXTURE (RACEMATE)

If you have an equal amount of the two enantiomers of a chiral compound, we describe this as a racemic mixture, or a racemate.

In terms of the optical rotation, the two enantiomers cancel one another out, so that the optical rotation of a racemate is zero.

DEFINITION—OPTICALLY PURE

This is a slightly odd term, but you'll get used to it. A single enantiomer of a compound has the maximum possible optical rotation for that compound. 'Optically pure' is another way of saying that all of the molecules in a particular bottle of a compound have the same **configuration**. There is only one stereoisomer present.

DEFINITION—SCALEMIC MIXTURE

This is quite an obscure term, but you should at least be aware that it exists. We saw that a racemic mixture is a 50:50 mixture of enantiomers. A **scalemic mixture** of enantiomers is any mixture that is not 50:50 and not a single enantiomer.

SECTION 5

BONDS CAN ROTATE

INTRODUCTION

Of course bonds can rotate! You can see this with your molecular models. Make a model of ethane and spin that carbon–carbon bond. If you can do it, so can the molecule!

Molecular models are great! When you twist and turn them, you can get a lot of insight into what might happen in a real molecule. If you cannot twist the model into a particular orientation, then chances are the molecule will also find it hard to adopt that orientation.

What we need to do in this section is make the formal link between bond rotation and energy.

Once we have established the basics, we will consider how some compounds prefer to react in specific orientations, and that the ease of adopting a particular orientation (conformational isomer) will very often determine the reactivity of the compound.

There is no doubt that the most important compounds we will look at in this section are cyclohexanes. These are also the compounds that cause the most difficulty for students.

I don't say this to worry you. Quite the opposite. I want you to know that you are not alone in your struggles, but you will succeed if you persevere.

You need to work on these structures, and you need to reach the point where they cease to be difficult. We will look at a few problems in this section, and some more involved problems in **Section 7**.

There are three 'Applications' chapters in this section. This is a new chapter type, and the distinction is made because you can apply a small amount of knowledge and solve quite complex problems. But you need to be shown how to apply the knowledge. You will learn better if you are actively learning, rather than simply reading an explanation.

BASICS 30
Isomerism in Organic Chemistry—Conformational Isomers

I deliberately wanted to keep the drawing of structure separate (in the first instance) from energetics. For conformational isomers, the whole point is that they have different energies, so we need to talk about the energy differences between the conformers now!

CONFORMATIONAL ISOMERS

__Conformational isomers__ (or __conformers__) are a special (but very common) type of stereoisomer that are related simply by rotation around single bonds. Almost every structure has several conformational isomers, and much of the time it doesn't matter.[1]

We are going to start with a couple of relatively straightforward examples, along with some terminology. In Applications 1, we will add more detail, looking at a wider range of compounds. We will see applications of conformational isomerism in substitution reactions in Applications 2. The energy difference between conformers is fundamental to E2 elimination reactions, which we will explore in **Applications 4**.

Later, we will look at conformational isomers of cyclohexanes. These really matter, so it's important that you get the basic idea before we reach that point. We will apply the fundamental principles of conformational analysis in cyclohexanes to substitution and elimination reactions in Applications 3 and **Applications 5** respectively.

The point, once again, is that everything is connected. As we cover more ground, all we see are more connections, rather than more learning.

[1] Don't mix up conformation and configuration (see Basics 24). They are different things, so using the wrong word for the wrong thing will just make you sound silly. You'd never call an apple an orange, despite them both being round fruits!

CONFORMATIONAL ISOMERS OF ETHANE

Let's look at a flying wedge structure for ethane.

The C–C bond is a single bond. We can rotate around it. It doesn't take much energy to rotate around it, but it does take **some** energy.

Let's look at the two extreme **conformations**. We refer to them as 'staggered' and 'eclipsed'.

staggered　　　　eclipsed

Look at the structure on the right. Imagine you are looking directly along the carbon–carbon bond. You would see all of the hydrogen atoms on the front carbon atom directly in front of those on the rear carbon atom. They are eclipsing one another!

As we saw in Habit 4, the Newman projection for the staggered conformation of ethane looks like this. We just didn't talk about the conformations.

There is an important convention with this representation. The three blue lines that join in the centre represent bonds to the **front** carbon atom. So, the three 'H^a' are on the front carbon. The three shorter purple lines that join the circle are on the rear carbon atom. It is important that you know this convention. As with all conventions, if you use it lots, you will get used to it and you will get better at interpreting/drawing it.

Make a model of ethane, look down the bond and convince yourself that the Newman projection looks about right.

Now for the eclipsed conformation. We cannot show this properly eclipsed. It just looks too messy.

We draw the bonds offset by a small angle, and we accept that we know it is eclipsed in reality.

DEFINITION—DIHEDRAL ANGLE

The angle between hydrogen atoms on **adjacent** carbon atoms is known as the **dihedral angle**. The dihedral angle in staggered ethane is 60° (or 180° or 300°). The dihedral angle in eclipsed ethane is 0° (or 120° or 240°).

dihedral angle 60°

dihedral angle 0°

ENERGY DIFFERENCES BETWEEN CONFORMERS

The energy difference between the staggered and eclipsed conformers of ethane is only 12 kJ mol^{-1}, so that bond rotation is easy.[2] A full 360° rotation is shown below, with three staggered and three eclipsed conformers. Remember that 0° and 360° are the same.

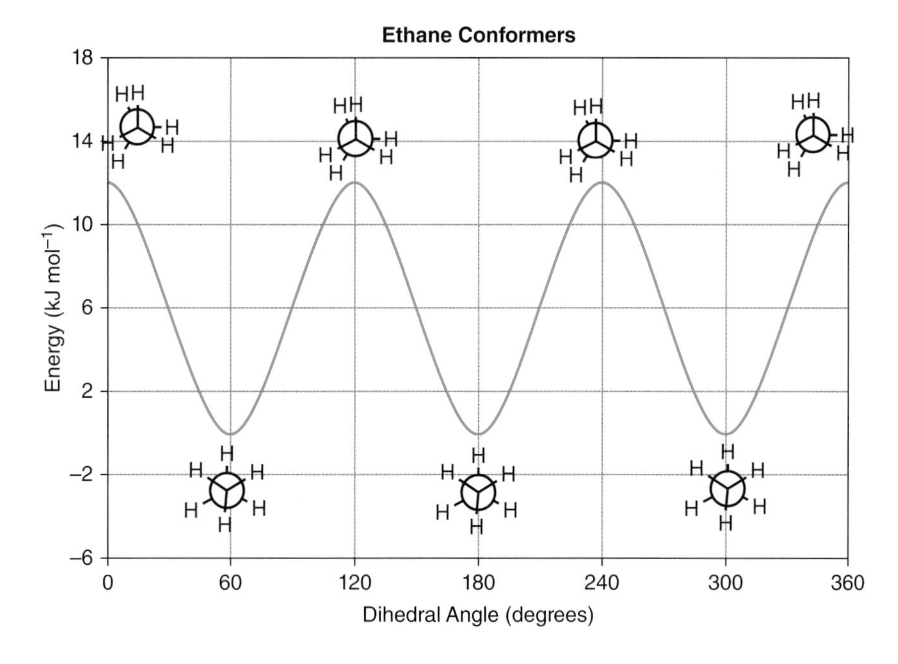

[2] Remember that the dissociation energy for a single bond is about 350 kJ mol^{-1}.

CONFORMATIONAL ISOMERS OF BUTANE

For butane, the situation is a little more complicated. Rotating around the central C–C bond gives four extreme conformers, two staggered and two eclipsed.

anti	*gauche*	Me-Me eclipsed	Me-H eclipsed
(0 kJ mol^{-1})	(+3.8 kJ mol^{-1})	(+19 kJ mol^{-1})	(+16 kJ mol^{-1})

The two staggered ones are referred to as ***anti*** and ***gauche***. The *anti* isomer is the lowest energy isomer. The energies above are relative, so we call this the zero point. The *gauche* conformer is about 4 kJ mol^{-1} higher in energy. Then there are two eclipsed conformers, depending on whether the methyl group is eclipsed with another methyl group (+19 kJ mol^{-1}) or with a hydrogen atom (+16 kJ mol^{-1}). Of course, there are other conformers in between.

TORSIONAL STRAIN

The energy difference between the conformers is due to two factors. If the substituents on the ends of the relevant bonds are large, there is a 'simple' steric effect. The methyl–methyl eclipsed conformer of butane is less stable because of an unfavourable steric interaction between the methyl groups.

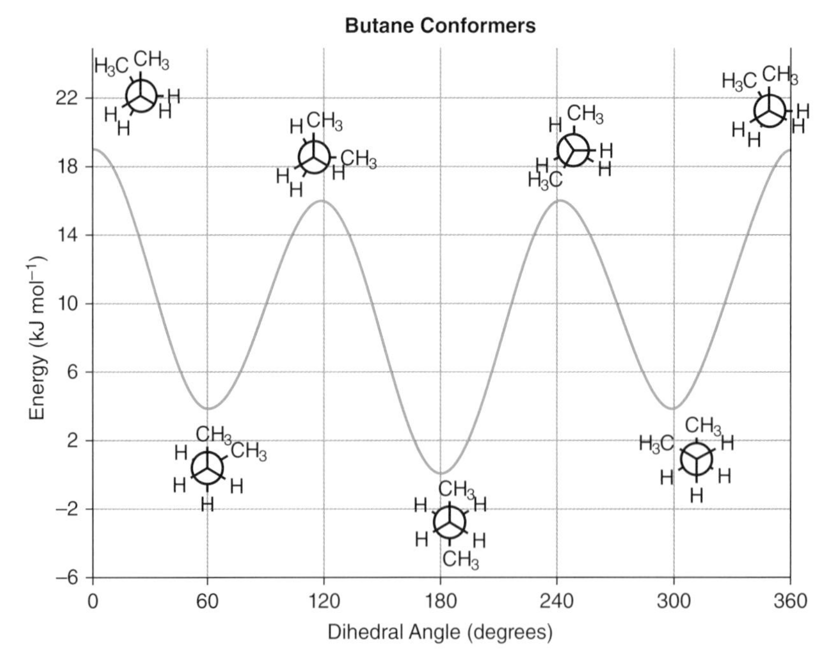

There is also a direct interaction between the electrons in the bonds as they become eclipsed. This is an electrostatic repulsion known as **torsional strain**.

You need to be familiar with both aspects of the strain, steric and torsional.

WHAT DO YOU NEED TO DO WITH THIS INFORMATION?

You can imagine that if we have longer chains, with more bonds to rotate, the number of conformers increases very rapidly indeed. With butane, we only considered rotation of the central C–C bond, but we could add conformers due to rotation of the terminal C–C bonds as well. It isn't easy to show all of this on a graph.

Make a model of butane and consider[3] the conformer in which **all** the bonds are fully eclipsed—it will have some very strange bond angles!

As you build your confidence, you should be able to predict which conformer is most stable for simple molecules. You need to be able to extrapolate from these simple structures to more complicated structures (*e.g.* with bigger or just different R groups or chain lengths) and be able to identify the unfavourable interactions—you need to know which specific interaction(s) in a structure are destabilizing it.

Moving forward, you will find that the outcome of many reactions depends on the level of conformational bias in a structure. If you don't grasp (and retain!) the basics, you won't be able to understand the more complicated stuff.

[3] I am asking you to *consider* it, because you won't be able to twist your molecular model enough to make it.

PRACTICE 11
Conformational Analysis

THE POINT OF THE EXERCISE

A flexible molecule will spend more time in its lowest energy conformation. Sometimes, a molecule has to adopt a particular conformation in order to undergo reaction. You need to be able to identify the most stable conformer of a given structure, as well as the less stable conformers. You must also develop a gut feeling for the energy differences between conformers.

QUESTION 1

Draw and label the conformers of ethane and butane without looking back at Basics 30. Indicate **approximate** energy differences between the conformers.

Use Newman and flying wedge projections.

You should know which is more stable, and the general region of the energy differences. At some point, you will need to consider whether a reaction which requires a particular conformer is feasible energetically, so you need to know how these energies relate to bond dissociation energies, activation energies, etc.

QUESTION 2

Look at the following three structures and consider the relative stability of the possible conformers.

The key question is 'what will the OH/NH_2/Cl group change, if anything?'

Cl–CH₂CH₂–Cl   HO–CH₂CH₂–OH H_2N–CH₂CH₂–NH_2

You should ask the question 'if Cl and O have the same electronegativity, why include both?' Is there something else that OH can do?

The answers to this question are in Applications 1. Don't be tempted to read ahead to the answers until you have had a good go at this.

QUESTION 3

Draw Newman projections for the following flying wedge projections.

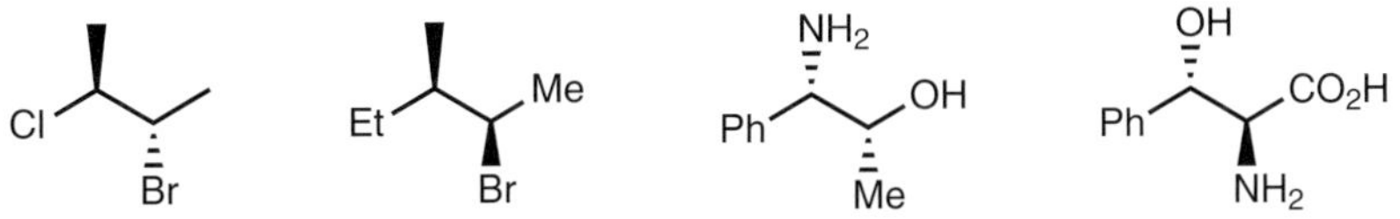

There isn't a single correct answer. You could draw them as different conformers or in different orientations. Explore this with your molecular models by rotating the whole model, or just one bond, and drawing what you see.

Be sure that, in your Newman projection, you draw the same stereoisomer with the same configuration (*R* or *S*) at all stereogenic centres.

QUESTION 4

Draw flying wedge projections for the following Newman projections.

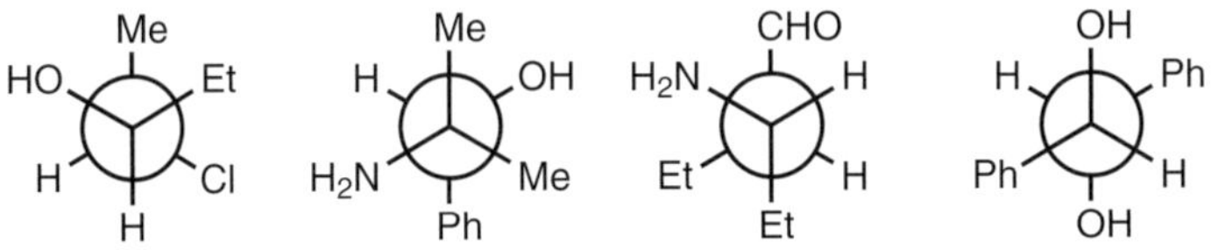

QUESTION 5

For the structures in QUESTION 3 and QUESTION 4, draw all possible staggered and eclipsed conformers. There should be three of each. Try to rank the conformers of each compound in order of stability. Consider the possible steric and electronic interactions.

Do I care if you get these questions right? Not really!

Perhaps you think that is a strange view for an educator to have.

It's not that I don't care. It's more a case that if you are thinking in the right way, you **will** get them right in time. I don't want you to care, or (more properly) I don't want you to worry about it. I want you to just let it happen.

You will try this once, and perhaps you won't be confident. You will work through Applications 1 and come back to this and you will do a bit better. You will find a few more examples, and eventually you will come back to this and not only be able to put the conformers in order—you will 'guess' how much more/less stable they are.

Oh, and I want you to focus on the important stuff—the questions you need to answer before you can confidently do this! 'How strong is a hydrogen bond between X and Y?', 'How electron-withdrawing is Z?'

So, it really doesn't matter if you answer these questions correctly. What matters far more is that you learn to ask yourself the right questions that will allow you to get these questions right.

ADDITIONAL EXERCISE

For the structures in QUESTION 3 and QUESTION 4, assign the stereochemistry of all stereogenic centres. Can you do this directly from the Newman projections?

APPLICATIONS 1

Conformational Isomers 2

This is an applications chapter. There won't really be any additional information. It just looks like there is. We will be taking what we learned in Basics 30 and adding some information from **Basics 5**, plus the occasional hydrogen bond, and we will see where this takes us.

This approach is really important. Whenever you see a 'new' example, you have to ask yourself some key questions in order to determine what is different about it.

Go back and read Basics 30 one more time, to make sure you are familiar with the idea of conformers and their energy differences. Make models of ethane and butane and rotate around the bonds to make sure you appreciate the interactions that lead to the energy differences.

BACKGROUND

I am going to run through the energy differences between the conformers of three compounds in this chapter. We will include butane as a reference point, and we will add 1,2-dichloroethane and 1,2-dihydroxyethane into the mix.

We established that the most stable conformer of butane is the *anti* conformer. This is because the methyl groups can get as far apart as possible. Here are the main conformers again.

anti	*gauche*	Me-Me eclipsed	Me-H eclipsed
(0 kJ mol^{-1})	(+4.2 kJ mol^{-1})	(+24.4 kJ mol^{-1})	(+13.7 kJ mol^{-1})

You may have noticed that these numbers are different to the ones we had before. In fact, the relative differences are also different! But the most stable conformer is still the most stable, and the least stable is still the least stable.

In science, it is important that we are able to compare data. Therefore, in order to illustrate the points in this chapter, I have generated computational data for the conformational energy differences.

The experimental data that we would need does not all exist, at least as far as I can tell, under identical conditions. What does exist depends on a range of factors, not least of which is solvent. Using gas phase calculations is a simplification that is worth pursuing.

Despite this, there are some discrepancies between calculated and experimental data for the level of theory I have used.[4] These will not materially affect the conclusions we can draw.

Calculated numbers won't be exactly the same as experimental numbers, but the same trends apply!

1,2-DICHLOROETHANE

Here is the structure of 1,2-dichloroethane. If we start with butane and replace the methyl groups on the ends, this is what we will get.

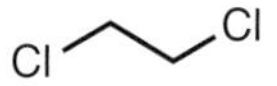

The key question is 'what effect does changing a CH_3 into a Cl have?'

In order to answer this, we have to ask what is different about Cl. Is it smaller or larger than CH_3? A methyl group is bigger than a Cl atom, but not by much.

Let's look at the conformers, and their relative energies, calculated in the same way as those above.

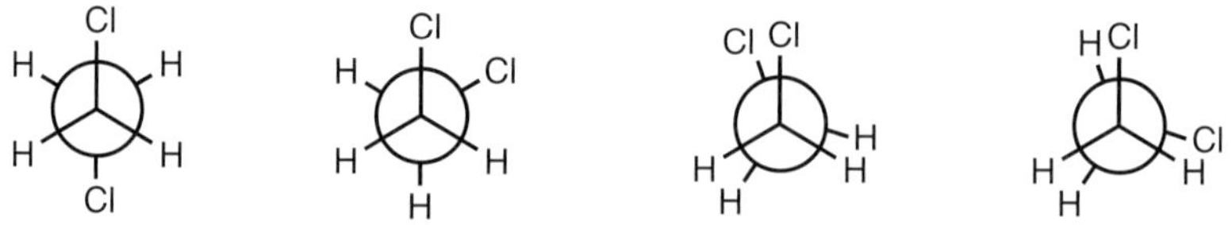

anti (0 kJ mol^{-1}) *gauche* (+8.9 kJ mol^{-1}) Cl-Cl eclipsed (+38.4 kJ mol^{-1}) Cl-H eclipsed (+19.2 kJ mol^{-1})

[4] For those who are interested, these are DFT (Density Functional Theory) calculations at the B3LYP/6-31+G* level. You won't know what this means yet, but when you do learn, you will be able to assess how reliable the calculations are likely to be.

Experimentally, the gauche conformer is only about 4.4 kJ mol^{-1} less stable than the anti conformer. This is only marginally more than we see for butane, and the reasons for the discrepancy are too complicated to discuss here.

The *anti* conformer is still the most stable. In fact, the order is still identical, but the numbers are significantly different.

The Cl–Cl eclipsed conformer is 38.4 kJ mol^{-1} less stable, which is a big increase over the corresponding situation in butane, despite Cl being smaller than CH_3. What could be causing this increase? Clearly the two Cl atoms want to be as far apart as possible, but it's not just due to their size.

It's the electronegativity. There isn't much else it can be. If it ain't steric, it's electronic!

We have strong permanent dipoles in the C–Cl bonds, and we can minimize the dipole moment of the molecule by having these dipoles pointing in opposite directions.

Of course, there ***is*** *a steric contribution to this. It just happens to be reinforced by an electronic contribution. It isn't easy to determine the relative amount of each type of contribution.*

We can delve a little deeper. The *gauche* conformer is destabilized relative to that in butane as well. The reason is the same. There isn't such a large dipole moment in the molecule, but it's still there. It is present in the Cl–H eclipsed conformer as well.

1,2-DIHYDROXYETHANE

Let's complicate the situation a bit more. Oxygen and chlorine have almost identical electronegativity. This makes it a good comparison. I have kept the *anti* conformer as the reference point.

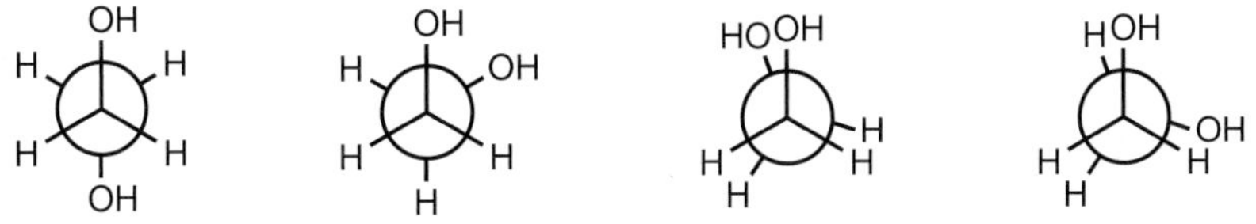

anti (0 kJ mol^{-1}) | *gauche* (–12.1 kJ mol^{-1}) | OH-OH eclipsed (+14.1 kJ mol^{-1}) | OH-H eclipsed (+9.1 kJ mol^{-1})

Here, we can see that the *gauche* conformer is actually the most stable. The fully eclipsed conformer is still the least stable, but not by anywhere near as large a margin as we have seen before. The only number that is broadly comparable is the OH–H eclipsed conformer.

So, there must be something stabilizing the *gauche* conformer in particular, and the OH–OH eclipsed conformer to some extent. Otherwise the two OH groups would get as far apart as possible.

What could possibly be holding two OH groups close together?

Well, since I put it like that, there can only really be one answer—hydrogen bonding. Here is a 3D model of the *gauche* conformer. The hydrogen bond in question is about 2.3 Å long.

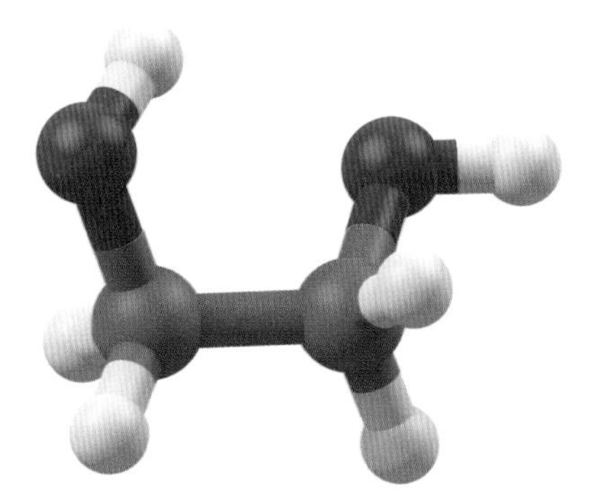

How strong are hydrogen bonds? The short answer is that it depends on the hydrogen bond donor and hydrogen bond acceptor. For OH groups, a value in the region of 20 kJ mol^{-1} is typical. The *gauche* conformer of 1,2-hydroxyethane would appear to be stabilized by about this amount, compared to that in butane.

Here is another important question. Did we just stabilize the *gauche* conformer or destabilize the eclipsed conformer? Probably a bit of both, and it depends on whether you are comparing to butane or to 1,2-dichloroethane.

We are starting to build a more complete picture of the interactions between atoms in organic molecules. The key skill that you need to develop is to analyse problems at this level when the structures become more complex. You need to see through the complexity and identify the fundamental processes. In order to do this, you need to be able to visualize the shapes of molecules. Another key skill is to learn to make valid comparisons.

Draw the corresponding conformers for a few more derivatives of ethane. Pick substituents at random and try to identify the factors that will lead to one conformer being stabilized (or destabilized) relative to another. As with everything else, you will find you get luckier with practice!

CONFORMATIONAL ISOMERS OF ETHENE?

In eth**a**ne, we have established that the barrier to rotation around the C–C bond is 12 kJ mol^{-1}. We have also established that eth**e**ne is planar, because any other geometry would not permit overlap of the p orbitals to form a π bond.

H H
C=C
H H

There is another way to look at this. The barrier to rotation about the C=C double bond in ethene is about 300 kJ mol^{-1}.

We say 'about' because we didn't precisely quantify the relative contributions of the σ and π bonding in **Basics 6**.

Looking at it this way, all bonds can rotate. Some just have higher barriers than others. A barrier of 300 kJ mol^{-1} means that there is no rotation. A barrier of 12 kJ mol^{-1} means there is rapid rotation.

We have seen various things that can stabilize one conformer or destabilize another conformer. These include steric or electrostatic repulsion, torsional interactions, hydrogen bonding and now π bonding.

As long as we understand fundamental principles, we could anticipate whether any new interaction has the potential to stabilize or destabilize a conformer.

WHERE IS THIS GOING?

Some reactions require a specific conformer for reaction to occur. If that conformer is not the most stable conformer, the activation energy for the reaction will be increased by the energy difference between the most stable and the required conformer.

In this case, the reaction will be slower.

So, conformational equilibria directly affect rates of reaction.

We will see lots of examples of this, for substitution in Applications 2 and Applications 3, and for elimination in **Applications 4** and **Applications 5**.

APPLICATIONS 2

S_N2 Substitution Reactions Forming Three-Membered Rings

As we progress further into the 'Applications' chapters, you will find that we need to draw on an increasing number of principles from various sections in order to rationalize or predict the outcome of reactions. This is very common in organic chemistry. In fact, it is rare that you can explain a reaction using only one idea.

Three-membered rings are very interesting. The tetrahedral bond angle is 109.5°. The corresponding bond angle in a three-membered ring is 60°. This deviation from the ideal bond angle represents a considerable amount of strain.[5] Compounds with three-membered rings are quite reactive. They are also surprisingly easy to form.

First of all, we will look at why this is. Then we will see some applications of this chemistry.

APPLYING NEWMAN PROJECTIONS AND ELECTRONEGATIVITY

Take a look at the following compound. Oxygen is electronegative. Bromine is electronegative.

Identify the most stable **conformation** (Basics 30, Applications 1) of this compound.

Let's make sure you understand what you need to do in order to work this out. You have a single stereoisomer of this compound. You might want to make a model of it. You should check that you have made the correct stereoisomer. Newman projections are great for showing conformers. Perhaps you can draw a Newman projection directly

[5] Although actually not as much as you might think.

from the structure above. Perhaps you need to make the model and draw the Newman projection from that.

*It doesn't matter **how** you do it. The point is that you have options, and you need to take the time to find out what works for you.*

I drew the Newman projection first of all in the conformer shown above. Here it is.

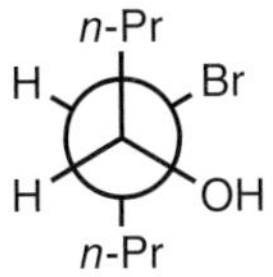

Now we need to think about stable conformers. It isn't always about the size of substituents (although this will have an impact). If we can get the OH and Br as far apart as possible, we will reduce the overall dipole moment in the molecule. Here is a reasonable guess at a stable conformer.

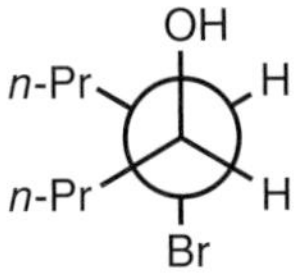

*This isn't a million miles from the examples we looked at in Applications 1. Of course, that's **why** we looked at them. We are now seeing an example of how the conformation of a compound makes a reaction more favourable.*

Okay, now we are in a position to think about reactivity. What happens if we treat this compound with a strong base? We will deprotonate the OH group, and this will favour the above conformer even more.

We didn't consider whether we might form a hydrogen bond between OH and Br, but we can be certain that by removing the proton, we will not!

Let's now ask a fundamental question.

In an S_N2 substitution reaction, where does the nucleophile (in this case a negatively charged oxygen atom) attack relative to the leaving group (Br)?

It attacks from the back, in order to overlap with the σ* orbital. We saw this in Reaction Detail 2. In this case, the nucleophile is **exactly** there! That makes it a great S_N2 substitution.

Finally, here is the reaction. We form a three-membered ring with oxygen—an epoxide.

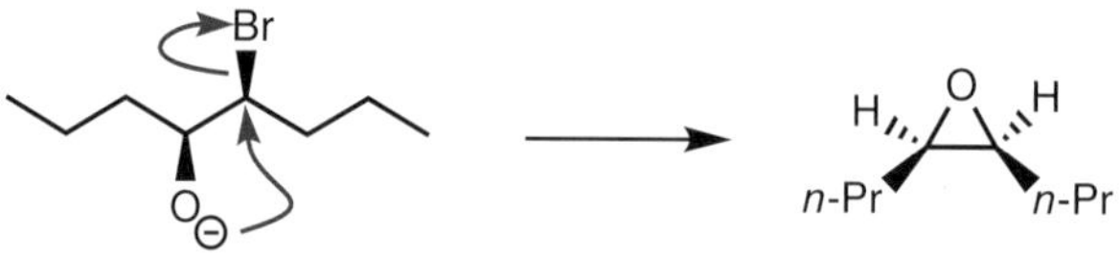

In this case, we form the *cis* epoxide. We don't need to worry about the absolute stereochemistry of the product.

Why don't we need to worry about the absolute stereochemistry of the product?[6]

In order to work through this problem, we needed to apply quite a lot of the basic principles. If there was any one aspect that you were not comfortable with, you would probably have found the whole problem very challenging, if not impossible. This is the problem with learning organic chemistry—you need to be familiar with all of the basic principles.

Underlying all of this, there is one absolute. ***Don't Draw Garbage!*** *As long as you can apply some of the principles, you will get part way through a problem. You will know if you are 'fudging it'! When you feel that you are drawing something that you don't understand, in order to get to the end of a problem, go back to the basics and make sure you understand the approach you need to take. If you do this over and over again, you'll find that the problems become easier!*

NEIGHBOURING GROUP PARTICIPATION (ANCHIMERIC ASSISTANCE)

In the above example, the nucleophile was in the same molecule as the leaving group. We can describe this as an **intramolecular reaction**. Now, we need to apply the same principles to reactions that involve a three-membered ring as an intermediate which is not ultimately retained in the product.

In the following examples, stereochemistry and regiochemistry are used as markers that allow us to follow what has happened.

Here is a really neat example. If we react compound **1** with hydroxide, we get compound **2** and not compound **3**.

[6] Hint—have another look at Habit 7.

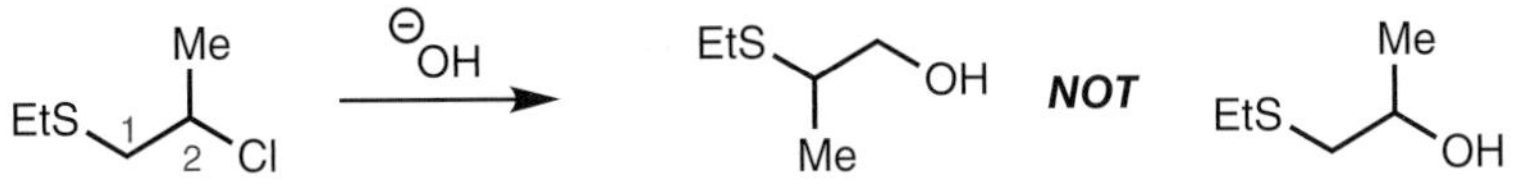

5.2

Compound **3** would be expected on the basis of attack of hydroxide at carbon atom 2. This has obviously not happened.

Let's analyse the problem, and work through the possibilities. Labelling carbon atoms is a useful tool that will allow you to keep track of what is going on. In this case, we can identify two possibilities. These are shown below.

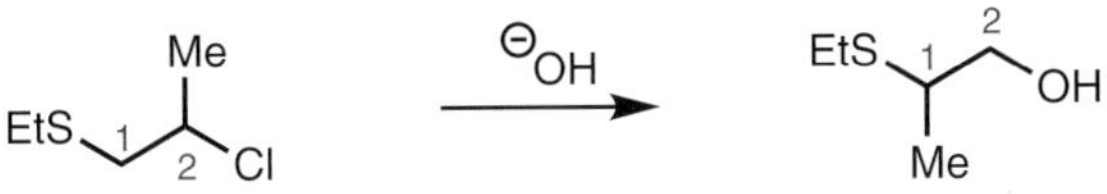

In the first case, the methyl group has moved from carbon 2 to carbon 1. In the second case, the sulfur has moved from carbon 1 to carbon 2. When we express it like this, it isn't easy to see which is the most likely.

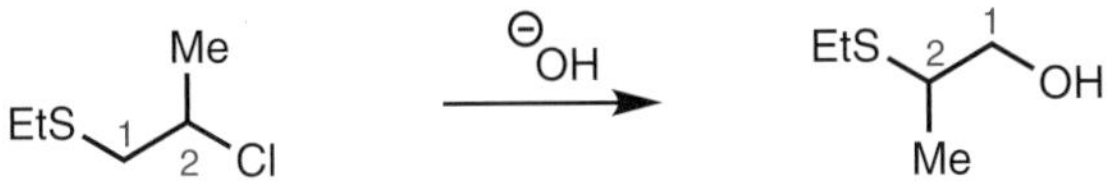

There is little doubt that a substitution reaction has taken place—just not the one we expected. We know that sulfur is a great nucleophile in substitution reactions. We have seen, above, that we can form three-membered rings quite easily in substitution reactions.[7]

Since this is a favourable process, we should draw it and see what the consequences are. Here it is! We get compound **4**. Clearly this is not the product (**2**), but it is a sensible reaction.

Sensible is a good start! But does it actually help us?

[7] The title of the chapter is also a bit of a clue!

It is only productive if we can go from compound **4** to compound **2**. We can make a stronger statement than this. Reaction of compound **4** should give compound **2** in preference to compound **3**. If it doesn't, then you might have to go back to the drawing board.

Here are two possible outcomes, shown using blue and purple curly arrows.

In either case, these would be S_N2 reactions.[8] *The question we must ask is 'which one is more favourable?'*

Compound **2** is formed if the hydroxide attacks at a *primary* carbon centre. Compound **3** is formed if hydroxide attacks at a *secondary* carbon centre.

Steric or electronic?

2

Make sure you consider both options. Sterically, we would expect attack at carbon 1, since it is less hindered. While this is correct, it is not the whole story. Don't be tempted to stop as soon as you find one factor that favours the observed outcome.

If the hydroxide was to attack at carbon 2, the build-up of negative charge in the S_N2 transition state would be destabilized by the electron-releasing methyl group. Remember that methyl groups stabilize positive charge and destabilize negative charge. Therefore, attack at carbon 1 is favoured on both steric and electronic grounds.

The SEt group acts as a nucleophile. It **participates** in the reaction as a result of its proximity to the site of substitution. It is a **neighbouring group**. This process is therefore referred to as **neighbouring group participation**. Alternatively, it can be described by the more exotic term, **anchimeric assistance**.

Don't let 'what it is called' bother you. We worked out what is happening by a careful consideration of principles.

[8] Make sure you understand why this is, by considering the alternative S_N1 pathway.

ANOTHER NICE EXAMPLE

We have more cases where stereochemistry can be used to give information about reaction mechanisms.

Consider the following reaction.

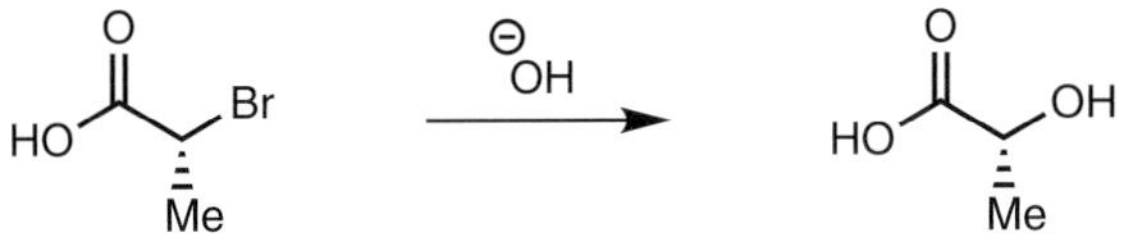

The stereochemical outcome here is retention of configuration. If this was an S_N1 reaction, we would get racemization, while if it was S_N2 we would get inversion.

There are only two possible ways to get overall retention of configuration. Either the reaction genuinely occurs with retention of configuration, in which case we need to propose a mechanism to account for this, or the stereogenic centre is inverted ***twice*** *(and only twice!).*

Instead of trying to picture the entire reaction sequence, we should break it down into elementary steps. Hydroxide is a nucleophile, but it clearly cannot simply be acting as a nucleophile in this case, or we would get the enantiomeric product. It could also be acting as a base. Our substrate is a carboxylic acid, so deprotonation is not only likely—it is inevitable!

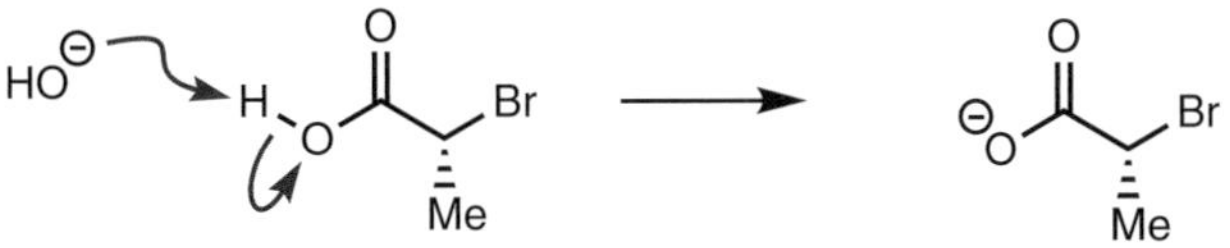

From here, we have a similar situation to those above. We have a nucleophile (not a great one, admittedly) close to the leaving group.

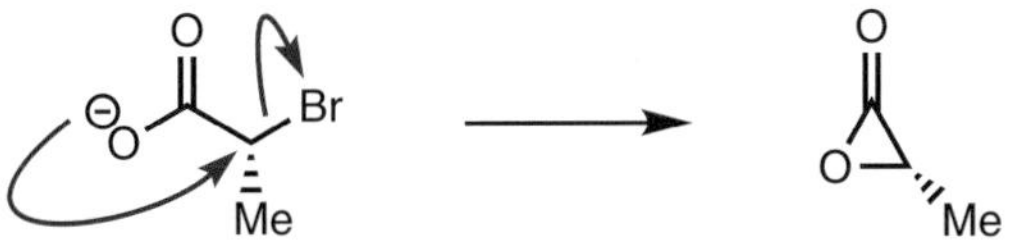

Now, I don't find the stereochemical outcome of this reaction particularly easy to see. But I know, with absolute certainty, that this reaction is an S_N2 substitution and S_N2 substitution reactions proceed with inversion of configuration.

We can now finish the job, with another S_N2 substitution reaction. We are breaking open the strained ring.

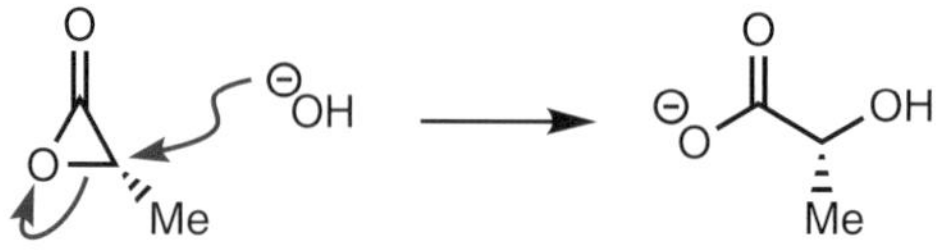

Hopefully it is now quite easy to see that this compound has the same sense of stereochemistry as the starting material. Make sure you know how to 'simplify' the problem. You know that the reaction proceeds with inversion of stereochemistry **twice**. Therefore, you know what the outcome **must** be. Even if you draw the intermediate stage with the wrong stereochemistry, you will still draw the final stage correctly.

In this context, three-membered rings appear to be 'special'. In fact, they are not. You will also find examples of anchimeric assistance in which a four-membered ring is formed (and broken). You won't find many examples of anchimeric assistance involving five- or six-membered rings, as these will usually be the final stable product.

2

BASICS 31
Introduction to Cyclohexanes

INTRODUCTION

This is cyclohexane. The problem is, it's not actually a hexagonal molecule. Each carbon atom is tetrahedral, so there is no way all six carbon atoms can be in the same plane.

Here is a top-down 3D representation of cyclohexane.

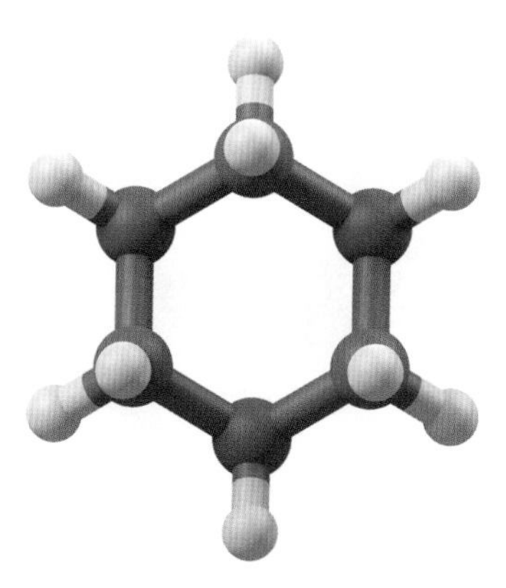

You can see six hydrogen atoms around the periphery (we might call it the **equator**!) of the molecule. Then we can see three hydrogen atoms that look as though they are directly on top of carbon atoms. You won't be surprised if I tell you that there are three more that are hidden behind the other three carbon atoms.

If you are surprised, make a model. Oh, just make a model of it anyway!

Here's a side view, so you can see all the hydrogen atoms that are pointing directly up or down.

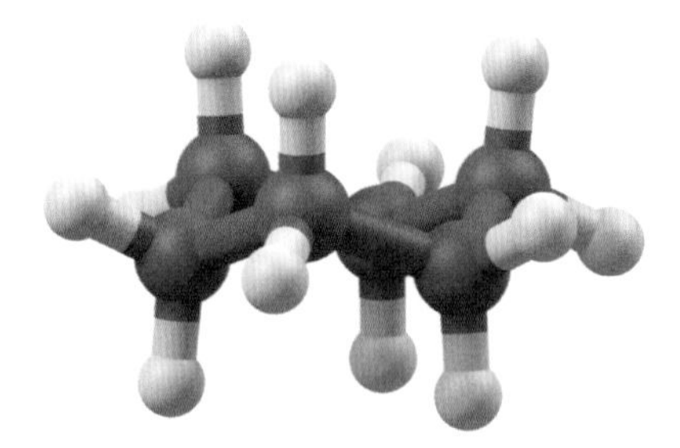

In this view, you can also see that, of the hydrogen atoms around the equator, some are pointing slightly up and some are pointing slightly down.

We need to get to grips with these structures, as they (and the principles that underpin them) are everywhere in organic chemistry.

CYCLOHEXANE CHAIR STRUCTURES

We refer to the structure above as a 'chair'. If I show it to you in this orientation, it is probably easier to see the analogy.

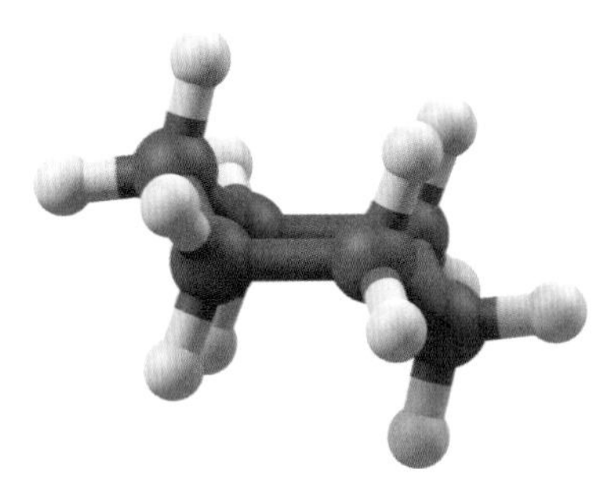

Ultimately, we need to get you to the point where you stop worrying about the appropriateness of the word 'chair', and start associating it naturally with the structures. You need to accept the term 'chair' for cyclohexanes to the point where you stop asking 'why does it look like a chair?'

We are now going to leave the 3D structures behind and start drawing the structures as you will see them in the chemical literature and as you will draw them yourself. Here is the 'flat' structure again, and here is the chair as it would normally be drawn. Note that by convention **we do not draw wedges and dashes for cyclohexanes**, either for the shape of the molecule or for the substituents attached to it.

Let's start adding the hydrogen atoms. We refer to the two 'different' locations for substituents as **axial** and **equatorial**. We are going to add the **axial** hydrogen atoms first. These are either straight up or straight down in the above orientation. Here's how to remember which way round they are. **If the C-C bonds going towards a particular carbon are going up, then the axial H is up**. Here is the structure with the three 'up' axial hydrogen atoms drawn in.

If the C-C bonds going towards a particular carbon are going down, then the axial H is down. We will add these next.

Now we know that the tetrahedral bond angle is 109.5°, so we can draw the **equatorial** substituents at a 109.5° angle to the axial substituents. Here's something important to remember. **If the carbon atom is to the left of the centre, then the equatorial substituent is pointing to the left. If the carbon atom is to the right of centre, then the equatorial substituent is pointing to the right**. Here is the structure with all of the hydrogen atoms drawn in.

It's starting to look a bit messy. Now you see why we often omit the hydrogen atoms! Here it is with only the equatorial hydrogen atoms shown.

Here it is a bit bigger, with the axial bonds/hydrogens in blue and the equatorial bonds/hydrogens in red.

5.3

CYCLOHEXANE CONFORMERS—FLIPPING THE CHAIR

We are going to do something funny with this structure now. We are going to take hold of the carbon atom on the far left and pull it downwards. We are then going to take hold of the carbon atom on the right and pull it upwards. This is what we get. It's a chair conformation again, but now, every hydrogen atom that was equatorial has become axial. Every hydrogen atom that was axial has become equatorial.

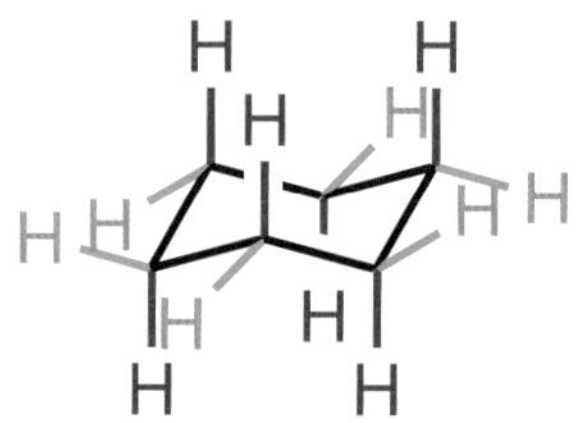

This is a very general phenomenon. Axial and equatorial are interchanged when we 'flip' (or **invert**) the chair.

*The **only** way to get used to this idea is to make a model of cyclohexane with your model kit and invert the chair form. Note that you aren't breaking bonds. This is just bond rotation, but you are rotating two bonds at the same time. These are still **conformational isomers**.*

We are going to explore this in more detail, with added substituents, and then we will fill in some more details.

EQUATORIAL IS BETTER!

As we saw with conformers of ethane and butane, we cannot consider these shapes without a brief look at their energies.

If you look at the diagrams above, you may be able to see that the three axial hydrogen atoms on the top face (or the bottom face) are quite close together. They are much closer than the equatorial substituents. Therefore, when we start putting substituents on the cyclohexane, axial is a bad place to be. Substituents prefer to be equatorial. We can show this by the following 'unequal' equilibrium, where the conformer with the equatorial methyl group is favoured.

The structure on the right is destabilized because the axial methyl group is close to the axial hydrogen atoms shown. Because these are two carbon atoms apart, this is referred to as a **1,3-diaxial interaction**. We are not going to worry too much about the position of the equilibrium (yet!), but the larger the group, the less likely it is for the group to be axial.

We will quantify the position of these equilibria in Basics 32.

When you get to a *t*-Bu group ($(CH_3)_3C$) it is almost impossible for this group to be axial.

the only stable conformer

really destabilised

The *t*-butyl group effectively 'locks' the conformation of the cyclohexane ring. This is really useful, as it means we can think about where any other substituents are relative to the *t*-butyl group. We will see how to use this in Applications 3.

A KEY SKILL—INTERCONVERTING FLYING WEDGE AND CHAIR CYCLOHEXANES

Sometimes you will encounter a cyclohexane structure in a flying wedge projection, and you need to work out where the substituents are in space. You need to be able to interconvert the structures confidently. You already know by now that I think this simply needs practice.

Look at the following two structures, a *cis*- and *trans*- disubstituted (two substituents) cyclohexane.[9] In this representation we need the wedge and dash bonds to show where the substituents are in space.

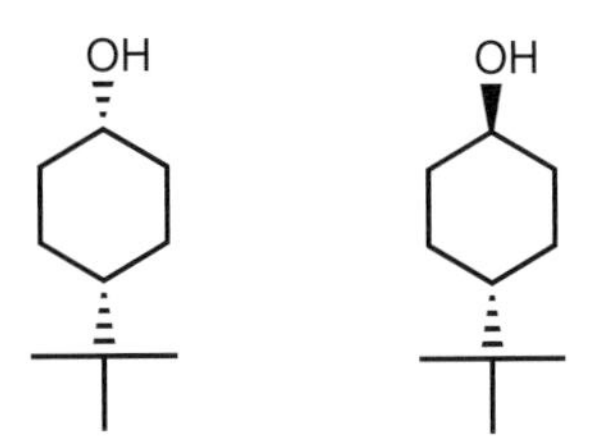

We can draw these in the chair as follows. It is easier at first to show the hydrogen atoms on the carbons with the substituents. The left-hand structure has *t*-Bu **down** (the

[9] If two substituents are on the same side, they are said to be *cis*; on the opposite side they are said to be *trans*. We can use this terminology for alkenes, although the *E* and *Z* nomenclature introduced in **Basics 3** is preferred.

H is above the *t*-Bu) and the OH **down**. Therefore, this is *cis*. The structure on the right has the *t*-Bu **down** and the OH **up**.

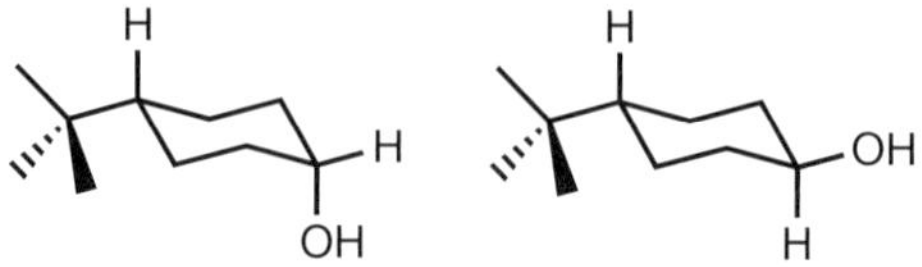

There's something else I did with these structures. I could have drawn these in a conformation that had the *t*-Bu axial. I didn't, because I would rather draw the conformer of the structure that is likely to be more stable.

With practice, it becomes easier to consider the most stable conformational isomer as well as the gross structure.

As you'd expect, you will find it quicker to consider all aspects of the structure together and only need to draw one structure. As you learn, make sure you draw **a** correct structure. Then you can consider whether it is the **best** correct structure.

Here's a really important point that causes a lot of problems.

There is no direct connection between up/down and axial/equatorial. It will depend on the conformation of the ring that you draw. We can see this in the structures above. The left-hand structure has both substituents ***down*** *and yet one is axial and the other equatorial.*

Let's have a more detailed look at the conformational equilibria for these two compounds, in order to reinforce this point.

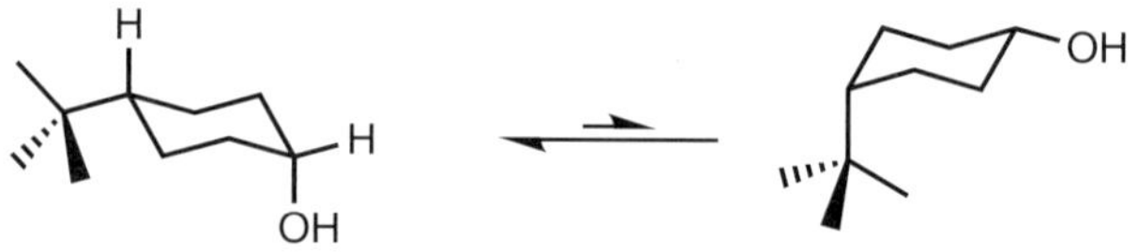

cis *cis*

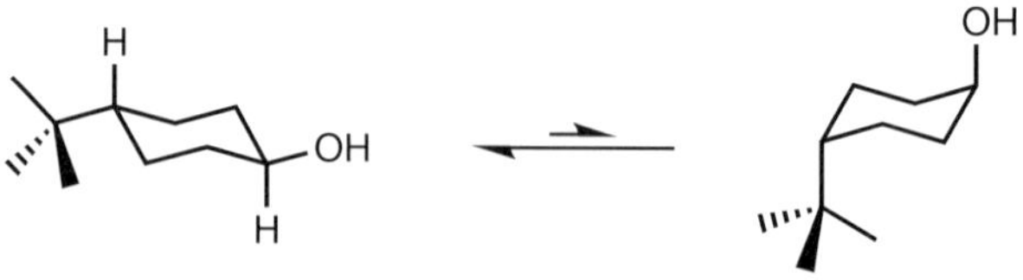

trans *trans*

We can see that there are two chair conformers for each compound. If we have two substituents on the same side, they are said to be *cis*. We cannot move one of them to the opposite side of the ring by rotating bonds. Therefore, the *cis* and *trans* isomers are different compounds.

The structures on the left are much more stable in both cases than the structures on the right. Therefore, the *cis* isomer must have an axial OH group. It's not great, but it's better than having an axial *t*-butyl group. This is useful, as these compounds allow us to explore the different reactivity of axial and equatorial groups.

When asked to draw the most stable cyclohexane conformer for a given structure, the common error is to draw the structure that has every substituent equatorial. ***This misses the point.*** *If you have two substituents, they are either both on the same side, or are on opposite sides. This alone will determine whether they are equatorial or axial. Once you have done this, you are then in a position to determine which conformer is favoured.*

PROBLEM-SOLVING CYCLOHEXANES

So, how do we determine the most stable chair conformation of a cyclohexane? The simple answer is that you should draw both conformers, and should identify the one which has either the largest substituent equatorial or the most substituents equatorial.

For example, it is better to have one *t*-Bu group equatorial and two methyl groups axial than to have the *t*-Bu group axial and two methyl groups equatorial. We will look at a more refined way to consider the conformational equilibrium in Basics 32.

You learn this sort of thing with experience. For now, we need to get used to drawing good cyclohexane chairs and interconverting them.

SOME EXAMPLES AND POINTERS

The important point is what you need to be able to do, and what pitfalls you can avoid. Converting a flying wedge projection of a cyclohexane into a chair is important. To do this, you need to work out which of the substituents will be equatorial and which will be axial. We will run through the method with a worked example. Let's look at the following molecule. I have numbered the carbon atoms so we can follow them through the method.

t-Bu
1
Cl 2 3
CH_3

We need to convert this into a chair form, but we could start with either of the following two chair forms. I've put the numbers on, and we can see that going from 1 to 3 is anticlockwise. It won't matter which we start with.[10] This is important, as you can be confident that you won't make a mistake at this stage.

The *t*-Bu group is 'up' (a wedge bond). If we put this on the structure on the left, it will be axial, but if we put this on the structure on the right, it will be equatorial. I've added the H on carbon 1 to show this more clearly.

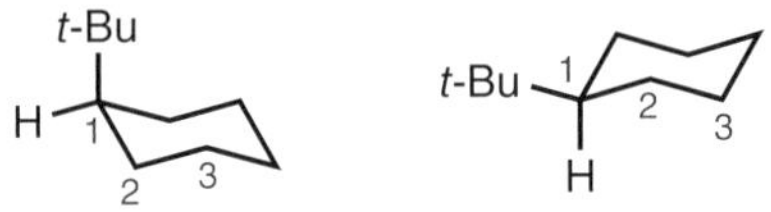

On carbon 2, the Cl is 'up'. On the left, this will be equatorial. On the right, it will be axial. Here it is.

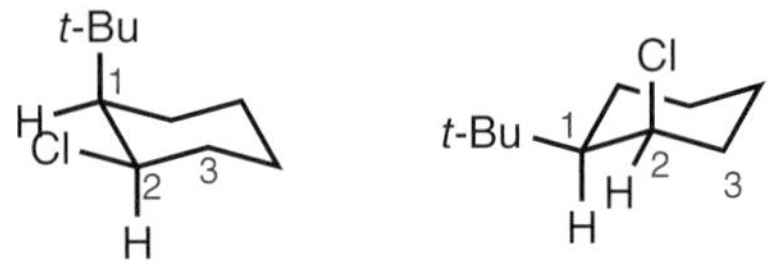

Finally, the methyl group on carbon 3 is 'down', which is equatorial on the left, and axial on the right. Here is the final structure.

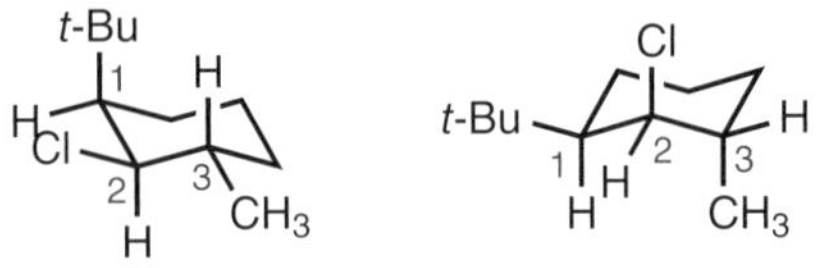

So, the two structures above are conformers of the same stereoisomer. You can prove this to yourself by making a model of the compound and interconverting the chairs.

Now we can think about which conformer is the most stable. On the left, we have two substituents that are equatorial and one axial. However, the bulky *t*-Bu group is axial, which is really bad. On the right, we have one substituent that is equatorial and two

[10] I always start with the structure on the right. I start drawing the right-hand side of it and work downwards and to the left. I don't really understand why I always do this, but to draw the left-hand structure feels more difficult. That's the benefit of practice and habit.

that are axial. But at least the really bulky substituent is equatorial. In this case, the conformer on the right is favoured.

There are a couple of key points we need to make.

- It isn't always possible to have all substituents equatorial. I know I said this before. It's worth repeating!
- When you convert one structural representation to another, always keep the stereochemistry the same.
- It isn't always easy to determine which conformer is going to be the most stable, particularly if you have one equatorial and one axial substituent of similar size. In this case, it is likely that they will be in equilibrium where both conformers are accessible.

BAD CYCLOHEXANES!

What sort of mistakes will you make? The most common one seems to be getting the proportions wrong. Structures like the one below are common.

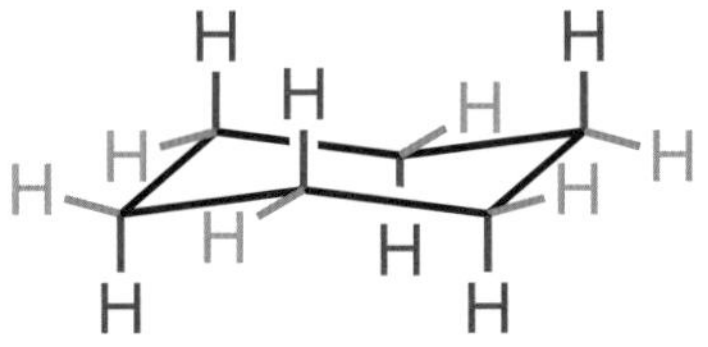

The axial hydrogen atoms look miles apart. This is because the C–C bonds are drawn much longer than the C–H bonds. This is unrealistic. You need to see this so that you can understand why it is wrong.

In an ideal world, I would never ever see it again!

There is also a tendency to sometimes draw the chair form with two horizontal bonds, as shown in the structure below on the left.

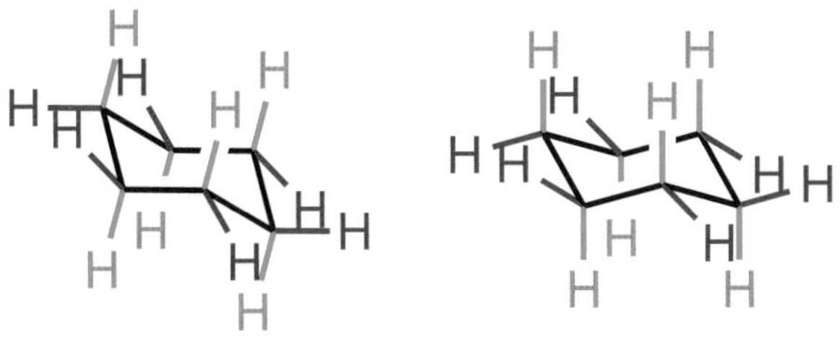

While this is not strictly incorrect, it does make it harder to draw the axial and equatorial bonds properly. This all comes down to developing good habits. The structure on the right is just 'nicer'! We are going to work on this with some examples in Practice 12.

The structure on the right isn't harder to draw than the structure on the left. You just need to know why it is worth making the effort!

NEWMAN PROJECTION OF CYCLOHEXANE CHAIR

I now want to show you what the cyclohexane chair looks like in a Newman projection. If you look at the structure on the left in the direction indicated by the arrow, you should see something like the Newman projection shown on the right. We are looking along two bonds simultaneously, which is why it looks like two Newman projections joined in the middle.

You can see something 'special' in this representation. All of the bonds are fully staggered. We have almost perfect tetrahedral bond angles of 109.5° throughout, and dihedral angles of 60° and 120°. The shape is quite predictable, which is what makes cyclohexane an interesting model system with which to test our understanding of reaction mechanisms.

Again, get your model kits working and try to see this for yourself. It will be much easier to remember that way.

WHERE DO WE GO FROM HERE?

There are lots of things we can do with cyclohexanes. They are useful molecules in their own right, since they can have useful biological properties. The fact that their shapes are (with practice) predictable is useful in this respect.

Moving forward, you can have structures with multiple cyclohexane rings fused together. We will see some examples of these in Basics 33 and some problems in Practice 13.

Looking even further forward, some reactions proceed through intermediates and transition states that have the same general shape as a cyclohexane chair. When you eventually encounter these, I want you to have the necessary key skills at the ready.

If you don't get to grips with the fundamentals of cyclohexane drawing, you'll struggle to cope with the more challenging applications. But if you practise and internalize the basics, the challenging applications won't be all that challenging.

We are going to add some more detail about cyclohexane shape and energetics in Basics 32 and Basics 33, but for now we need to work through a lot of problems dealing with the basics.

PRACTICE 12
Drawing Cyclohexanes

THE POINT OF THE EXERCISE

Six-membered rings are really stable, and that doesn't just apply to benzene. Nature loves six-membered rings, and they are found in many biologically active natural products. You need to become good at drawing them.

*There are lots of instances where you will need to draw something that **looks like** a cyclohexane chair. If you get good at drawing cyclohexanes, without really worrying about why you are doing it, the other stuff will take care of itself.*

We are going to practise the basics now, and then come back to more complex examples in Practice 13.

QUESTION 1

12

Draw the most stable chair conformer for each of the following cyclohexanes.

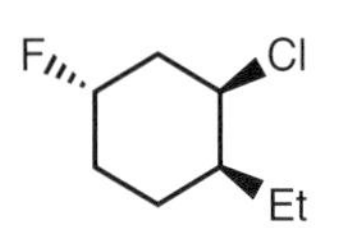

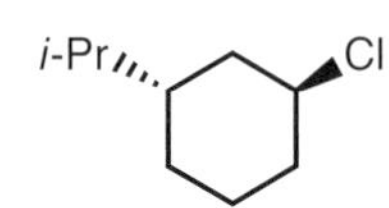

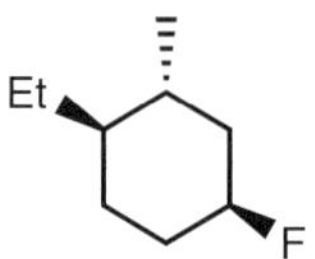

Right, we need to get a common problem with this question out of the way right now.

Did you draw, in each case, a structure which has all substituents equatorial?

This would be understandable. After all, you know that a substituent 'prefers' to be equatorial rather than axial.

The problem is, everything is connected. In some cases, if you have one substituent equatorial, another substituent **must be** axial.

Go back and have another go! Check that you didn't make this mistake.

The methodology is key here. Maybe you can see it. Maybe you cannot. Perhaps you need to make a model of the compound.

By now you should be 'internalizing' your recognition of wedges and dashes (i.e. you shouldn't need to think too hard about which is up and which is down).

Hopefully by now for each compound you have a lowest energy conformer which has one substituent axial.

There's another common mistake!

Does the chair structure you have drawn represent the same enantiomer of the compound as the flying wedge structure above? It needs to!

This may seem like a lot to take on board. Trust me when I tell you that with practice you will be able to do this without thinking about it. All you need to do is consider this point every time and practise lots!

Now draw the less stable chair conformer for each compound, and make sure you have drawn the same enantiomer of that one as well. If you just swapped each equatorial substituent for an axial one (and vice versa) then you have probably drawn the mirror image. You need to flip the chair.

QUESTION 2

Draw flying wedge projections for the following cyclohexanes.

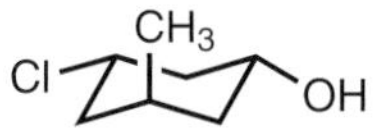

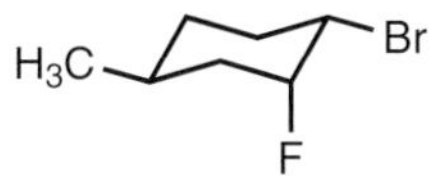

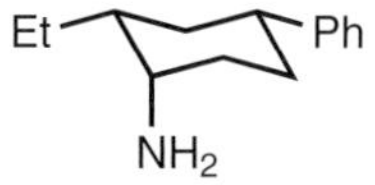

Perhaps you should add the hydrogen atoms on the carbons atoms with the substituents, to make sure you can distinguish between 'up' and 'down'.

The issue I find here is that students draw the bonds at very strange angles, so that 'up' and 'down' are not at all clear. Make sure you look back at the representations in Basics 31. Everything is there. You can do this. You need to learn to spot the mistakes, so you can stop making them. Use your molecular models. I don't know a better way.

QUESTION 3

You can do more of this on your own. Draw five cyclohexanes in flying wedge representations and convert them into chair forms. Draw five cyclohexanes in chair forms and convert them into flying wedge representations. Do a few of these problems every week until you can do it without thinking about it. You don't need me to give you more problems to solve. You can do this for yourself. There isn't really a good or a bad example. You can learn from all of them.

QUESTION 4

Which of the following structures is chiral? The point here is to try to recognize symmetry, rather than formally assigning stereogenic centres (Habit 6). In each case draw the structure in its most stable chair conformation.

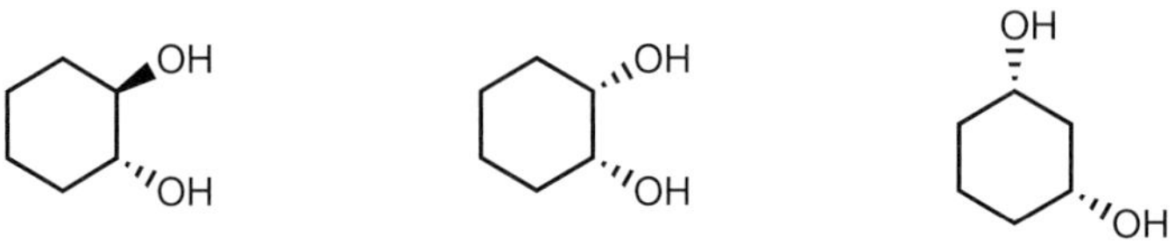

For structures with a plane of symmetry, is this more apparent in the chair form or in the flying wedge form? Sometimes one representation is easier to see than another.

This is a recap of material we looked at in Habit 7, but we are now considering the shape of the compound.

It turns out that this example is (at least in principle) rather complicated. We will use this as a case study in **Worked Problem 6**. I will use this to show you how you can analyse a simple problem on a range of levels.

APPLICATIONS 3

Substitution Reactions of Cyclohexanes

This is another 'Applications' chapter, so you need to do the work, rather than just reading through and convincing yourself that you understand everything. We are applying what we have discussed about substitution reactions to cyclohexane systems.

WHY ARE CYCLOHEXANES SPECIAL?

The short answer to this question is that they are not. They just make really good examples, since we can readily predict the shape of the molecules and the bond angles. Recall that a cyclohexane chair has more or less perfectly tetrahedral carbon atoms, locked into a staggered conformation. Every dihedral angle is 60° or 180°. We can add particular substituents that 'lock' the chair into one conformation, so that we only have to consider the reactivity of one structure.

Go back to your molecular model of cyclohexane and check this, one more time!

In this chapter, we will look at the substitution reactions of cyclohexanes. We will consider steric and electronic effects, and we will determine what the outcome is likely to be for either of the two possible mechanisms. We will consider which product(s) will be formed, and how quickly they will be formed. We are really starting to build things up now.

Of course, the arguments that we will use in this chapter don't just apply to substitution. The principles can be applied to any reaction. And that is the most important point! We will apply them to elimination reactions in ***Applications 5****.*

THE S_N1 MECHANISM

Consider the following two cyclohexanes. Because we have the *t*-butyl group locking the chair, one structure has the bromine in an equatorial position; in the other it is axial.

Let's get one thing out of the way. Neither of these compounds will actually react by an S_N1 mechanism. Loss of bromide anion will give a secondary carbocation, which is not that stable. This doesn't undermine the logic we are going to use.

We know that these are the only conformers of these compounds that we need to consider. If we flip the chair in the left-hand structure, we will have two axial substituents. This would be very unstable. In the right-hand case, flipping the chair would place the Br equatorial, at the expense of the *t*-butyl group becoming axial. Again, this is unfavourable. So, the structure on the left has an equatorial Br. The structure on the right has an axial Br, and there's nothing you can do about it!

Make models of the compounds.

Check that you have followed the arguments. Never miss an opportunity to test your understanding of the basics.

In order that you internalize this information you need to do the work, and I will guide you.

Draw a reaction profile.

The compound with an axial Br is less stable (higher energy) than the compound with the equatorial Br. However, when they lose bromide to give a carbocation, they will give exactly the same carbocation. Formation of the carbocation will be *endothermic* in both cases. It will be preceded by a transition state that is (a little) bit higher in energy again.

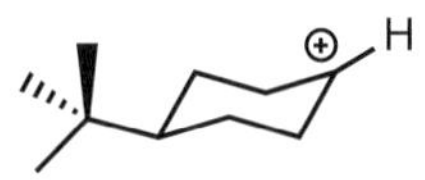

What does this mean about the activation energy for the process (use the Hammond postulate to relate the energy of the carbocation to the energy of the transition state that precedes it)?

Hopefully you came to the conclusion that the compound with axial Br will react faster in an S_N1 reaction. If you didn't, work through it again and see if you can work out why this is the case.

I'm not going to give you the reaction profile here. I am confident that you can work it out for yourself.

We should remind ourselves why the compound with the axial Br is less stable. This group undergoes unfavourable interactions with the axial hydrogen atoms shown.

If we lose the bromide anion, then both compounds will form the same carbocation, shown below. The crowded axial group is leaving. What I am aiming for here is that you 'feel' the relief of strain in this molecule as the axial Br leaves. Perhaps that is a strange way to express it, but most good organic chemists will develop an emotional response to certain structures and reactions. You just need to know which one is more favourable. If you can 'feel' the steric strain in compounds, this will become more natural.

I explicitly drew the hydrogen atom in the structure above, because it is neither axial nor equatorial—the carbocation is planar!

Let's finish the job. Quite a lot of this section is focusing on application of principles, as there are relatively few clear-cut examples of S_N1 mechanisms. You might want to view it as 'this is how it would work *if* it was S_N1'.

THE SECOND STEP

In Reaction Detail 2, we talked about the stereochemical outcome in S_N1 substitution reactions, and made the point that the common term used (racemization) only applies in the strictest sense if we have a substrate that only contains one stereogenic centre, and is present only as a single enantiomer. Something that is already racemic (or simply not chiral) cannot undergo racemization. Go back and have another look at this explanation before proceeding.

In the present case, if we consider the attack of a nucleophile onto the carbocation above (completing the S_N1 mechanism), the nucleophile could attack either side of the planar carbocation.

(after subsequent loss of $H^{\oplus}$ from the nucleophilic atom)

In one case you will get an equatorial product, in the other case axial.

If you get a different product from attack on either side, the sides must be different!

Since the two sides are different, there is the possibility that one will be more favoured than the other. This is a stereoselective process (Basics 28).

Here is an example that is as close to S_N1 as you are likely to find.[11] Only the product with the axial methoxy group is formed. Clearly, axial attack on the carbocation intermediate is favoured. There is no doubt that the product formed, with the large phenyl group equatorial, will be more stable. This doesn't necessarily mean it will be formed fastest.

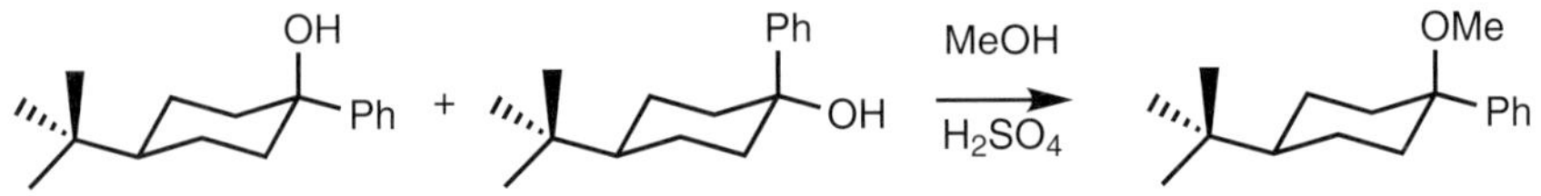

To form this product, the nucleophile must get past the axial hydrogen atoms.

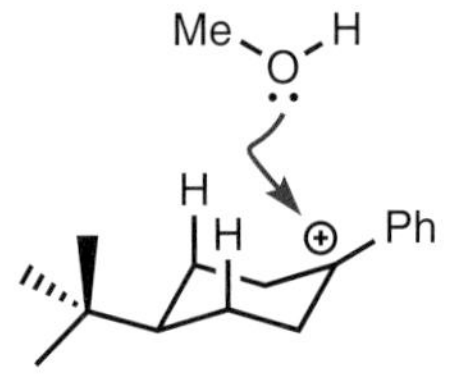

However, to get the alternative product, we would have to compress the bulky phenyl group against the same hydrogen atoms, and we would have **torsional strain**[12] between the incoming nucleophile and the axial C–H bonds pointing down in the structure below.

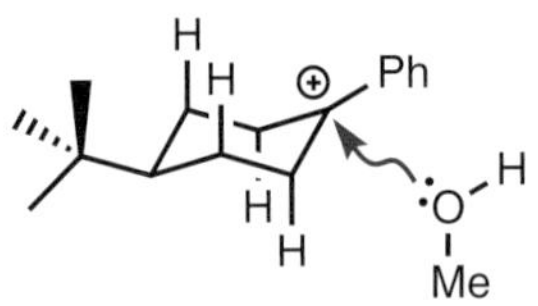

The exact outcome of any given reaction will depend on the substituents. It is not easy to predict whether the interaction of the incoming nucleophile with the axial hydrogens will be more or less severe than the compression of the phenyl group against the same hydrogens.

[11] Make sure you understand why this is!

[12] Torsional strain is the increase in energy in a structure as a result of eclipsing of bonds. We saw this in Basics 30.

Now you have seen the principles, you should be able to predict where to add a substituent to increase one or another of the effects, or what the difference will be if you use a larger or smaller nucleophile.

There is one more complication in this case. You could protonate the product, and then break the C–O bond to form the same intermediate carbocation—the outcome of the reaction could be the result of an equilibrium favouring the more stable product!

 Draw a curly arrow mechanism that shows this possibility!

S_N2 REACTIONS OF CYCLOHEXANES

Let's now consider the S_N2 mechanism. For the compound with equatorial Br, the nucleophile must get past the axial hydrogen atoms indicated below. This is a significant amount of steric hindrance (it hinders approach of the nucleophile). This steric hindrance increases the energy of the transition state, which in turn increases the activation energy. Therefore, displacement of an equatorial bromide is slower than displacement of an axial bromide.

There are a couple of key points here. First of all, as this is an S_N2 substitution, it still proceeds with inversion of configuration. The carbon atom at which the substitution takes place is not a stereogenic centre (at least, not in this case). However, there are still clear stereochemical consequences in this reaction.

This example emphasizes the importance of drawing structures correctly and realistically. If you get the C–C and C–H bonds out of proportion, you will convince yourself that the hydrogen atoms are either too small or too far away to have any influence.

The discussion up to this point won't allow you to predict how much faster or slower one stereoisomer will react. But perhaps you have a cyclohexane, or similar structure, with one axial and one equatorial Br—you need to know whether you can react one selectively in the presence of the other (can you get chemoselectivity—**Basics 26**?).

For now, just get used to the fact that the two isomers will react at different rates, understand why this is and be ready to apply it when you need it.

BASICS 32

Quantifying Conformers of Cyclohexanes

We covered the basics of cyclohexanes in Basics 31. Most of what you do with cyclohexanes concerns the relative stability of two conformational isomers. In Applications 1, we saw that it isn't just the size of a substituent that affects conformational equilibria. If this applies to ethane and butane derivatives, there is no reason why the same principles would not apply to cyclohexanes.

In this chapter, we will look at the same ideas we saw in Applications 1, but here we will apply them to cyclohexanes.

Finally, we will formalize the conformational equilibria of cyclohexanes by using a table of 'A values' to help us decide which conformer will be favoured in more complex situations.

I'll be honest. I don't use these 'A values' myself, but I think it is educationally useful to present them so that you can access the information. What I want is for you to be able to look at two conformers of a cyclohexane, and to determine which will be most stable. Or conversely, to be able to say that it is pretty close, and you cannot be confident which will be more stable.

Here are the three examples we considered for ethane/butane in Applications 1. Of course, now we are talking about cyclohexanes, we only need to consider two conformers—the chairs. I would argue that this simplifies the problem. It all depends on how good you are at drawing chairs.

32

CH_3 CH_3 ⇌ CH_3 CH_3

Cl Cl ⇌ Cl Cl

OH OH ⇌ OH OH

The rules from Applications 1 have not changed. We will consider the size and electronegativity of the substituents, as well as the potential for hydrogen bonding in the case of OH.

Once again, I will reinforce the point that there isn't that much to do, but you need to know how to apply basics in a range of situations.

We are going to do this in two stages. First of all, we will compare the equatorial and axial cyclohexanes with one another. Then we will compare them with the ethane/butane conformers from Applications 1.

DIRECT COMPARISON OF CYCLOHEXANES

trans-1,2-Dimethylcyclohexane[13] is pretty easy. It's steric effects all the way. You have two substituents that will either both be equatorial or both axial. The conformational equilibrium will strongly favour the conformer on the left, in which they are equatorial.[14]

CH_3 CH_3 ⇌ CH_3 CH_3

13.3 kJ mol^{-1} more stable

32

For *trans*-1,2-dichlorocyclohexane, we can draw on what we did for 1,2-dichloroethane in Applications 1. We would expect the dipole repulsion in the diaxial conformer to be favoured. However, cyclohexanes have 1,3-diaxial steric interactions. When we take these interactions into account, it turns out that the diaxial conformer of *trans*-1,2-dichlorocyclohexane is favoured, but only just.

On purely steric grounds, substituents prefer to be equatorial. Therefore, a situation in which a preferred conformer has two axial substituents is significant.

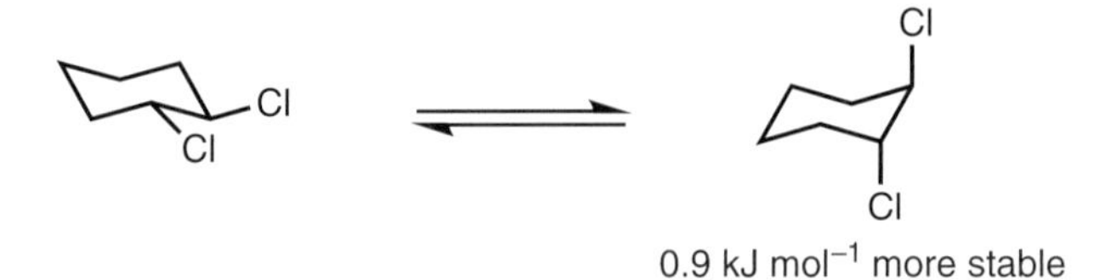

0.9 kJ mol^{-1} more stable

[13] You may have noticed that in this sentence, the capital letter is the 'D' of 'Dimethylcyclohexane' and not the 't' of '*trans*'. It is worth pointing this out, so you can start to get used to it.

[14] Don't rush this. If you are still getting used to which one is axial and which one is equatorial, or whether axial or equatorial are normally preferred, work through this slowly and carefully, re-drawing the structures and making notes. Use this as an opportunity to reinforce the basics.

Notice that I am talking about 1,3-diaxial interactions in the right-hand conformer, but the relevant axial hydrogen atoms are not shown explicitly. They are still there!

For the 1,2-diol, we would expect the diequatorial conformer to be preferred. Hydrogen bonding is significant, and the dichloro compound above only just favours the diaxial conformer. Add the strength of a hydrogen bond and it is pretty clear-cut.

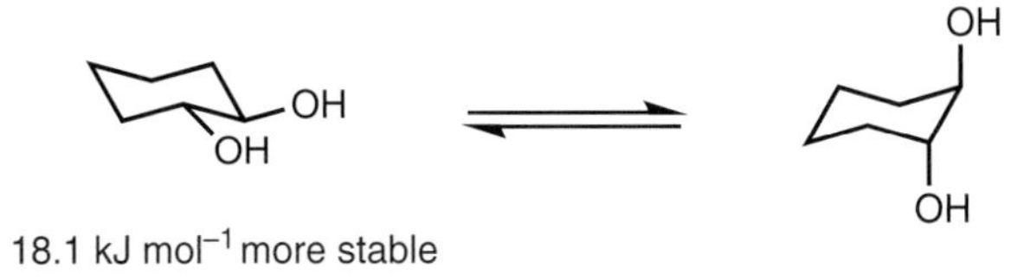
OH
OH
OH
OH
18.1 kJ mol−1 more stable

COMPARING CYCLIC AND ACYCLIC CONFORMERS—QUANTIFYING AXIAL INTERACTIONS

Now we should compare the cyclohexane data above with the ethane/butane conformer data from Applications 1. All of this is data calculated at the same level of theory using the same software, so it is directly comparable.

Here are the butane conformers alongside the 1,2-dimethylcyclohexane conformers. The *anti* conformer corresponds to the diaxial conformer. The *gauche* conformer corresponds to the diequatorial conformer. They are drawn in a different orientation here, to make this comparison clearer.

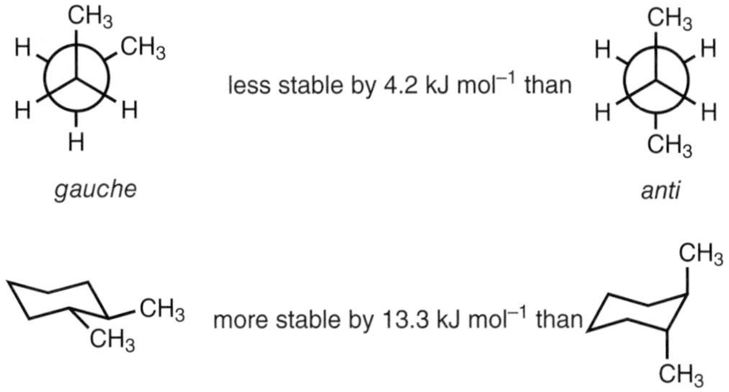
CH3
H
CH3
H
H
H
gauche
less stable by 4.2 kJ mol−1 than
CH3
H
H
H
H
CH3
anti
CH3
CH3
more stable by 13.3 kJ mol−1 than
CH3
CH3

So, a 4.2 kJ mol^{-1} preference for the *anti* conformer is over-ridden by the 1,3-diaxial interactions. The extent of this can be estimated as 4.2 + 13.3 = 17.5 kJ mol^{-1}.

Remember that this number is based on computational data and is an estimate. The comparison is sensible, but ignores some interactions present in the cyclohexane. Try to identify these.

Here is the same comparison for the 1,2-dichloro compounds.

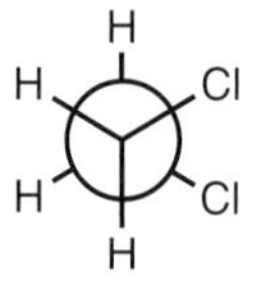

less stable by 8.9 kJ mol^{-1} than

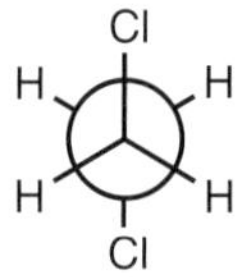

gauche *anti*

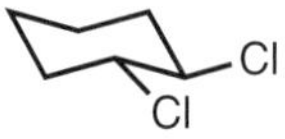

less stable by 0.9 kJ mol^{-1} than

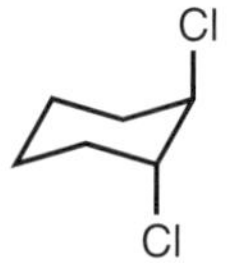

In this case, in the cyclohexane the diaxial interactions (which are steric rather than electronic) mitigate against the dipole repulsion. The *trans*-diaxial cyclohexane conformer is still more favoured, but not by as much as we might have expected.

Finally, the dihydroxy compounds.

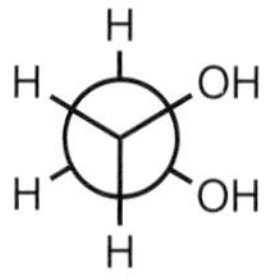

more stable by 12.1 kJ mol^{-1} than

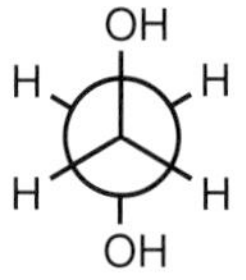

gauche *anti*

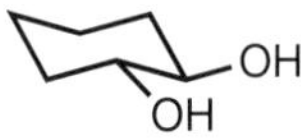

more stable by 18.1 kJ mol^{-1} than

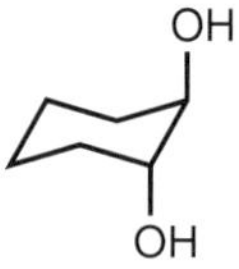

Both the *gauche* ethane conformer and the diequatorial cyclohexane conformer have the potential for intramolecular hydrogen bonding. The net effect is that with the dichloro compounds, the diaxial cyclohexane was only just favoured. Add a hydrogen bond to the diequatorial conformer, and it is very much favoured. Comparing with the ethane-1,2-diol, we get a stronger preference for the diequatorial conformer because the 1,3-diaxial steric interaction further disfavours the diaxial conformer.

TAKING STOCK

We have considered conformational equilibria for ethane/butane and cyclohexane systems. We have seen the similarities and the differences.

Applying these principles, you should be able to predict which conformer will be favoured for a range of compounds. The principles apply to all ring sizes and chain lengths. Of course, there are added complexities, mainly associated with specific interactions in the conformers. However, the principles won't change.

QUANTIFYING THE CONFORMATIONAL EQUILIBRIA

Consider the following equilibrium.

The conformer on the left makes up 95 per cent of the compound at equilibrium. The conformer on the right is only 5 per cent of the equilibrium mixture. From this equilibrium, we can calculate that the conformer on the left is approximately 7.2 kJ mol^{-1} more stable. The 'A value' is the difference in free energy between the conformers in which a given substituent is equatorial or axial.[15]

A larger substituent/A value will favour the equatorial conformer even more strongly.

Here is a sample of A values for a small selection of substituents. In effect, we can consider the A value to represent the steric impact of a given substituent. We will look at a few of these in more detail.

Substituent	A value/kJ mol^{-1}
Me	7.2
Et	7.4
i-Pr	9.1
t-Bu	20.7
Cl	1.8
Br	1.6
Me_3Si	10.6

First of all, an ethyl group or an isopropyl group don't have much more steric impact than a methyl group, despite being quite a lot bigger. We can see why on the following structures.

[15] 'A values' are generally quoted in kcal mol^{-1}. Since all other energies in this book are in kJ mol^{-1}, I have converted the numbers for consistency.

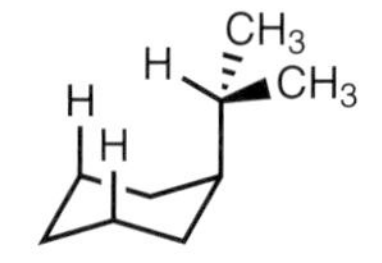

Here, one or even two methyl groups can be rotated out of the way. Therefore, the key interaction is an H with an H in all three cases.

This is a really important point. The stereochemical outcome of many reactions depends on steric interactions between the small hydrogen atoms. These interactions are generally 'rigged' by the presence of other substituents. This may well be the most important principle in this chapter, but the rest is still important!

There is then a big jump in A value for a *t*-butyl group. Here, there must be a steric interaction of a methyl group.

A trimethylsilyl group is very much like a *t*-butyl group, but with silicon instead of carbon. Its A value is much smaller, despite the group being bigger. This is because a C–Si bond is longer than a C–C bond, so the bigger group is further away and therefore it has less impact.

The same applies for Br—the long C–Br bond reduces the impact of the substituent.

These numbers have significant limitations. They do not take into account the interplay of substituents—you cannot simply use the numbers additively, unless you are confident that the substituents cannot interact in either conformer.

We would expect the trans-1,2-dimethylcyclohexane to favour the diequatorial conformer by 14.4 kJ mol^{-1} on this basis. In fact, the calculated number above was 13.3 kJ mol^{-1}. It isn't easy to tell whether the discrepancy relates to inaccuracy in the calculation, or a small steric interaction in the diequatorial conformer. Have a think about it!

The numbers do not take electronegativity or hydrogen bonding into account.

I think it is a good idea to show you these numbers. They are instructive in emphasizing key points about stability of conformers. I think it is useful to be aware of the key trends. If you genuinely understand them, the amount of learning is minimal. I would not advise you to learn the numbers. There are too many limitations. It is far better to develop the skill of spotting when steric and electronic effects apply, and determining which will predominate in any given case.

WHERE IS THIS GOING?

Once again, some reactions require specific conformers. For the reaction to take place, the conformer needs to be energetically accessible.

You need to be able to understand the factors that determine the position of the conformational equilibria in cyclic compounds as well as in acyclic compounds.

We already looked at the substitution reactions of cyclohexanes in Applications 3. We will look at the corresponding elimination reactions of cyclohexanes in **Applications 5**, and we will see how the conformational equilibria affect the rate and outcome of the reactions.

*Whenever you encounter a reaction in which two cyclohexane conformers could react, you need to ask the questions '**which conformer will react faster?**' and '**which conformer will predominate?**' This must become a habit.*

32

BASICS 33

More Conformers of Cyclohexanes and Related Compounds

Now you are getting used to cyclohexane chairs, we need to look at the other possible conformers.

5.5

CYCLOHEXANE BOATS

It turns out that a chair is not the only shape that a cyclohexane can adopt. Remember how we inverted the chair forms by pulling one side of the molecule up and the other down? Do you think that both things happen at the same time with molecules?

Try doing it at the same time with your molecular models. It isn't easy.

Perhaps they happen one at a time! If we just flip one side of a cyclohexane, we will get something like this. We describe the structure on the right as a 'boat'.

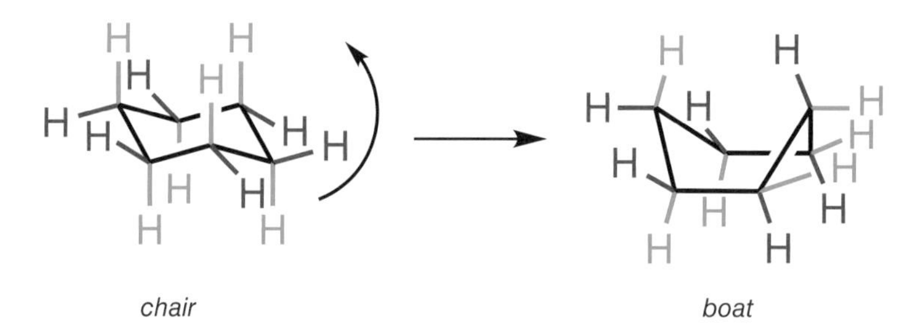

What I want you to see from those structures is that if a hydrogen atom is 'up' on the left, it is still 'up' on the right. If it is 'down' on the left, it is 'down' on the right.

The boat is half way between the two chair forms. When you interconvert the cyclohexane chair forms using your molecular models, you will usually find you go through a boat form.

The boat conformer is less stable than the chair structure. This is because there is a significant level of eclipsing of bonds, and there is a particularly unfavourable **flagpole** interaction as shown below.

flagpole interaction

There's a really important point to make here. You could have already predicted all of the destabilizing factors in the cyclohexane boat structure. All (?) you have to do is look very carefully at the structure and examine the dihedral angles. If you try to apply the basic principles every time, you will get better at it, and you will find yourself doing it almost subconsciously.

CYCLOHEXENE

There's another important molecule I'd like us to look at—cyclohexene. Here is the structure, and I've added an arrow to indicate the direction in which we should look. Remember that this is a double bond, so that the two carbon atoms making the double bond are in a plane, as are the two carbon atoms attached to them. Therefore, we have four carbon atoms in a plane, and the remaining two carbon atoms out of the plane.

Generally, one of these carbon atoms will be above the plane and one will be below it, so we have two conformers that look like this.

The terms 'axial' and 'equatorial' don't quite apply here. Instead we would use '*pseudo*-axial' and '*pseudo*-equatorial', particularly for substituents on the carbon atoms that are out-of-plane. It won't surprise you to find that *pseudo*-equatorial is better than *pseudo*-axial!

This isn't the easiest thing in the world to see. Make a model of it. Once you get the idea, you will see it more clearly in the diagrams.

We will look at some of the implications of these conformers in **Worked Problem 7**.

CYCLOPENTANE

We won't do much with cyclopentanes for now, but you should at least have some idea. The preferred conformer of a cyclopentane is described as an **envelope**. Essentially, you have four carbon atoms in a plane, and one out of the plane.

The complicating factor is which of the carbon atoms is out of the plane (the flap of the envelope). This tends to be an equilibrium, so it is all of them at various times. Play around with your molecular models and you will see this.

If you don't look at this with your molecular models, it will just be another fact to learn.

cis- AND *trans*-DECALINS

'Decalin' is the trivial name given to a compound in which two cyclohexane rings are fused as follows.[16] The systematic name is bicyclo[4.4.0]decane. The ring fusion can either be *cis* or *trans* as follows.

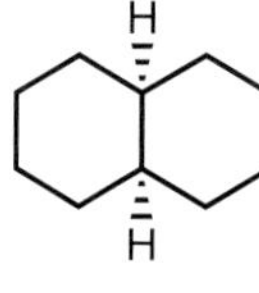

cis-decalin

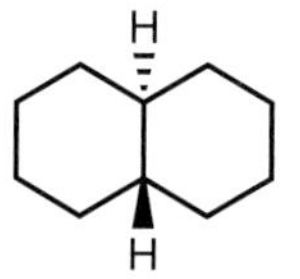

trans-decalin

5.6

So, how do you draw these? Well, each six-membered ring can be a chair. The point is not to panic, and to start with what you know.

In this case, draw the left-hand ring as a chair, and then view the parts of the right-hand ring as substituents. So, for the *cis*-decalin, we can expect one of the carbon atoms of the right-hand ring to be equatorial and one to be axial. In the *trans*-decalin they can both be equatorial. I'll start this with the following, where the wiggly lines across bonds indicate that there is more to come.

[16] Deca = 10. Ten carbon atoms. It does make sense!

If you then continue, you will make the second ring to be a chair, and you will see what the structure looks like. Your molecular models will help with this. They should look something like this.

The most common mistake when drawing cis-decalins is to start with a fused-ring structure like the one on the right, and then add hydrogen atoms at very strange angles to 'force' it to be cis-fused. When students do this, they do know it is wrong, but aren't sure how to correct it.

There's just one more thing to do with these structures.

Take your molecular model of both of them (you have made one, right?) and invert the chair of the left-hand ring (or the right-hand ring).

You should find that you can do this for the *cis*-decalin, but not for the *trans*-decalin. If you do this for the *trans*-decalin, both of the substituents that make up the right-hand ring will have to become axial, and there is no way that can happen with a fused six-membered ring. It's easier to see this with molecular models than it is to explain it.

These sorts of motifs are present in lots of important compounds. All the steroid compounds contain a ring system with three six-membered rings and one five-membered ring. Testosterone, the male sex hormone, is one such compound.

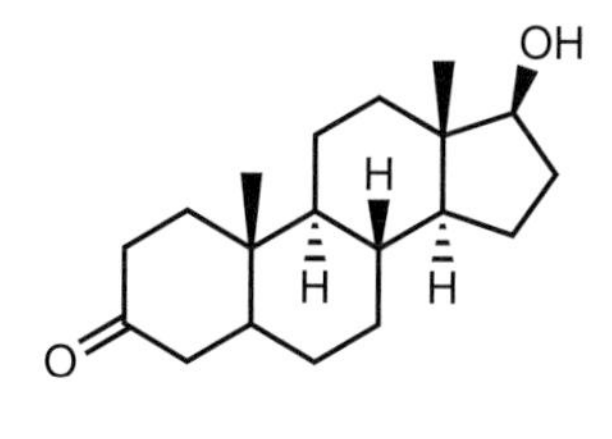

testosterone

Work out the shape of this compound! Make a model of it—there is little conformational flexibility. Practise drawing it as a series of fused chair conformers.

ONE MORE SUBTLE POINT

Chair and boat forms of cyclohexane are 'ideal' structures.

Get your molecular model kit and make a cyclohexane boat. Identify the C–H bonds that are eclipsed, and the flagpole interaction.

Now the key question. What can the molecule do about it?

Bonds can rotate!

It is possible to alleviate some of the eclipsing and flagpole interaction by twisting the boat slightly. Organic chemists lack imagination. We call such a conformer a **twist boat**! You will see this conformation again in **Worked Problem 6** and **Worked Problem 7**.

This is a great example of how your knowledge and understanding builds in layers. You learn the cyclohexane chair form and get used to drawing it. You then learn about the boat form, and you get used to it. Finally, you recognize that these are just two extreme forms, and there are lots of conformers in between. You internalize the reasons for cyclohexane forming these conformers, and you gradually learn to identify a 'bad' cyclohexane.

Make a model of *cis*-1,4-di-*t*-butylcyclohexane.[17] What shape is it?

You would initially expect it to have one equatorial *t*-Bu group and one axial *t*-Bu group. You will never have *t*-Bu axial, so you have to 'naturally' find the conformer that 'feels' best by understanding what an unfavourable steric interaction is. You have gone beyond the basic chair and boat.

5.7

I deliberately discussed this without drawing a structure. You need to see it for yourself.

[17] You have a name, with stereochemistry and positioning of substituents. You need to convert it to a chair structure in the first instance, and then see what (if anything) is bad about it. And then you follow your instincts.

PRACTICE 13

Drawing More Complex Cyclohexanes

THE POINT OF THE EXERCISE

We drew quite a lot of cyclohexanes in Practice 12, but when you get right down to it, those examples were quite easy. There was only one ring, so you just had to work out whether there were more/bigger substituents in the equatorial position.

Most of the examples you might encounter in research are more complicated, and they will have more than one ring. We need to practise these as well.

We are going to use some of the structures, or parts of structures, that we first encountered in **Practice 3**.

I will warn you up front. Some of these are truly horrible! Do not allow this to deter you. For the level that this book addresses, you will almost certainly never see examples as difficult as these on an exam. But by learning to cope with these challenging examples, anything you do see on an exam will be a walk in the park! Train hard, fight easy!

5.8

QUESTION 1

Let's start with a fused system that isn't too bad (it's all relative!). The two rings are *trans* fused.[18]

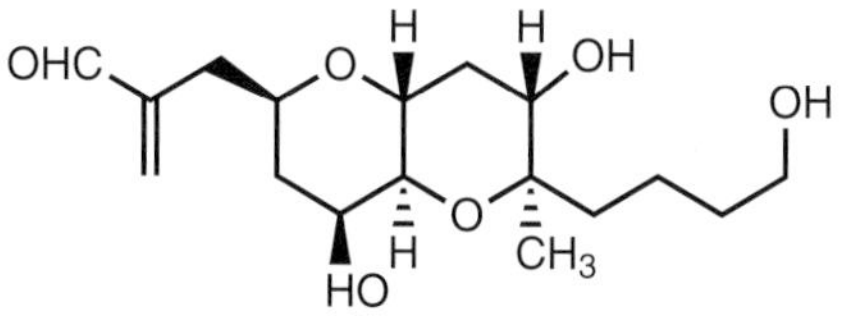

Make sure you have drawn the same enantiomer as well as a sensible conformer with all the right groups axial/equatorial.

[18] I am showing you what to look for!

QUESTION 2

Here is the next one. There are two distinct challenges here. There are two reasonably long alkyl chains that will take time to draw for no added benefit. Just call them 'R^{1}' and 'R^{2}'.[19]

Pseudomonic acid C

The second challenge is that one of the bonds in the ring is a dashed bond. Think about what this means for the stereochemistry of the hydrogen atom attached to the carbon, and then think about the stereochemistry of the 'R' group.

It's not a question of whether this is easy or difficult. It's just a question of 'internalizing' your ability to visualize the shapes of molecules, **no matter how they are drawn**.

QUESTION 3

The structure below is what you would get if you cleave the O–O bond in artemisinin.

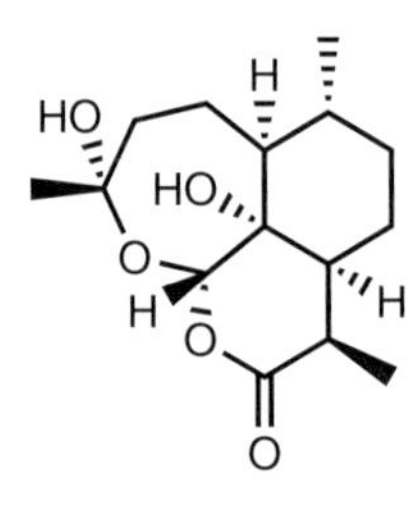

There are two fused cyclohexane rings (well, one of them has an oxygen and a carbonyl, but that's the point—it doesn't change all that much). There is also a seven-membered ring fused onto both of them. This one is a bit more flexible.

Start with the two six-membered rings, and ignore the seven-membered ring at first. Then add this on.

[19] But make sure the number is a superscript, as a subscript would tell you how many of the groups you have!

This is **not** an easy problem, but it is a common problem. I made a model of it and twisted it to look for the steric interactions in the different conformers.

You then have to decide which orientation to draw it in.

*When I said it wasn't easy, I meant it is **horrible**. But if you can do this, you can do anything. If you can't do this at first, just keep coming back to it. When you manage to draw something that looks okay, you will know you have this sorted!*

QUESTION 4

Find lots more structures of natural products which contain saturated six-membered rings and draw them in the best chair conformer. Make sure you draw the same enantiomer where applicable.

*What happens if you choose a really difficult one? That's up to you. Make a model, persevere, put it to one side and come back to it. Whatever works and does not cause you too much stress. Don't let anything knock your confidence, and make sure you **do** come back to the difficult problems.*

COMMON ERROR 8

Cyclohexanes

The errors commonly made when drawing cyclohexanes have been mentioned already, so I just want to summarize them briefly here.

BAD DRAWING

Ultimately, this is the main problem. Everything else follows on from this. We have already looked at 'squashed' chairs with bonds out of proportion (Basics 31). We looked at chair forms in the 'wrong' orientation, so that the axial bonds are not exactly vertical.

There is one more error that I see a lot—substituents at incorrect angles. Here is cyclohexane with one of the equatorial hydrogen atoms drawn explicitly. On the left, the angle is correct. On the right is a common mistake.

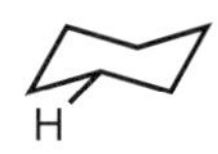

correct

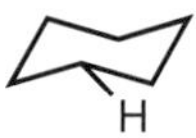

incorrect

You might think this doesn't matter. It all comes down to being able to visualize the shape of a molecule correctly. Perhaps if you do this, you won't make any 'big' mistakes. If you have fused cyclohexanes, such as the decalins we looked at in Basics 33, they will certainly look strange.

It isn't any more difficult to get the angles correct.

As with so much else we have discussed, you won't draw cyclohexanes badly because you don't understand. You will do it because you haven't developed the habit of drawing them correctly.

WEDGES AND DASHES

By convention, we do not need wedges and dashes in cyclohexane chair structures. Drawing them doesn't add anything. You **must** get used to the representation.

DRAWING ALL THE HYDROGEN ATOMS

Here is the structure of 4-*t*-butylcyclohexanol. On the left, we have the structure I would draw. On the right we have the same compound but with all the hydrogen atoms shown.

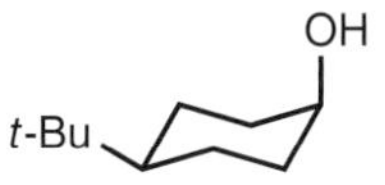

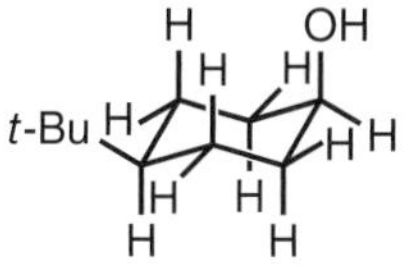

Which is clearest?

I am really hoping you think it is the one on the left. In the structure on the right, it's harder to actually spot the substituents.

Of course, the real issue is that when you need to 'do something' with the structure on the left, you can visualize where the hydrogen atoms are if you need them.

AXIAL AND EQUATORIAL

In the early stages, students sometimes mix up axial and equatorial. If you are having problems, slow down and check. As always, making a mistake doesn't mean you lack the ability to succeed. But if you don't correct this problem early on, you will keep reinforcing the bad habits until they are hard to break.

I'm sure you have been told that practice makes perfect. It doesn't! Only perfect practice makes perfect!

The other common mistake is assuming that all substituents can be equatorial. Depending on the number and stereochemistry of substituents, it may be that some are axial and there's nothing you can do about it!

Oh, and finally—spelling it 'equitorial'. Just don't!

*You might think that's being ridiculously pedantic. Equ**a**torial relates to the equator. It has meaning, and the meaning will ensure that you do **not** forget which one is which.*

cis-DECALINS

Here is a common error when drawing *cis*-decalin in the chair form. I mentioned the error in Basics 33. Because the *trans*-decalin is easier to draw, it is common to instinctively draw this when you are given *any* decalin structure. In order to 'make' the hydrogen atoms *cis*, you then draw one of the hydrogen atoms at a very strange angle. The chair form shown below **must** be *trans*-decalin. The bond in red cannot possibly be correct.

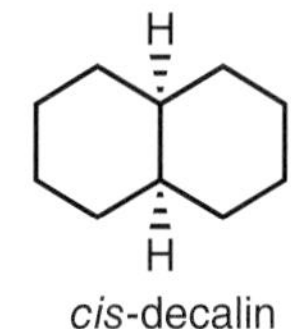

cis-decalin

H
H

drawn incorrectly

SECTION 6

ELIMINATING THE LEARNING

INTRODUCTION

Forgive the pun! But the point is a good one. If you understand everything, you won't have as much to learn! As you would expect, this section is **mostly** about elimination reactions. But it isn't just about elimination reactions. There are some fundamental ideas here as well. Then again, there always are!

REACTION DETAIL 4
Elimination Reactions

It isn't easy to split elimination reactions into mechanisms and stereochemistry, although **Applications 4** does focus on the latter aspect. We will still see the stereochemical implications of the E2 mechanism in this chapter as we look at the molecular orbitals.

For each of the elimination mechanisms, we need to consider all the following aspects.

- Which orbitals are involved?
- Which product(s) will we get (regioselectivity, stereoselectivity)?
- How fast is the reaction?
- How do changes to substrate structure affect the outcome?

Before we start, we need to recap one key point.

MORE SUBSTITUTED ALKENES ARE MORE STABLE

An alkene with more alkyl groups will tend to be more stable. This is true right up to the point where the attached groups are so bulky that there are unfavourable steric interactions.

> *You can always find an exception to a rule, but that doesn't make the rule less valid.*

The reason why a more substituted alkene is more stable is not trivial, and is explored in **Basics 34**.

THE E1 ELIMINATION REACTION IN MORE DETAIL

Here is the overall mechanism of an E1 elimination reaction.

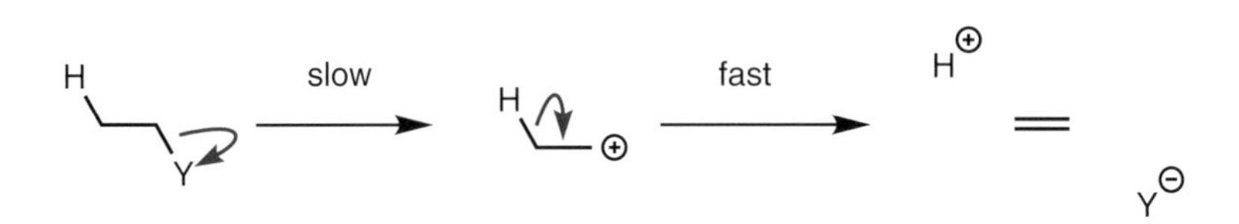

The first step is slow. It is the formation of a carbocation. The example above is a little misleading, because it shows a *primary* carbocation, which would never form (it would be **really** slow!). Once you have formed the carbocation, loss of a proton is faster.

Since only the substrate is involved in or before the rate-determining step, the rate only depends on the substrate. There **may** be a base that removes the proton, but it would only be a weak base, and it doesn't affect the rate.

An E1 elimination has:

rate = k[substrate]

We did the rate stuff for substitution in **Fundamental Reaction Type 1**. I deliberately left this aspect until a bit later for elimination reactions, so you had time to think about it.

If you form a carbocation, you can either add a nucleophile (S_N1) or lose a proton (E1). Whether you add a nucleophile or lose a proton will often depend on something as simple as 'do you actually have a nucleophile present?'

There is one additional consideration. In order to lose the proton, the corresponding C–H bond orbital must be properly aligned to overlap with the empty p orbital of the carbocation. This is shown, in a very simple case, below.

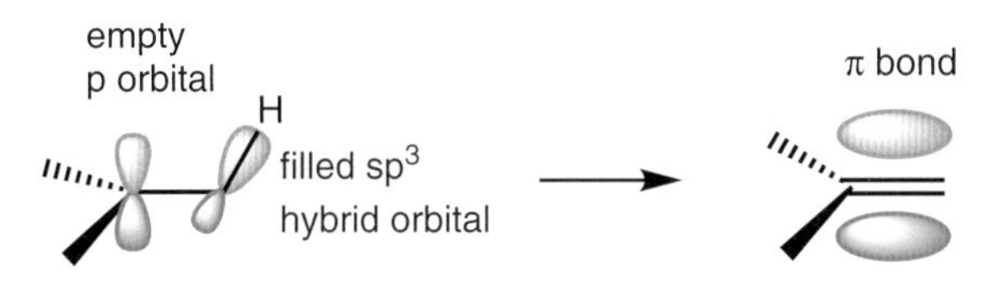

The reason for the need for proper orbital alignment is pretty straightforward—the π bond cannot develop unless the relevant orbitals start to overlap.

How do you learn to tell whether this is possible? You make models of compounds and build your intuitive understanding of the structures. I'm going to keep adding these reminders. If you've got to this point, chances are you have already been building models and drawing lots of structures. If you haven't, I'm guessing this is the point where you are starting to struggle. As always, the answer is simple—go back to basics and practise some more!

We are going to include some examples to work through in **Applications 4** and **Applications 5** that will help you develop this skill.

THE E2 ELIMINATION REACTION IN MORE DETAIL

We have seen so far that in the E2 elimination, the proton and the Y group leave at the same time, and that a base is required. This only just scratches the surface of this reaction.

transition state

Since there is only one step, and both reactants are involved in this step, an E2 elimination has:

rate = k[substrate][base]

We are forming the new double bond at the same time as we break the C–H and C–Y bonds. Remember that bonds are formed by overlap of orbitals, and so we must consider the shape and symmetry of the orbitals involved.

We are breaking the C–H bond, to leave the electrons with the organic product. Therefore, we will be using the electrons in this bond (an sp^3 hybridized C–H bond) to form the double bond.

We need the bonding orbital of the C–H bond.

We need to think about where the electrons in the C–H bond go.

They are being shared with the 'other' carbon, which already has a share in eight outer electrons.

If we are going to 'give' it some more electrons, we have to take some away. These electrons go to 'Y'.

*But recall (**Basics 9**) that one way to think about breaking a bond is to put electrons into its antibonding orbital.*

We can consider the overall process to be overlap of the C–H bonding orbital with the C–Y antibonding orbital.

When we draw these orbitals, it looks like this!

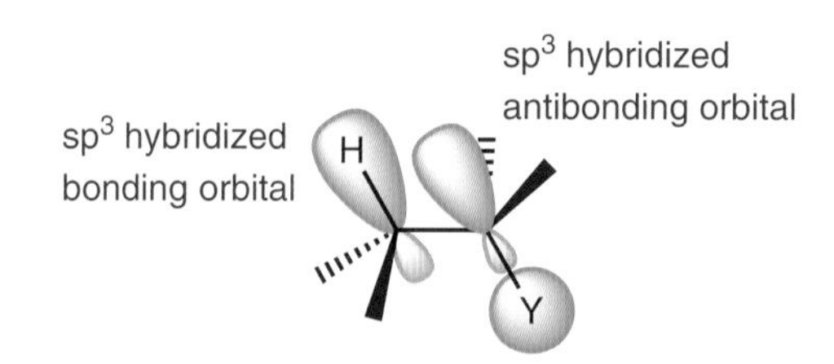

When we look at the orbitals like this, it is already starting to look like a double bond. As the reaction proceeds, these orbitals overlap and distort, and eventually become the π bonding orbital of the alkene. This results in a 'rehybridization' of the carbon atoms to become planar, sp^2.

However, in order for the reaction to begin, the C–H and C–Y bond orbitals must be able to overlap. In fact, they must be able to be aligned exactly as above, where the H and Y groups are on opposite sides of the central C–C bond. We refer to this arrangement as ***anti* periplanar**. This just means that H–C–C–Y are all in the same plane, but H and Y are on opposite sides.

You might imagine that orbital overlap would still be possible if H and Y were on the same side (syn). It is, but this arrangement is higher in energy, and so less favoured. There are some examples of syn eliminations, but they require high temperatures.

Let's think about the consequences of this for a minute. To do this, we will look at a Newman projection. There is no such thing as the right or wrong representation for a particular structure. However, if you don't at least consider using a Newman projection to work out the stereochemistry of E2 eliminations, it suggests you haven't got comfortable with them. They are perfect!

Now we can see that since H and Y are *anti* periplanar, if we have four other different groups attached to the two carbon atoms as shown, then R^1 and R^2 will end up on the same side of the double bond, as will R^3 and R^4.

Imagine taking off the H and the Y, and squashing the molecule!

Here are two hypothetical reactions. They give different double bond isomers.

Each of the starting materials has two stereogenic centres. They are diastereoisomers. You get a different alkene product depending on which diastereoisomer of the precursor you have. Therefore, if you want to make a particular alkene geometry, you need to know how to make the correct diastereoisomer of the precursor.

That's enough stereochemistry for now. We will come back to this in **Applications 4**.

THE E1cB MECHANISM IN MORE DETAIL

The E1 and E2 mechanisms are the most common mechanisms for elimination reactions. There is a third mechanism, which we mentioned briefly in **Fundamental Reaction Type 2**. We are only going to cover one example here, and this example serves to illustrate several key points.

Here is the first step in the reaction. We will use piperidine as the base.

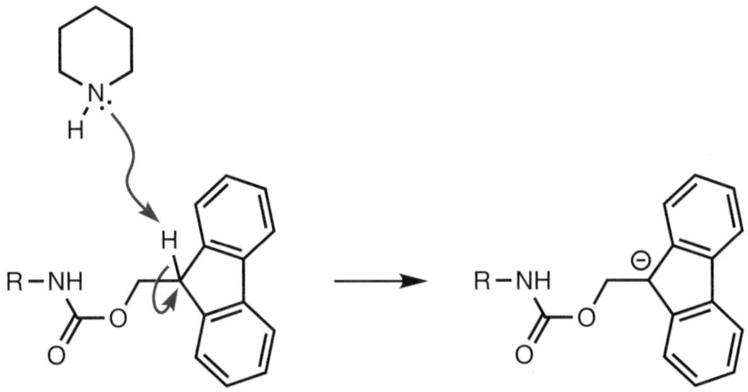

We are forming a carbanion, and this will only happen if the carbanion is (relatively) stable. Remember, piperidine isn't a very strong base. It's just an amine.

*In this case the carbanion is aromatic. We saw Hückel's rule in **Basics 10**. This carbanion has 14 π electrons, corresponding to $4n + 2$ where $n = 3$.*

So far this isn't an elimination reaction. The next step fixes this. We lose an oxygen leaving group, to form an alkene bond.

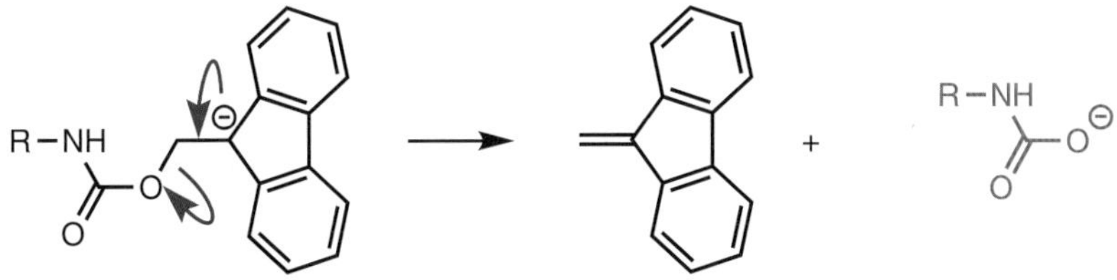

As far as the E1cB reaction is concerned, we are finished. We have an alkene product. However, we should look briefly at the 'by-product' as it is more important in this case.

We have a carboxylate group attached to nitrogen, and this isn't very stable. It can lose CO_2. In principle, this gives us an 'N minus', which would be very unstable, but we get around this by protonating the nitrogen as it leaves. The mechanism as drawn below doesn't imply that we have free protons in the presence of the base (remember, this reaction takes place under basic conditions). It just means we aren't going to worry too much about where the proton is coming from.

$$R{-}NH{-}CO_2^- + H^+ \longrightarrow R{-}NH_2 + CO_2$$

Of course, if you keep track of everything, you will note that we have protonated piperidine in the first step. Perhaps the protonated piperidine can give back the proton?

We just have one more thing to consider—the rate equation. You might have imagined that this would therefore be a 1st order reaction. Since the base and substrate are both involved prior to the rate-determining step, they are involved in the rate equation.

rate = k[substrate][base]

It is E1cB because the rate-determining step is unimolecular!

Look at it this way. If you add more base, you will increase the concentration of the carbanion intermediate, so that there will be more of it to react.

REGIOSELECTIVITY IN E2 ELIMINATIONS

This is going to get a little complicated, so we will start with the generalization and work towards specific examples.

In general, if you can get more than one possible alkene product in an elimination reaction, you will tend to form the more stable product (the more substituted alkene) as the major *regioisomer*.

Here is a simple example.

base

major

minor

This happens because the more substituted alkene is formed fastest under these conditions. If it is formed fastest, it must have the lowest activation energy.

The transition state for elimination is stabilized by the developing overlap of the C–H bond and the C–I bond in the transition state. This is an electronic effect.

REGIOSELECTIVITY IN E2 ELIMINATIONS—EXTREME SUBSTITUENT EFFECTS

I'm going to be honest about this next bit. It's not easy. If you don't understand it the first time, move on from it, and come back to it later!

Extreme situations are really instructive. Sometimes they exhibit slightly different reactivity or selectivity. How we analyse them is important.

The reaction below is almost identical to the example above, and yet it gives a different regiochemical outcome.

We have just established a 'rule' that where we can get two different regiochemical outcomes, the one that leads to the more substituted alkene is favoured. We are immediately 'breaking' that rule.

F

NaOEt

It would be very hard to argue that this is a steric effect. Fluorine is much smaller than iodine, and yet the base removes the more hindered proton in the compound with iodine. Surely if this was steric, it would be the other way around!

It must be an electronic effect!

With the iodine as leaving group, we don't have to work too hard. This is a clear E2 elimination and gives the most stable alkene product (most substituted). As the hydrogen and iodine are removed, there is considerable double bond character in the transition state, so that factors which stabilize the double bond will also stabilize the transition state.

What happens with fluorine? Well, fluorine is extremely electronegative, and so polarizes all of the C–H bonds as shown below. But one of them is more polarized than the other. We need to work out which one this is.

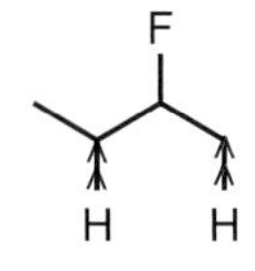

The most polarized C–H bond will be the one with the greater δ+ charge on the hydrogen, so the greater δ− charge on the carbon. If the carbon can stabilize the δ− charge, this would be more favourable.

If we put a negative charge on the carbon on the right (after removing the H), it will be a *primary* carbanion, and will therefore be the more stable of the two possibilities. Therefore, with an extremely electronegative substituent, the hydrogen on this carbon will be removed preferentially, giving the observed outcome.

*I am **not** saying that the hydrogen **is** removed first. I am relating the stability of the full anion to the partial negative charge that results from polarization of the C–F bond.*

It does take a little while to get your head around this, but bear with it. In fact, this reaction is almost becoming an E1cB reaction.

In **Perspective 4** we will explore the idea that rather than having clearly defined E1, E2, and E1cB mechanisms, most reactions sit on a continuous 'scale of mechanism' which can be defined by how much bond-breaking and bond-forming takes place at each stage. In the above case, there is considerable lengthening of the C–H bond, but not enough to consider it to be broken to form a carbanion intermediate.

*Again, we refer back to the comment towards the end of **Basics 16**. Everything we know about the stabilization of a full negative charge also applies to the stabilization of a partial negative charge.*

REGIOSELECTIVITY IN ELIMINATION REACTIONS—HOFMANN VERSUS ZAITSEV

We are now in a position to explore some ancient history!

In 1875, Alexander Zaitsev[1] formulated a rule that if an elimination reaction can give two regioisomeric products, the more substituted alkene product predominates.

This is the 'general' rule we saw above.[2]

In direct contradiction to this, August Wilhelm von Hofmann found that elimination from quaternary ammonium salts favoured the less-substituted alkene product.

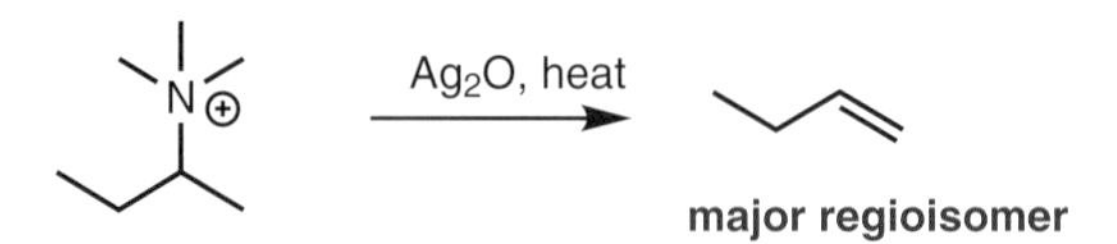

[1] You will also see this written as 'Saytzeff' or 'Saytzev'.
[2] General in the sense that the majority of E2 elimination reactions follow this trend.

Can they both be right?

These were experimental observations, so clearly they **are** both right. The two leaving groups in these processes are very different. It might be argued that the quaternary ammonium salt in the Hofmann example is very electron-withdrawing, and it is more like the 2-fluorobutane example above. However, it can also be argued that the quaternary ammonium salt is more hindered, so that the less hindered proton is removed. Both arguments have been presented over the years, and current wisdom favours the steric interpretation.

Of course, this cannot be applied to 2-fluorobutane, so there clearly can be a case where the electronic effect dominates.

As I said, it's complicated!

PERSPECTIVE 4
A Continuum of Mechanisms

So far, much of the discussion of substitution and elimination mechanisms has focused on whether a reaction is S_N1, S_N2, E1, E2, or E1cB. This encourages you to think in absolute terms about particular mechanisms.

The reality is at once more complicated, yet simpler.

The molecules don't need labels. They do whatever is most energetically favourable.

A DEEPER LOOK AT SUBSTITUTION REACTIONS

We always state that S_N1 substitution reaction of single enantiomer chiral substrates proceeds with racemization.[3] We saw, at the end of Reaction Detail 2, that this isn't quite true.

In many cases, there is slightly more inversion of configuration than retention of configuration. Remember that racemization would require an equal amount of each outcome.

We could take the 'simple' view that some of the reaction is proceeding through an S_N1 mechanism (racemization) and some by an S_N2 mechanism (inversion).

This is getting closer to the truth, but it is not quite correct.

Let's conduct a 'thought experiment'. We have an alkyl halide (R–X) that can form a relatively stable carbocation. We break the C–X bond by gradually increasing the bond distance.

*Here's the key question. **When** does the nucleophile attack?*

If the nucleophile doesn't attack until R and X are completely separated, solvated and all information about the chirality in R has been lost, there is no doubt that you will get a racemic product.

What happens if the R–X bond is 80 per cent broken (whatever that might actually mean!) and the nucleophile 'can't wait'? Where will it attack? At this point, even though

[3] Actually, we are usually less precise, and most of this definition is only implied.

there is significant S_N1 'character' to the reaction, it will be easier for the nucleophile to attack opposite to the leaving group—you will get an S_N2 outcome.

A reaction that is 'pure S_N1' is pretty rare—it needs a stable carbocation intermediate and the right solvent. A reaction that gives a 'pure S_N2 outcome' (notice the subtly different wording here) is pretty common, as any individual molecular process that isn't pure S_N1 will give the S_N2 outcome.

*In a reaction that is **mostly** S_N1 (almost racemization), the majority of the molecules will react to form a planar carbocation, but for those that do not (because the nucleophile cannot wait!), the inversion will take place with considerable 'S_N1 character'.*

Let's look at this in more detail with some computational data.[4] What I have done here is ask the question 'what would the transition state for the given S_N2 substitution reaction look like?' This doesn't mean that the reactions will all be S_N2. The substitution of *t*-Bu-Br will definitely not be! These are just calculations of what the transition state for a substitution reaction would look like **if** it was S_N2.

The data in the table are key interatomic distances in the **transition states** for a range of alkyl bromides being substituted with chloride as shown below.

$$R{-}Br + Cl^{\ominus} \longrightarrow R{-}Cl + Br^{\ominus}$$

R	Cl–C Distance/Å	C–Br Distance/Å	Electrostatic Charge
CH_3	2.435	2.472	-0.395
CH_2CH_3	2.460	2.587	+0.390
$CH(CH_3)_2$	2.578	2.674	+0.888
$C(CH_3)_3$	2.859	3.057	+1.441

Now let's make an assumption. For the purposes of the argument, it won't matter if the assumption is true or not. We are going to be looking for trends. Bear with me!

Let's assume that the substitution of iodomethane is a pure S_N2 reaction. Therefore, 2.435 Å corresponds to a situation in which the C–Cl bond is 50 per cent formed. Similarly, 2.472 Å corresponds to a situation in which the C–Br bond is 50 per cent broken.

The first thing to note is that since we are breaking a C–Br bond and forming a C–Cl bond, we wouldn't expect the C–Br and C–Cl distances to be the same. They are

[4] As with calculations described in other chapters, I have used DFT calculations with the B3LYP hybrid functional and a 6-31+G* basis set. It still won't mean anything to you, and it still shouldn't worry you.

different halogens. The C–Br and C–Cl distances in the starting materials/products won't be the same, so we shouldn't expect them to be the same in the transition states.

We can also look at the charges on atoms in the transition states. The electrostatic charge is a calculated property, and we shouldn't try to read too much into the tabsolute numbers. Nevertheless, we expect this particular reaction to be S_N2, and we would definitely expect a build-up of negative charge in the transition state.

Now let's look at the final entry in the table, substitution of *t*-butyl bromide. Here, both the C–Cl and C–Br distances are much longer. In the transition state, the C–Br bond is more than 50 per cent broken, and the C–Cl bond is less than 50 per cent formed.

This makes a lot of sense. The methyl groups in the *t*-butyl group add steric crowding and electronic destabilization to the S_N2 transition state. Having more C–Br bond breaking and less C–Cl bond forming allows a positive charge to build on this carbon atom (see the electrostatic charge in the table—more than a full positive charge!). The methyl group stabilizes this positive charge, and the longer bonds relieve the steric crowding.

Although we have calculated the S_N2 transition state, there is considerable S_N1 character. Since we expect the substitution of tertiary alkyl halides to be S_N1 (depending on reaction conditions!), we should not be surprised by this.

Now have a look at the two other rows in the table—ethyl bromide and isopropyl bromide. There is actually a pretty smooth trend in both the interatomic distances and the charges in this series.

What we are seeing within this trend is a gradual increase in 'S_N1 character' to substitution reactions that are S_N2. The nucleophile is still attacking from the back. If this was a single enantiomer substrate, we would still see inversion of configuration.

While we should continue to ask key questions such as 'is this substitution S_N1 or S_N2?', we should begin to develop the idea that mechanisms are a continuum rather than an absolute. Pure S_N1 substitution reactions are extremely rare, because they need a very stable carbocation intermediate. A substitution reaction may well have all the features of an S_N2 mechanism (one step, no intermediate, inversion of configuration) and yet there may be more bond-breaking and less bond-forming in the transition state—the presence of substituents may impart some S_N1 'character' to this reaction.

Don't get too hung up about this. These are subtle ideas, and they take a little getting used to. If you don't like this, or don't get it, move on and then come back to it later. With experience, you will find that this is a natural way of thinking about changes in mechanism.

MORE O'FERRALL–JENCKS DIAGRAMS

There is a neat way of representing the smooth transition between mechanisms—a More O'Ferrall–Jencks diagram. We aren't going to do too much with them, but I want you to be aware that they exist, and what they show you.

Here is a More O'Ferrall–Jencks diagram for S_N1 and S_N2 substitution. The blue diagonal line represents a 'perfect' S_N2 substitution reaction. The R–Br bond is being broken and the R–Cl bond is being formed simultaneously. The red line represents an S_N1 substitution, in which the R–Br bond breaks, and then the R–Cl bond forms.

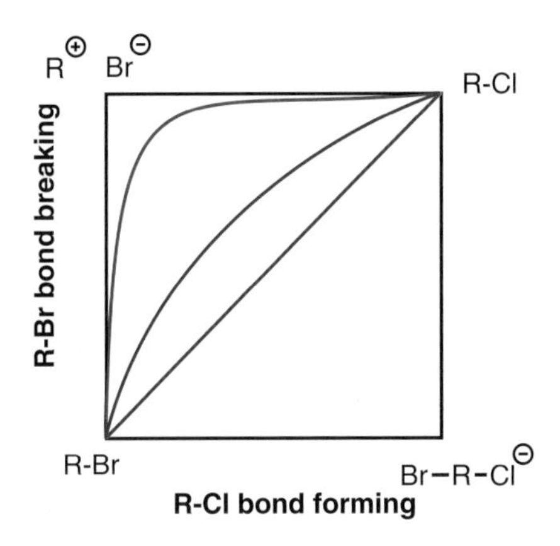

The purple line is close to S_N2, but it has some S_N1 character. The R–Br bond breaking is ahead of the R–Cl bond forming, but only slightly.

Of course, the bottom right of this diagram is empty, because we cannot form the anionic intermediate shown. You can have this type of (associative) mechanism for third row elements (*e.g.* phosphorus) as they have available d orbitals.

Draw a More O'Ferrall–Jencks diagram for elimination reaction mechanisms. Use the diagram above as your starting point. What mechanism is represented at the bottom right?

BASICS 34

More Substituted Alkenes Are More Stable

INTRODUCTION

It is an accepted fact that a more substituted alkene is more stable. That is to say, ΔH for the formation of the more substituted alkene from the elements is more favourable.

We will look at why this is the case.

But, it is more important to see the bigger picture. If a reaction can give two different products, it will very often give more of the most stable product. We saw this in **Reaction Detail 4** for elimination reactions. We also saw that sometimes this does not happen, and we can propose an explanation for why this is the case.

Finally, we should consider the reactivity of differently substituted alkenes. Just because one alkene is more stable than another, it does not necessarily follow that the more stable alkene is less reactive.

We will use a simple reaction, protonation of an alkene, to allow us to discuss this.

WHY ARE MORE SUBSTITUTED ALKENES MORE STABLE?

We will only consider simply alkyl substituents. In fact, for now we will only consider methyl groups.

Here are seven alkenes.

H H H H (**1**)   H H H_3C H (**2**) H_3C H H_3C H (**3**)   H H H_3C CH_3 (**4**)

H CH_3 H_3C H (**5**) H_3C CH_3 H_3C H (**6**) H_3C CH_3 H_3C CH_3 (**7**)

It isn't a trivial task to directly compare the 'stability' of these alkenes. We could directly compare the heat of formation for **3**, **4** and **5**. These are isomers, so we can determine which is most stable with respect to the elements. However, comparing the heat of formation of **5** with that of **7** is meaningless.

Instead, we compare the ΔH for the hydrogenation of these compounds. We did that in **Basics 20** to provide evidence for the stability associated with conjugation.

I am not going to provide the numbers here, but they demonstrate that **1** is the least stable and **7** is the most stable.

Each additional alkyl group stabilizes the alkene bond.

Structure **2** is a little bit more stable than **1**. Structure **6** is a little bit less stable than **7**.

Let's look at some numbers. Structure **2** is about 10 kJ mol^{-1} more stable than structure **1**. Structure **7** is about 2 kJ mol^{-1} more stable than structure **6**.

If it ain't steric, it's electronic!

The stabilization is an electronic effect. We saw in **Basics 16** that a methyl group can be electron-releasing. We also saw in **Common Error 3** that it depends whether the methyl group has anywhere to release electron density to.

In this case it does. The π bond of the alkene has a corresponding π* orbital which is empty. Donation of electron density from the C–H bond of a methyl group into the π* orbital leads to a net lowering in energy of the system. We can show this in two ways. We can sketch the orbitals, or we can draw an orbital energy diagram.

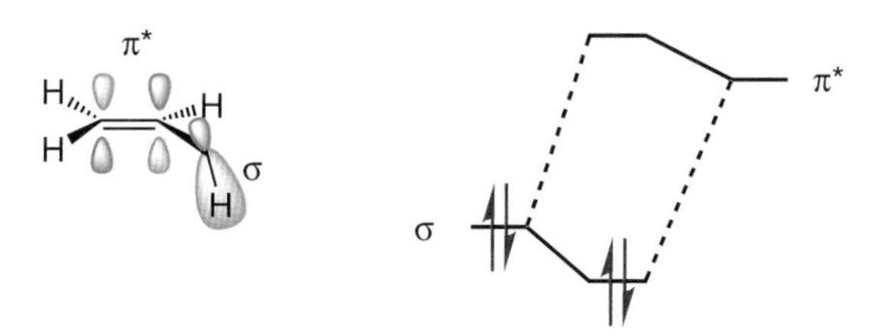

There would be no point drawing an orbital energy diagram for an interaction between two orbitals that did not have the correct symmetry to overlap.

In this case they do. Similarly, there would be no point drawing a representation of the orbitals if it was not going to lead to stabilization. Both aspects are equally important.

If it ain't electronic, it's steric!

Why do we gain 10 kJ mol^{-1} from the first methyl group but only 2 kJ mol^{-1} from the fourth? This is steric. By the time you add the fourth methyl group, it is getting pretty crowded. You do still gain some energy, but not as much as you would have expected.

When you start to add really bulky substituents, it doesn't follow that more substituted is more stable. It is important that you understand principles rather than blindly following rules.

Now we can consider the relative stability of the three disubstituted alkenes **3**, **4** and **5**. In alkenes **4** and **5** the two methyl groups are on different carbon atoms, but in the case of **4**, they are on the same side (*cis* or *Z*). This is more crowded, so that **4** is less stable than **5**, by about 4 kJ mol^{-1}. The 1,1-disubstituted alkene **3** is about the same stability as **4**.

Note that these energy differences are relatively small, but they can make a difference.

ARE MORE STABLE ALKENES LESS REACTIVE?

Follow this next bit carefully! I would say this is more important than the stability, as the energy differences we are now going to encounter are larger.

We will use the same arguments we encountered when looking at the inherent reactivity of conjugated systems compared to non-conjugated systems (**Basics 21**).

Consider the two protonation reactions shown below.

Draw a curly arrow mechanism for each reaction. Explain the regiochemical outcome (**Basics 27**) for each process (*i.e.* why do you get *this* carbocation and not the alternative one?).

$H^{\oplus}$

6 → **8**

$H^{\oplus}$

9 → **10**

Alkene **6** is more stable than alkene **9**. It has two additional alkyl substituents directly attached to the alkene bond, so we might estimate an additional 15 kJ mol^{-1} of stability. However, carbocation **8** is *tertiary*, which is considerably more stable than the *secondary* carbocation **10**. The energy difference is about 60 kJ mol^{-1}, as we saw in **Perspective 2**. Remember that this number is based on a calculation in the gas phase. Perhaps the real difference will be smaller, but it is still a lot more than 15 kJ mol^{-1}.

*We don't have to go 'uphill' quite as far in energy terms for the reaction of alkene **6**.*

Draw an energy profile for this. Start by putting the alkenes and carbocations on, and then consider the height of the barriers using the Hammond postulate (**Basics 19**).

I am not going to give you the profile. From it, you should be able to deduce that alkene **6** is more reactive.

Does this make sense? We could form alkenes **6** and **9** from a single elimination reaction, and we expect to get more of alkene **6**.

Draw an example of such an elimination reaction.

IS THIS TRUE FOR EVERY REACTION?

The short answer is 'no'! It only applies to the reactions of alkenes that produce carbocation intermediates. Many alkene reactions do involve carbocation intermediates, so this clearly applies.

Reactions such as hydrogenation,[5] that do not involve carbocation intermediates, can be more directly affected by steric factors. In this case, a more substituted alkene is not necessarily more reactive.

In every case, we must consider the overall reaction profile, not just the relative stability of the two possible starting materials.

[5] Look it up!

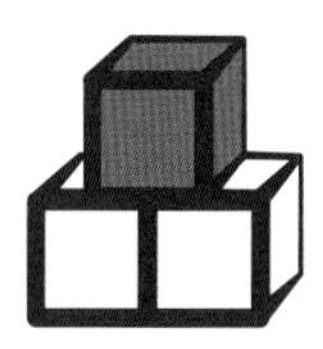

BASICS 35

Enthalpy Changes for Reactions Involving Anionic Species

A STRANGE STATEMENT

The two statements below appear to be direct contradictions. How can they possibly both be correct? The devil is in the details.

Addition of HBr across an alkene can be exothermic. Elimination of HBr to form an alkene can also be exothermic.

LOOKING AT THE REACTIONS

Here is the addition of HBr to ethene.

 Use the data in **Basics 13** to calculate the enthalpy change for this reaction.

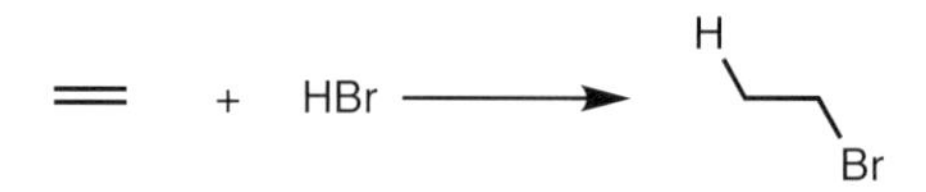

You should find that it is exothermic by 53 kJ mol^{-1}.

Did you do the calculation? I hope so! You must have got the point of all this by now!

Now, here is the elimination reaction. We will use ethoxide as base.

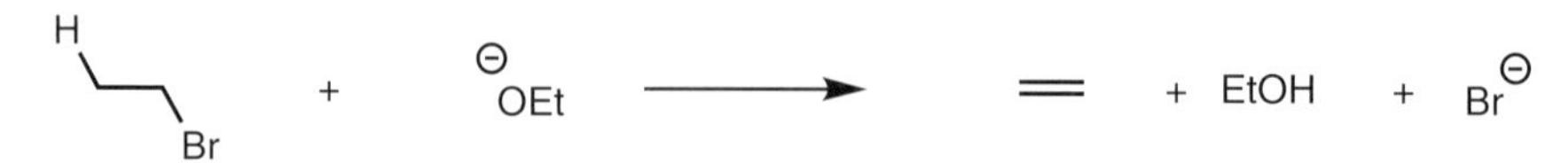

Here, the bond-forming reactions are slightly different, but there is a bigger issue. Bromide is more stable than ethoxide. How do we quantify the stability of anions?

We have a reliable measure for this—the pK_a values.

CONSIDERATION OF pK_a IN ENTHALPY CALCULATIONS

In the above elimination reaction, we are breaking a C–H bond, a C–C bond, and a C–Br bond. We are forming a C=C bond, and an O–H bond. But we are exchanging ethoxide for bromide.

Here are acid–base equilibria for HBr and ethanol, along with the respective pK_a values.

$$\text{HBr} \rightleftharpoons \text{H}^{\oplus} + \text{Br}^{\ominus} \qquad \text{p}K_a\ -9$$

$$\text{EtOH} \rightleftharpoons \text{H}^{\oplus} + \text{EtO}^{\ominus} \qquad \text{p}K_a\ 16$$

Bromide is more stable than ethoxide. Much more stable! There are 25 pK_a units between them!

Let's put these two equilibria together.

$$\text{EtO}^{\ominus} + \text{HBr} \rightleftharpoons \text{EtOH} + \text{Br}^{\ominus}$$

For this equilibrium, we would expect the equilibrium constant, $K = 10^{25}$.

This is because pK_a is a logarithmic scale. It is defined as

$$\text{p}K_a = -\log_{10} K_a$$

Equilibrium constant and free energy are related by the following equation.

$$\Delta G = -RT \ln K$$

Since this equilibrium has two species on each 'side', we can (as an approximation) assume that ΔS will be close to zero. Therefore, $\Delta H \approx \Delta G$.

$$\Delta H = -RT \ln K = -RT \ln (10^{25}) = -143 \text{ kJ mol}^{-1} \text{ at } 298 \text{ K}$$

Note that we have $\log_{10}$ and ln (natural logarithm) in these calculations. As a chemist, you need some appreciation of what these mean.

REALITY CHECK!

Whenever you do a calculation, it's always a good idea to make sure the result is reasonable. In this case, the answer is in the same order as bond dissociation energies. If it was a few joules per mole, it wouldn't make a difference. If it was mega joules per mole, nothing else would matter. The answer here is sensible!

PUTTING IT ALL TOGETHER

Here's the elimination reaction again.

H, Br + $^{\ominus}$OEt ⟶ = + EtOH + $Br^{\ominus}$

We are breaking C–H, C–C, and C–Br bonds. We form C=C and O–H bonds, but we also gain the extra 143 kJ mol^{-1} for the pK_a. The net effect of this is

$$\Delta H = (410 + 350 + 270) - (611 + 460 + \mathbf{143}) = -184 \text{ kJ mol}^{-1}.$$

Make sure you go back to **Basics 13** to check that you understand where each number comes from.

You might be tempted to look at this and feel that you don't even need the additional 143 kJ mol^{-1}. Mathematically, this is true. However, you wouldn't get the EtOH formed without this aspect, so you need to see the bigger picture.

Repeat this calculation using a range of different bases,[6] and see how the pK_a affects the predicted enthalpy change for the reaction.

HINT You can simplify the maths by using

$$\ln(10^x) = x \ln 10 = 2.3x$$

[6] Look up some pK_a values, but make sure they make sense.

APPLICATIONS 4

Stereochemistry of Elimination Reactions

We already considered some aspects of the stereochemistry of elimination reactions in **Reaction Detail 4**. As these were stereochemical aspects that were dictated by electronic factors (orbital overlap), we call these 'stereoelectronic aspects'. It's another hybrid word that means what it sounds like.

This chapter is 'applications'. The corresponding chapter for substitution reactions was classed as 'Reaction Detail', because new information was presented there. You already know everything you need to predict and understand the stereochemistry of elimination reactions. You just need to know how to apply it.

There are many types of compound that we could use to illustrate the key points. Here, we are going to focus on one type, bridgehead alkenes, in the first instance, before we generalize the discussion.

BRIDGEHEAD ALKENES

A bridged compound has a ring that is 'bridged' by one or more atoms. It is distinct from a fused compound. Here are two examples in which you can see the distinction. The so-called 'bridgehead' carbon atoms are indicated.

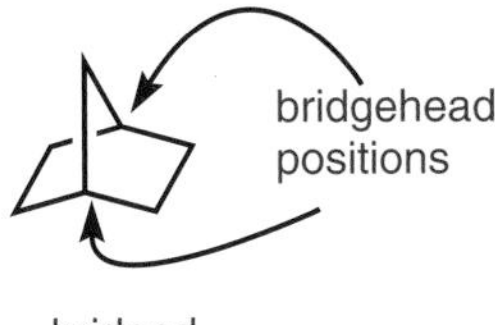

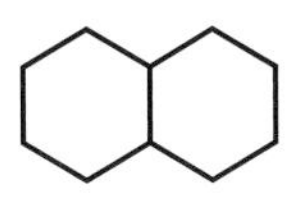

4

Right, now to the exercise.

Get your molecular model kit out, and make (or try to make) models of the following compounds which contain an alkene at the bridgehead.

6.1

Remember that alkenes are sp^2 hybridized and should be planar. Make sure you think about **WHY** this is the case!

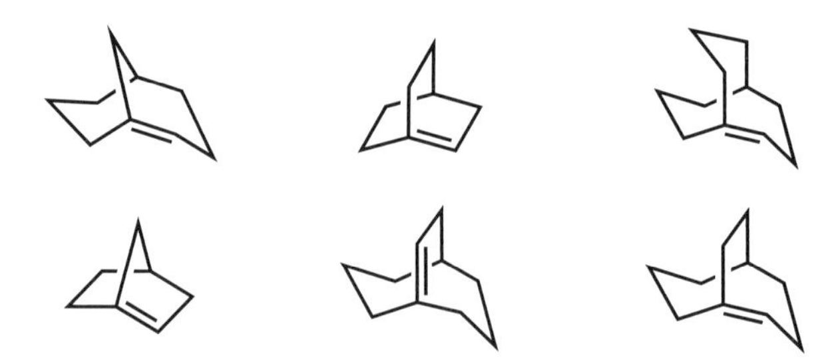

Don't try too hard with all of these. I'm not responsible if you break your molecular model kit. If you cannot make the molecule, think about what this might be telling you about its stability.

BRIDGEHEAD ALKENES AND E1 ELIMINATION

Now let's consider this from an E1 elimination perspective. Here are the carbocations that we might form in a hypothetical E1 elimination reaction. They are all *tertiary* carbocations, so they should all be stable—if they can become planar (or close to planar) of course.

This makes an assumption about the precursor we are using to try to make the alkene, but it is a reasonable assumption. If we want the best chance of an E1 elimination, we need a tertiary carbocation.

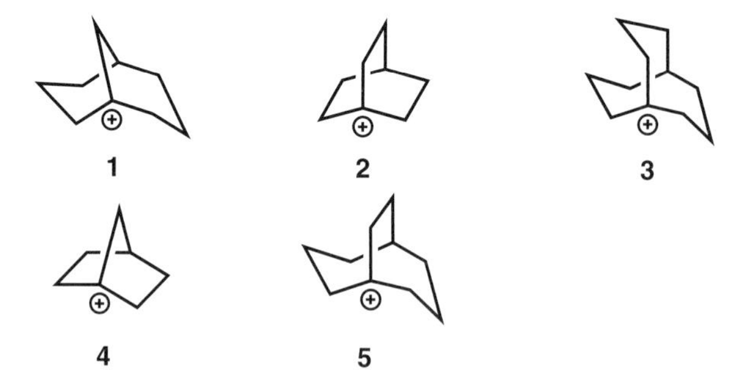

4

We have seen one of these carbocations (**4**) in **Basics 17** (in which we established that it cannot become planar because of strain). We encountered carbocation **2** in **Perspective 2** and found that it is actually reasonably stable. Based on this, you are in a position to anticipate which of the other carbocations might be stable.

But can you have an elimination reaction?

We saw in **Reaction Detail 4** that in order to have E1 elimination, we need to have a C–H bond that is aligned with the empty p orbital of the carbocation, so that you could get effective overlap to form a double bond.

This is the same overlap that provides stabilization to the carbocation, so that naturally you should reach the same conclusions.[7]

Make models. Consider where the empty p orbital is and identify whether you have any C–H bonds than can be in the same plane. I'm not going to give you the answers just yet.

BRIDGEHEAD ALKENES AND E2 ELIMINATION

Next, we can consider an E2 elimination mechanism. In **Reaction Detail 4** we saw that the two leaving groups (let's call them 'H' and 'X') must be *anti* periplanar for the orbitals to overlap in order to permit an E2 elimination reaction to take place. Make models of the structures below and determine whether the C–Br bond can be *anti* periplanar with a C–H bond on an adjacent carbon atom (**any** C–H bond on **any** adjacent carbon!).

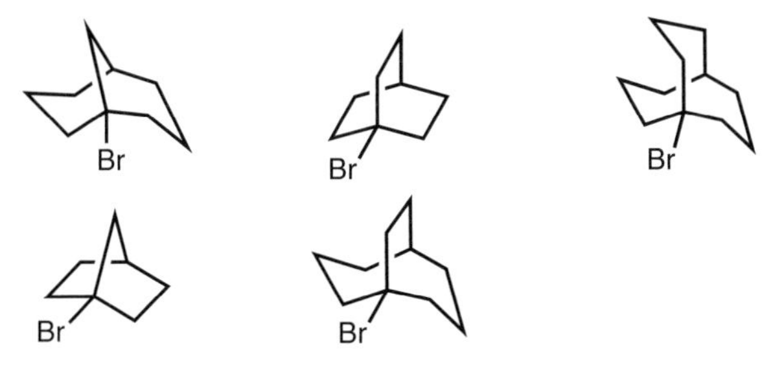

Again, this is really looking for the same thing, despite different mechanisms, because the one absolute in an elimination reaction is that you must form a double bond.

BRIDGEHEAD ALKENES—BREDT'S RULE

Now to the rule—**Bredt's rule**: You cannot make a compound with a bridgehead double bond in a bicyclic system, unless the system has seven or more **non-bridgehead** atoms in the ring system.

[7] There is a complication—an adjacent C–C bond, properly oriented, will also provide stabilization to a carbocation, but will not subsequently allow formation of a double bond.

Obviously, having made the model, you have realized, without needing a formalized rule, why this is the case. The systems would have to be too distorted. You would never get overlap of the p orbitals required to form a π bond in the first place.

Bredt's rule is great, but you must understand ***why*** *you cannot form these compounds—it is because they are unstable. You would not get the stability associated with the π bond. As with so many other things, once you really understand it, you stop worrying about labels. It doesn't matter what you call it.*

STEREOCHEMISTRY OF E2 ELIMINATION REACTIONS

Now let's look at one more aspect of the stereochemistry of elimination reactions. It is a direct consequence of the need for an *anti* periplanar transition state. If you know the basics, you can predict all of this.

Here is a specific example. Don't just read through this section. Draw the structures and see if you can anticipate the outcome.

The two bromides below give the following stereochemical outcome in their elimination reactions. Furthermore, the elimination reaction of compound **6** is slower than the reaction of compound **7**.

Me Ph Ph Br
NaOH
Ph Ph Me
6

Me Ph Ph Br
NaOH
Ph Me Ph
7

If our understanding of reaction mechanisms is sound, we shouldn't actually need too much information in order to explain these observations. If you can explain why you get the stereochemical outcome, but not the rate of reaction, your explanation may be incomplete (or wrong!). I firmly believe that the right explanation will 'feel' right. You will just know!

First of all, the mechanism cannot be E1 or E1cB, since both mechanisms would give the same stereochemical outcome. Try to work out why this is the case.

So, now we know that it is an E2 elimination, all we have to do is produce a stereochemical representation where the H and Br which are being eliminated are *anti* periplanar, and everything else should just follow from this.

With compound **6**, first of all, we will put on the missing hydrogen atoms.

Me Ph Ph Br = H Me Ph Ph H Br

6

Now, how you proceed from here depends on what you find the easiest. It is probably easier at first to make a model of the compound, then rotate the bonds and look at it to determine the correct Newman projection. However, let's do it the hard way for now.

We will look at this molecule from the left, with our eye slightly above the middle in order that we are looking straight down the C–C bond. This is what we get.

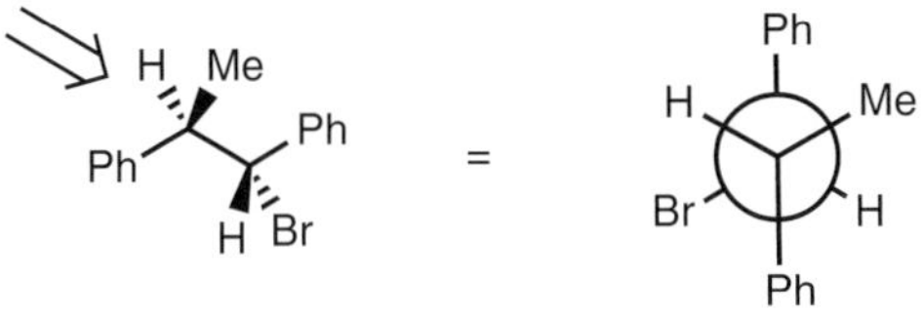

H Me Ph Ph H Br = Ph H Me Br H Ph

From the direction in which we are looking, the H on the front atom will be on our left, and the methyl group on our right, while the bromine on the rear atom will be on our left and the hydrogen on our right.

However, the H on the front atom and the Br on the rear atom are not *anti* periplanar. I personally find it easier to see if we have the H and the Br in a vertical plane, so we need to rotate the front atom 60° clockwise and the rear atom 60° anticlockwise. This is what we get.

It's much easier with a molecular model!

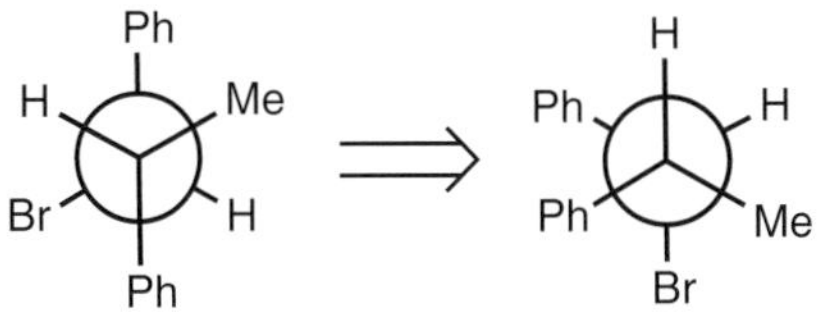

Ph H Me Br H Ph ⟹ H Ph H Ph Me Br

Now the H and the Br are finally *anti* periplanar, and the two phenyl groups are on the same side, so that these will end up *cis* on the double bond.

We can take a big short cut with compound **7** now. We know that the only difference between compounds **6** and **7** is the stereochemistry at the right-hand (rear) stereogenic centre. How do we invert just one stereogenic centre? We swap two substituents. So,

4

swapping the Ph and H on compound **6** as drawn above gives us compound **7** in a correct Newman projection.

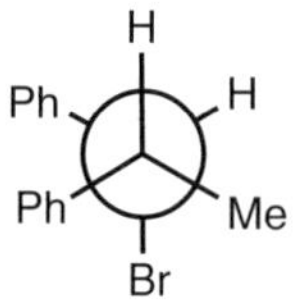

6 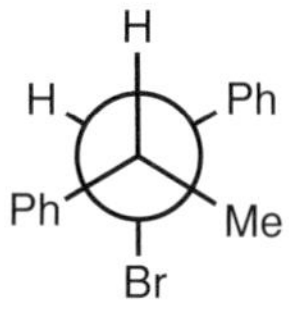 **7**

Now we can see that the Ph and the H will end up *cis*, as drawn above.

You can do a bit more with this. We are going to consider the rates of these reactions in **Worked Problem 3**.

MOVING FORWARD?

Good news! There won't be any further new ideas for this. If you continue to make models of the compounds and look at them very carefully, you will understand **all** the new examples. We are going to look at them in the context of cyclohexane chemistry in **Applications 5**. There's just one more idea we need to look at first!

BASICS 36
Stereospecificity

STEREOSPECIFICITY

This is a convenient point to formalize the idea of stereospecificity.

A stereospecific reaction can only produce one stereoisomer.

We have encountered several stereospecific reactions so far. An S_N2 substitution reaction proceeds by attack of the nucleophile opposite to the leaving group, which corresponds to inversion of configuration for a single enantiomer chiral compound (Reaction Detail 2). The nucleophile **must** attack from the back so that the desired orbital overlap can be achieved. An E2 elimination reaction has a transition state in which the two leaving groups are *anti* periplanar. As we have seen (**Applications 4**), this means we can very often only get one alkene double bond isomer as product.

We can formalize this by consideration of the reaction coordinate and possible alternative transition states.

In the case of an S_N2 substitution, any transition state in which the nucleophile could attack to give a stereochemical outcome other than inversion would have a very high energy, to the point where this doesn't happen.

In the case of an E2 elimination, the stereospecificity is a result of needing to develop orbital overlap to form the π bond. In fact, you could get **some** (but not as much) orbital overlap if the two leaving groups are *syn* periplanar (on the same side). You would expect a *syn* periplanar elimination to be possible, but it would have a higher barrier.

In fact, syn eliminations are known, and they are also known as pyrolytic elimination reactions. You really have to cook them!

STEREOSELECTIVITY

If a reaction has a choice of which of two (or more) stereoisomers can be formed, but it gives more of one stereoisomer than another, it is stereo**selective** (Basics 28), not stereospecific. In this case, one pathway is more favourable (lower energy) than the other(s).

AN IMPORTANT DISTINCTION!

If a reaction can potentially give two different stereoisomers, but actually (as far as can be detected!) only gives one, it is **very** stereoselective but not stereospecific. The term 'stereospecific' is reserved for those reactions that can, as a direct result of their mechanism, only give one stereoisomer.

APPLICATIONS 5
Elimination Reactions of Cyclohexanes

Just as we did with substitution reactions in Applications 3, we will take what we know about elimination reactions and apply them to cyclohexanes.

WHY ARE CYCLOHEXANES SPECIAL?

To reiterate the comment that I made last time, they are not. We should just remind ourselves that that a cyclohexane chair has more or less perfectly tetrahedral carbon atoms, locked into a staggered conformation. Every dihedral angle is 60° or 180°. We can add particular substituents that 'lock' the chair into one conformation, so that we only have to consider the reactivity of one structure.

In this chapter, we will look at the elimination reactions of cyclohexanes. We will consider steric and electronic effects, and determine what the outcome is likely to be for E1 and E2 mechanisms. In many respects, this chapter is a 'copy' of Applications 3. The same factors apply for the same reasons.

The conformational bias really affects E2 elimination more than E1 elimination. However, the fundamental principles apply to both mechanisms.

THE E1 MECHANISM

This discussion is exactly the same as we saw for the S_N1 substitution reaction. This shouldn't surprise you. After all, we have established that the rate-determining step in each case is the same, so the only difference will be 'what happens next'.

Consider the following two cyclohexanes, **1** and **2**. Because we have the *t*-butyl group locking the chair, structure **1** has the bromide in an equatorial position; in structure **2** it is axial.

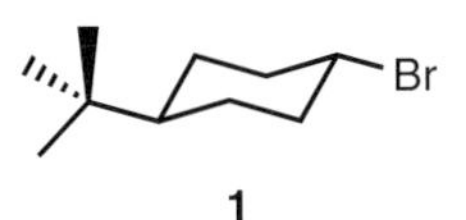

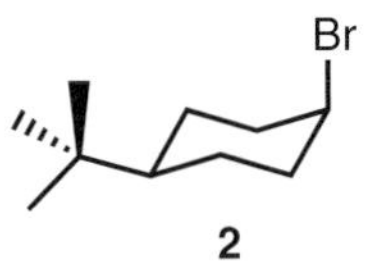

5

Let's get one thing out of the way. Neither of these compounds will actually react in an E1 process. Loss of bromide anion will give a secondary carbocation, which is not that stable.

We know that these are the only conformers of these compounds that we need to consider. If we flip the chair in structure **1**, we will have two axial substituents. This would be very unstable. In structure **2**, flipping the chair would place the Br equatorial, at the expense of the *t*-butyl group becoming axial. Again, this is unfavourable. So, structure **1** has an equatorial Br, structure **2** has an axial Br, and there's nothing you can do about it!

Make models of the compounds. Check that you have followed the arguments. Never miss an opportunity to test your understanding of the basics.

In order that you internalize this information you need to do the work, and I will guide you.

Draw a reaction profile. The compound with an axial Br is less stable (higher energy) than the compound with the equatorial Br. However, when they lose bromide to give a carbocation, they will give exactly the same carbocation.

Formation of the carbocation will be endothermic in both cases. It will be preceded by a transition state that is (a little bit) higher in energy again.

What does this mean about the activation energy for the process (use the Hammond postulate to relate the energy of the carbocation to the energy of the transition state that precedes it)?

Hopefully you came to the conclusion that the compound with axial Br will react faster in an E1 elimination reaction.

This is exactly the same as we saw for S_N1 substitution in ***Applications 3****.*

I am not going to repeat the rest of the discussion from that chapter. Go back to it and see how it applies here. Let's look at the second step. We just lose a proton from the carbon adjacent to the carbocation as shown below.

It is almost certainly the axial hydrogen that is lost, because the axial C–H bond is better able to overlap with the empty p orbital of the carbocation. It isn't easy to find examples that prove this unambiguously.

Try to design a substrate (or several substrates) that would help you determine whether this is true.

E2 ELIMINATION REACTIONS OF CYCLOHEXANES

Let's reiterate the one **absolute** for E2 elimination reactions. The H and (in this case) Br leaving groups **must** be *anti* periplanar. If the leaving group is equatorial, there can be no hydrogen atom *anti* periplanar.

1 **2**

The *anti* periplanar bonds for the equatorial Br in structure **1** are shown in red in the structure below.

Make a model of this compound and convince yourself that this is the case.

1

In structure **2**, with the axial Br, there are two hydrogen atoms (shown in red in the structure below) that are *anti* periplanar. Again, make a model of the compound and make sure you understand it. Don't take my word for it. Internalize it. Own it! Once you do this, you don't have to learn it or memorize it.

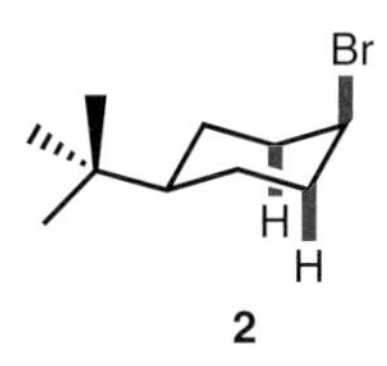

2

Therefore, for a cyclohexane to undergo a 1,2-elimination reaction, both leaving groups (H and Br in this case) ***must*** *be axial. Now you can see the importance of being able to work out which groups are axial, and which are equatorial, in any given cyclohexane or related compound.*

If the Br is equatorial, the only way for the compound to undergo E2 elimination of HBr is for the chair form to flip. In this case, this will place the *t*-butyl group axial, which is extremely disfavoured.

Br
ring flip
H
Br
eliminate
ring flip
1

Both the flipped chair form, and the subsequent transition state for elimination, are high in energy. This means that the energy of the transition state for elimination has effectively been increased by the difference in energy between the two conformers.

Draw this on a reaction profile. Draw the two different reactions on the same scale. Put the two substrates (one with axial and one with equatorial Br) with the correct *approximate* relative energies. Put the 'flipped' chair form for the equatorial Br on there, again with the approximately correct energy.[8] Put the transition states in. You can even put the products in, with the flipped chair form where appropriate.

Most importantly, don't 'connect' things that are not connected along the reaction coordinate. There is one path from the equatorial Br compound to the product, and a different path from the axial Br compound to the product.

Therefore, compound **1** undergoes elimination slowly or not at all, while compound **2** undergoes ready elimination.

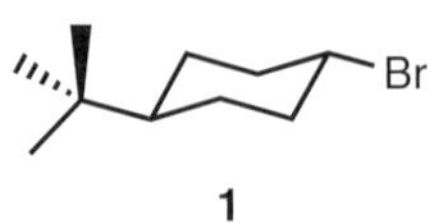

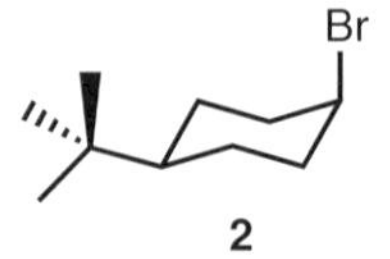

[8] Use the 'A values' in Basics 32.

Let's look at a slightly more complicated example. Compound **3** undergoes elimination to give a mixture of two products as shown. This is the favoured conformation of this compound, since the *t*-butyl group is equatorial. Since the tosylate group (see **Reaction Detail 1**) is axial, it can undergo ready elimination. There are two axial hydrogen atoms which are in an *anti* periplanar arrangement with respect to the leaving group (OTs). Therefore, we can observe elimination in two directions to give **regioisomeric** products. The major product has the **more stable** (more substituted) alkene.

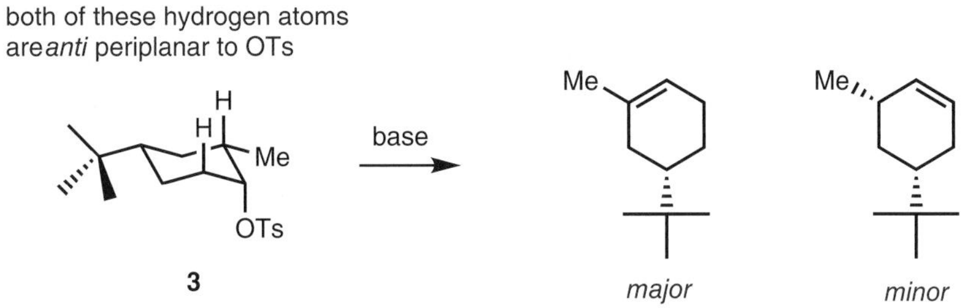

both of these hydrogen atoms areanti periplanar to OTs

H H Me OTs base Me Me

3 *major* *minor*

Compound **4** would undergo E2 elimination much more slowly, if at all. Certainly, if it did, it would only produce a single alkene product. In the left-hand structure below, all of the substituents in compound **4** are equatorial. There is no hydrogen atom *anti* periplanar to the OTs group. Therefore, elimination from this stable conformation is not favoured. The only way the OTs group and a hydrogen atom can eliminate is if the OTs group is axial, which requires the cyclohexane chair to 'flip' (all equatorial substituents become axial and vice versa). Therefore, the elimination process has a high activation barrier and is slow. Additionally, once the chair has flipped (structure **5**), there is only one hydrogen atom *anti* periplanar to OTs, so that only one elimination product is expected.

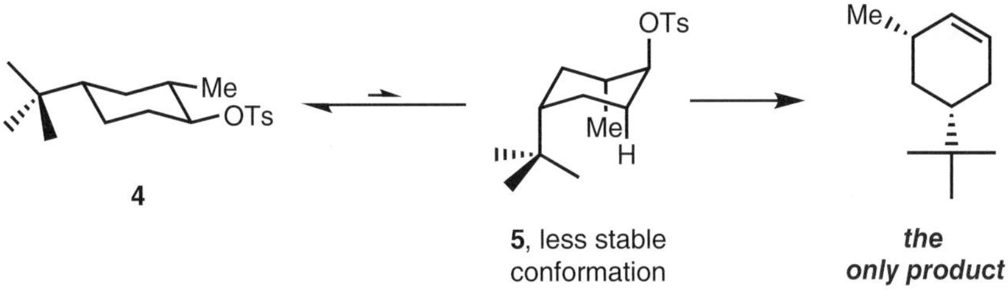

Me OTs OTs Me H Me

4

5, less stable conformation

the only product

I cannot find any evidence that this reaction has actually been done. I'm not particularly surprised by this!

This is great. If we understand the shapes of compounds, we can predict which will react faster, and what products are most likely to be formed. Remember what I said at the start of the chapter. The principles that we are using apply to all compounds, not just to cyclohexanes. It's just that using cyclohexanes at first gives us a clearer picture of the structures. Get used to these, and the rest will follow.

5

We can add a further level of complexity to this. Don't think I am adding complexity just for the sake of it. These types of system are commonly found in natural products and other important organic compounds. Compounds **6** and **7** are cyclohexane chairs with an additional ring as a result of two of the substituents being joined together. We can only form this additional ring if these two substituents are axial.

6.2

Don't take my word for it. Make a model of it.

The only difference between these two compounds is the stereochemistry of the Br atom. We have to ask the simple question 'which one is axial, and which one is equatorial?' You probably cannot see this at a glance, but you know how to work it out!

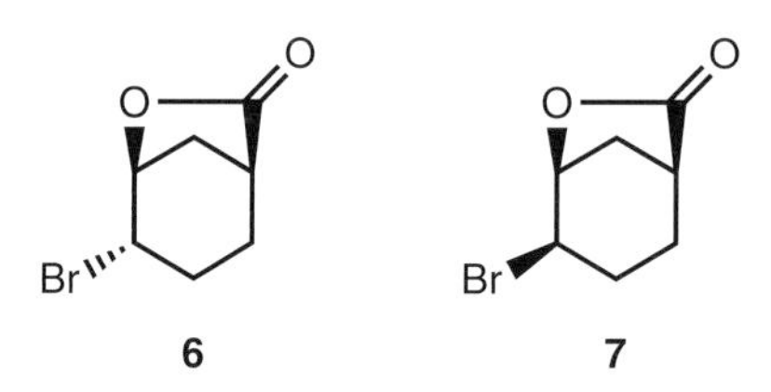

Hopefully you saw that the **only** accessible conformers are as follows:

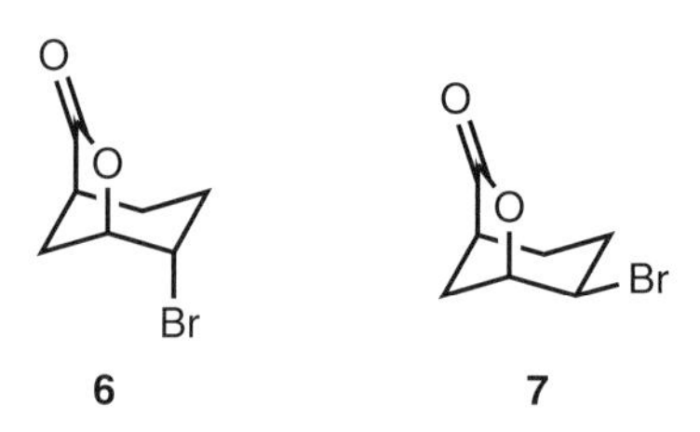

We have already seen that we can't eliminate an equatorial bromide. Look at the model you just made, and you will see why. The only bond *anti* periplanar to the C–Br bond in compound **7** is a C–C bond. That won't do!

So, compound **6** can eliminate, but only in one direction, while compound **7** can't eliminate at all. I'll leave you to work out what the product is from compound **6**. Again, it comes down to knowing where the hydrogen atoms (that we so often do not draw) are.

Draw the outcome of the elimination reactions of these compounds!

These examples might look difficult, but once you internalize the method, you can cope with anything. This is a far better learning strategy than memorizing specific examples and outcomes from your lecture notes.

A LITTLE MORE COMPLEXITY?

Actually, I'm not telling the whole truth when I say that compound **7** can't eliminate. It just can't eliminate HBr! Now you need to ask another fundamental question.

If it can eliminate, but not HBr, what else could it eliminate instead? Think about good leaving groups.

This is the point where you test yourself. If you understand the principles, you will draw sensible curly arrows and an outcome that makes thermodynamic sense. If you don't understand the principles, there is a significant chance of you drawing something that is very random!

Again, I'm not going to give you the answer. If I do, you'll look at it and say 'yeah, that makes sense'. When you get the right answer for yourself, you will **know** it makes sense! Discuss it with your friends or with your tutors. Alternatively, have a look at the video that covers this example!

That's all we need to say about substitution and elimination in cyclohexanes. What you need to do is apply the same principles to related systems, perhaps with different ring sizes. You need to be looking for the same ideas, even if the systems are a bit different.

COMMON ERROR 9

Elimination Reactions

WHAT IS ELIMINATION?

The first error, which is more common than you would probably expect, is confusing elimination with substitution.

Here's the problem. Compounds that undergo elimination reactions do also undergo substitution. Most of the same principles apply, so that the exam questions do look rather similar.

And you are 'eliminating' (or at least getting rid of) the leaving group!

I'm afraid the answer is to draw out the mechanisms so many times that you cannot become confused, even under pressure.

ERROR: ELIMINATION ALWAYS GIVES THE MOST STABLE ALKENE!

There is a common assumption that elimination will always give the most stable alkene. As we have seen, in an E2 elimination the transition state geometry defines the outcome. We will look at the rates of some of these reactions in **Worked Problem 3**.

As with everything else, it is a relatively small shift to always consider the transition state geometry. Newman projections are really helpful here.

I find that students either do this all the time, or never.

Nail it or fail it!

It's an adjustment that you need to make, and it will pay dividends in the long run. Everything will feel easier, in time.

REACTION DETAIL 5

Allylic Substitution

INTRODUCTION

An adjacent double bond can influence the outcome of a substitution reaction in several ways. The purpose of this chapter, as always, is to explore fundamental principles. Therefore, we will establish, and then build on, what we know.

THE S_N1 MECHANISM IN ALLYLIC SYSTEMS

Compound **1** is allyl alcohol. It's a trivial name for this specific compound. Compound **2** is *an* allylic alcohol, in which an OH group is on a carbon atom immediately adjacent to a double bond.

OH OH

1 **2**

We know that a carbocation is stabilized by an adjacent double bond. We call this an allylic carbocation. We have seen, in terms of resonance structures, why an allylic carbocation is stabilized. Here it is for the carbocation formed from compound **2**.

H_3C ⊕ ⟷ ⊕ H_3C

3 **4**

Draw a curly arrow mechanism for the formation of carbocation **3** from compound **2**, using an acid catalyst.

So, should we be drawing carbocation **3**, or carbocation **4**?

*The key point is that the positive charge is not localized on one end of the allylic system. When I see structure **3**, I am automatically aware of structure **4**. The two individual structures are limiting representations of the 'real' structure. Have another look at **Basics 17**.*

So, here's the key question. If we take compound **2** and treat it with HBr, what product(s) should we expect to form? Well, there is no doubt that we would protonate the alcohol group, and this would lead to carbocation formation. The HBr is acidic! We then have a carbocation, as drawn above, and a bromide anion/nucleophile.

So, the real question is 'where would a nucleophile attack?'

This should not depend on which resonance form we draw. We cannot make a statement such as 'the nucleophile will attack **there** because that's where the positive charge is'. A nucleophile could attack at either end, but not in the middle.

*We now need to determine what factors favour attack at each end. This is regioselectivity (**Basics 27**).*

You get a mixture of products, favouring compound **5**.

H_3C–CH=CH–CH_2–OH —HBr→ H_3C–CH=CH–CH_2–Br + H_3C–CH(Br)–CH=CH_2

2 **5** **6**

A more substituted alkene tends to be more stable (**Basics 34**). Alkene **5** is disubstituted. Alkene **6** is monosubstituted.[9] You would also expect the *secondary* alkyl halide **6** to be more crowded, and hence less stable, than the *primary* alkyl halide **5**. We saw this aspect in **Basics 16**.

*There are two reasons, one electronic and one steric, why **5** is more stable than **6**. It doesn't necessarily follow in all cases that the transition state leading to **5** will be lower in energy than the transition state leading to **6**, but it is true in most cases.*

 Draw a reaction profile (**Basics 14**).

Okay, that takes care of the S_N1 mechanism in allylic systems. There is a stabilized carbocation intermediate which favours the process (as long as the reaction conditions favour carbocation formation), but we need to add consideration of the site of nucleophilic attack. It is a minor refinement of the classic S_N1 mechanism.

*Did this look familiar? We already did it in **Common Error 4**.*

Now we need to look at S_N2 substitution in allylic systems, and we find that a new mechanism/outcome is possible.

[9] Remember, all we are talking about is how many substituents are directly attached to the double bond.

THE S_N2' MECHANISM

If we react bromide **7** with hydroxide, under conditions that do not favour an S_N1 mechanism, it turns out that we get product **8**, rather than product **9**.

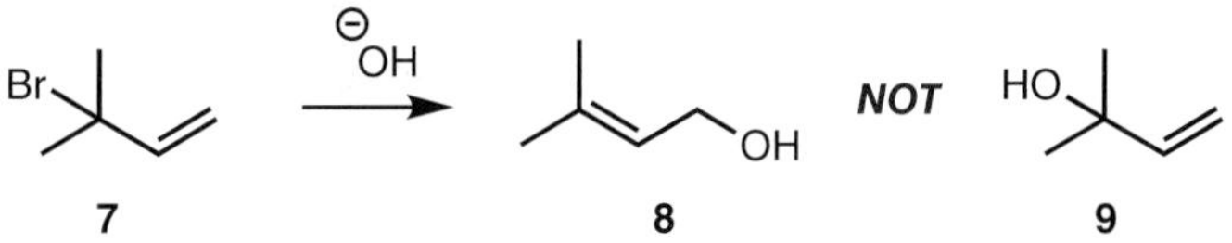

Br **7** $\xrightarrow{^{\ominus}OH}$ OH **8** ***NOT*** HO **9**

Formation of compound **9** would require attack of hydroxide at a very crowded *tertiary* centre, which we have seen (**Reaction Detail 1**) is not very favourable. Here are the curly arrows for the formation of compound **8**.

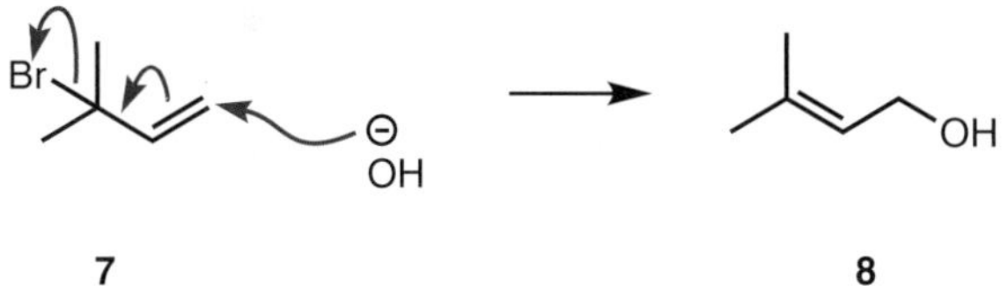

Br **7** $^{\ominus}$OH ⟶ OH **8**

If we focus on the curly arrows for a moment, some aspects look like substitution, but other aspects definitely look like elimination. The Br leaves, and we form a new double bond.

This is like an elimination reaction, but instead of 'getting the electrons' from a C–H bond, we are getting them from an alkene π bond.

On the other hand, we are adding a nucleophile and losing a leaving group, so overall it is definitely a substitution reaction.

In terms of kinetics, this is definitely an S_N2 reaction (substitution, nucleophilic, bimolecular) so we describe it as **S_N2'** (S_N2 prime) to distinguish it from a 'normal' S_N2 reaction.

5

INTERLUDE—ALKENE CHEMISTRY

The general trend in alkene chemistry is that alkenes are electron-rich, and so they use their electrons in reactions. Alkenes generally react with electrophiles, not with nucleophiles.

This reaction is an exception to the generalization.

We need to have a way of understanding why this exception is 'allowed'. The following discussion is far from rigorous, but it will serve the intended purpose.

BACK TO S_N2'

Recall that if we want to break the C=C π bond, we need to donate electron density into the antibonding (π*) orbital.

Giving electron density to something that is already electron-rich is not normally favourable. It will only work if the electrons have 'somewhere' to go.

In this case, the curly arrows show the π bond 'moving' to an adjacent position, with loss of bromide. This would represent donation of the π bonding electrons into the σ* orbital of the C–Br bond. It isn't easy to draw all of this on a single diagram. I've done my best, below, but I cannot show you the π and π* orbitals of the alkene at the same time.

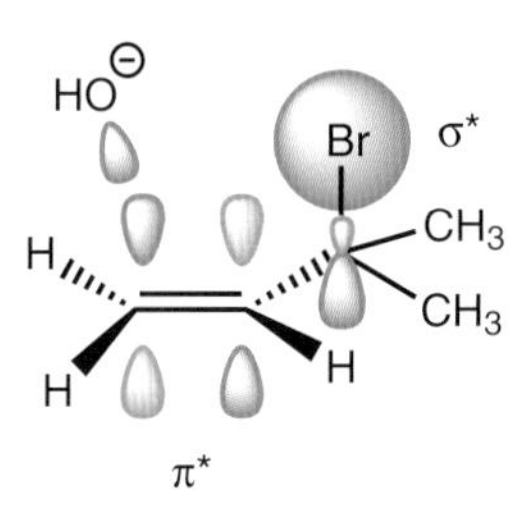

Perhaps this is a case where hybridized orbitals are too 'simple' to show the complete picture? This does not detract from their usefulness, and indeed I would argue that the diagram above might not be perfect, but it does give some useful insight.

5

This reaction can be really useful. Perhaps you need one regioisomer (Basics 27) of product, but you cannot access the correct starting material for a 'simple' S_N2 substitution? The outcome in the above example is primarily governed by steric factors.

You won't be surprised to find that some reagents 'prefer' to attack the alkene double bond.

All we are trying to do for now is establish what can potentially happen. We are not going to look at lots of examples in order to try to work out the preferred outcome in each case.

STEREOCHEMISTRY OF S_N2' REACTIONS

We know that S_N2 reactions proceed with inversion of configuration. In the above reaction, I have drawn the nucleophile attacking the same side of the double bond that the leaving group leaves from.

Sometimes you get nucleophilic attack from the same face as the leaving group. Sometimes you get attack on the opposite side.

It isn't all that easy to tell what happens. You have double bond geometry to contend with, as well as a rotatable bond. We are going to look at the complexity, from an entirely hypothetical level, in **Worked Problem 8**.

For now, I will simply state that one way to reduce the level of complexity is to use systems that only have one possible double bond geometry and limited potential for rotatable bonds.

Yes, those would be cyclohexenes!

Here is a commonly used example. Two different nucleophiles give different stereochemical outcomes. The general view, supported by computational studies, is that with the amine nucleophile a hydrogen bond between the nucleophile and the leaving group determines the outcome.

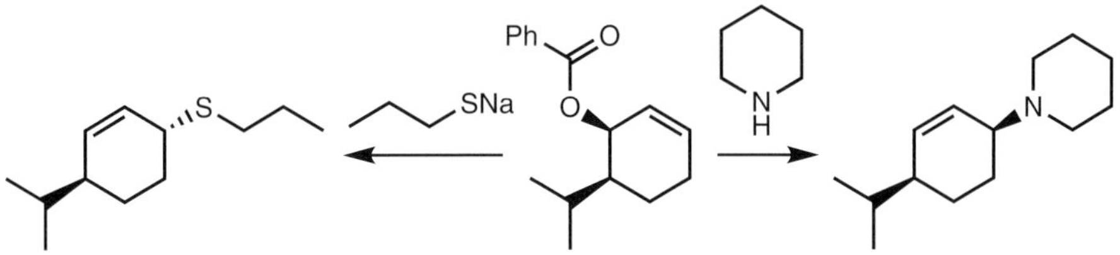

Draw the cyclohexene starting material (refer back to cyclohexene conformers in **Basics 33**), considering where the isopropyl group will be sited (*pseudo*-equatorial!). Draw the hydrogen bonding arrangement that leads to formation of the amine product.

As always, the point of the exercise is to get better at drawing increasingly complex structures. You will probably need to use your molecular models to get a 'feel' for the structures before you start drawing.

If you don't need to make a model, and you still get it right, good![10]

[10] But check that you drew the correct enantiomer of the starting material, **and** the correct diastereoisomer!

SECTION 7

BUILDING SKILLS

INTRODUCTION

Up to this point there has been quite a lot of information delivery, interspersed with 'Applications' chapters and 'Practice' chapters. All of this should have given you a solid base. You have practised drawing structures, curly arrows, and resonance forms. You have applied this (and practised some more) to a range of scenarios, and hopefully followed the relevant discussion.

Now it's time to step it up a gear. Here are some more involved examples that require you to pull together all of the aspects we have covered.

This is the real challenge when you are learning organic chemistry. Most of the time, you will need to think about shape, energy, reactivity, and selectivity all at the same time. You need to learn how to do this instinctively, so that you are not answering questions by rote learning or memory.

Although these chapters are 'Worked Problems', don't expect to find comprehensive detailed answers to every problem. I want you to have to discuss them with your friends. I also discuss the answers to the problems in the accompanying videos, as described in the *Additional resources* section at the start of the book.

The problems presented vary in scope and complexity. In all cases, one major point is to reinforce the basics. In some cases, it stops there. In other cases, you will be guided to think about problems in a more advanced way.

You can connect with the Facebook page and ask questions directly. If you find other problems that you would like me to apply this approach to, you can ask! I hope to build a considerable resource that will help many students who are struggling with organic chemistry.

For now, work through the problems in the following chapters. There are ten chapters with questions and guidance, followed by ten chapters with answers.

You will get more benefit if you have a go, even if you mess it up completely. Your most profound, and persistent, memories involve either pleasure or pain. It's the way we are 'wired'. Feeling the frustration of 'not getting it' is an unavoidable part of the educational journey. Don't allow yourself to believe that you are less able, just because something is a struggle.

Throughout this book, there are a great many cross-references to other chapters. I hope you have seen them as an opportunity to revise material you have already covered. Again, though, we will now step it up a gear. We are going to introduce a new text style to point out which aspects from other chapters we are using to solve problems.

> *Ultimately, we will be using bits from **all** chapters, but we will build slowly. When we use something from a previous chapter, you **must** go back to the original chapter. It is how the facts will make it from short-term memory to long-term memory, and the understanding will become instinctive.*

You've bought the book now! You may as well make the best possible use of it.

WORKED PROBLEM 1

Curly Arrows and Reaction Profiles

INTRODUCTION

Let's start with a statement.

Substitution reactions are often accompanied by elimination reactions as competing pathways.

This is standard textbook material, but we are going to do this as a series of worked problems so that you apply basic knowledge to predict the outcomes of reactions rather than simply reading the results of other people's work. There is quite a lot of guidance throughout, and some (partial) answers. Now over to you.

Make sure you have a go at the questions before looking at the answers. Otherwise, you will simply convince yourself that you would have got it right!

QUESTION 1 – CURLY ARROWS

For the simplified substrate below, draw correct curly arrows that show either a substitution or an elimination reaction. Assume S_N2 and E2 mechanisms.

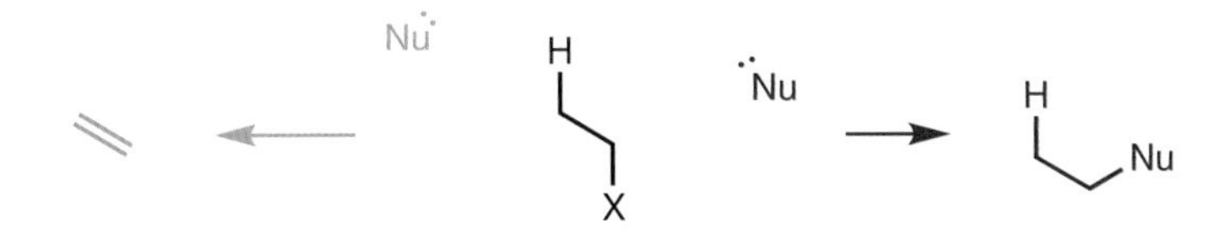

Have another look at **Basics 8** *but remember we didn't do the curly arrows for elimination there. Those were in* **Fundamental Reaction Type 2**.

Go back and double-check the mechanisms. Repetition is good.

Guidance 1

Don't over-complicate this. A base is a nucleophile. A nucleophile is a base.

1

QUESTION 2 – ENERGETICS

Considering only the energetics of the reactions, what factors determine whether substitution or elimination will be the more favoured process?

Don't over-think this. We are looking at Basics 14 *and* Basics 15.

Guidance 2

The wording of the question is clear—'considering only the energetics'. We don't want to get bogged down with 'if you put an extra methyl group here you will get more elimination'. We are currently at a much more fundamental level.

QUESTION 3

Ethanol is a good solvent to favour carbocation formation.

When you see this sort of statement in an exam, make sure you appreciate what it is telling you.

The results of the solvolysis of three different *tertiary* alkyl bromides, **1**, **4**, and **8**, are given below.[1]

Br EtOH → + OEt

1 **2** (19%) **3** (81%)

Br EtOH → + + OEt

4 **5** (29%) **6** (7%) **7** (64%)

Br EtOH → + + OEt

8 **9** (55%) **10** (7%) **11** (38%)

1

[1] This isn't *quite* true. One of the papers I used as a source gave ratios of elimination products and yields, but it did not explicitly state that the balance of material was substitution. I had to make a couple of reasonable extrapolations for the third reaction, as I could not find data generated under exactly comparable reaction conditions. It is easy with the benefit of hindsight to look back and consider the reactions that **should** have been done to provide proof of mechanisms. Much of the foundations of mechanistic organic chemistry were provided by a relatively small number of people and without the instruments we now take for granted. I hope you will forgive me for 'making up' data that provides a coherent story.

 Draw curly arrow mechanisms for the formation of all products in both reactions.

 **Guidance 3**

You just did this for S_N2/E2 in **QUESTION 1**. Make sure you draw the correct mechanisms in this case. You do need to be on autopilot to some extent. But it needs to be focused. The automatic response needs to be the correct mechanism.

It's okay to draw carbocation formation using one curly arrow from the C–Br bond going to the Br. The ethanol plays a role in this process, but once we accept that the conditions are reasonable for carbocation formation, we don't need to show this in the curly arrow mechanism.

Once again, this is a combination of **Basics 8** *and* **Fundamental Reaction Type 2**.

When you are drawing formation of the substitution products, don't 'randomly' form ethoxide from the ethanol. Use ethanol as the nucleophile, and **then** lose the proton to form the stable neutral product.

QUESTION 4

Let's focus on the first reaction (**1** → **2** + **3**) for a minute.

 Draw a reaction profile for the competing E1 elimination and S_N1 substitution reactions for the first of the two processes.

This is **Basics 14** *and* **Basics 15**.

 **Guidance 4**

You have seen reaction profiles for competing S_N1 and S_N2 substitution. Remember, selectivity is selectivity.

You know which is the major product and which is the minor product. Therefore, you know which process has a higher activation barrier.

The difficulty in these questions is to know how far to go. You cannot predict, with any level of confidence, the magnitude of the activation barrier in each case, but you ***can*** *state with confidence which process (substitution or elimination) has the higher barrier.*

> *You could even work out the enthalpy change for formation of the two products (***Basics 13***). You would have to assume that the other product in the elimination reaction is HBr, and that the ethanol is unaffected.*

QUESTION 5

Now draw reaction profiles for the reactions of compounds **4** and **8**.

The key difference is that we are now forming three products instead of two. You have the yields for each product, so you can put the activation energies for the second step in the correct energetic order.

WORKED PROBLEM 2

Competing S_N1 Substitution and E1 Elimination

INTRODUCTION

Here is the reaction scheme from the previous chapter. Now it's time to add more depth to our analysis.

Br EtOH OEt

1   **2** (19%)   **3** (81%)

Br EtOH OEt

4   **5** (29%)   **6** (7%)   **7** (64%)

Br EtOH OEt

8   **9** (55%)   **10** (7%)   **11** (38%)

QUESTION 1

Here are the three compounds, and their respective carbocations.

Which of the compounds (**1**, **4**, **8**) will form a carbocation more rapidly under these conditions?

7.2

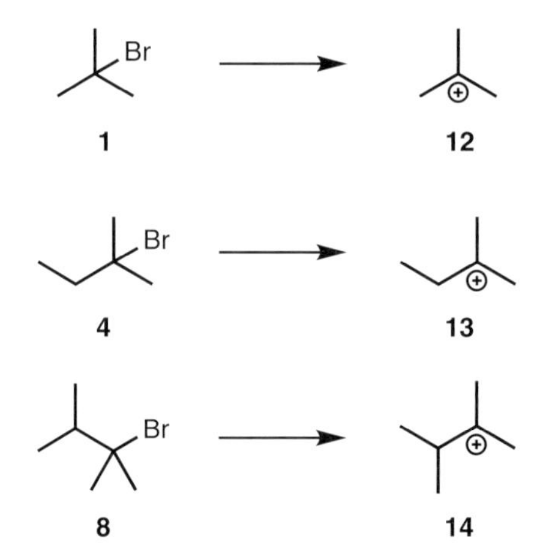

Guidance 1

All the carbocations are *tertiary*, so they will all be pretty stable.

*We know that carbocations are stabilized by hyperconjugation (**Basics 16**).*

There are **nine** C–H bonds that can participate in hyperconjugation in the first carbocation, **eight** in the second and **seven** in the third.

? Is it possible for all nine C–H bonds in the *t*-butyl carbocation to simultaneously overlap with the vacant p orbital?

Hopefully you said 'no'! Based on this, the difference between nine C–H bonds and eight or seven is insignificant.

? What about steric factors?

That's quite enough guidance. Perhaps you can answer this question by drawing a reaction profile. Perhaps you can just 'see it'.

QUESTION 2

? Why does the proportion of elimination product increase from the first to the second reaction, and again to the third reaction?

Guidance 2

There are really only three situations that would account for the data. As we go from the first to the second and then third reaction:

1. Substitution is slower; elimination is unaffected.
2. Elimination is faster; substitution is unaffected.
3. Substitution is slower and elimination is faster.

I really like reducing problems to the simplest possible level. It means we can analyse things properly.

The 'decision' about which product to form isn't made until the carbocation has been formed. Therefore, we don't need to consider which of the compounds will form the carbocation more readily.

QUESTION 3

What effect will increasing temperature have on the proportion of substitution and elimination?

Guidance 3

This is another of those questions that looks a little unfair, because we are spending most of our time (rightly) talking about molecular structures. This one is about basic physical chemistry principles.

Have another look at **Basics 11**.

WORKED PROBLEM 3

Competing S_N2 Substitution and E2 Elimination

INTRODUCTION

We have done a lot with S_N1 and E1 mechanisms, and it is all good! However, these are the less common mechanisms. Fortunately, we would draw almost identical conclusions about S_N2 and E2 mechanisms, and for almost identical reasons.

QUESTION 1

Compound **1** undergoes reaction in ethanol to give 62 per cent of the elimination products and 38 per cent of the substitution product. Upon treatment with neat DBU, only elimination is observed, with a 92:8 mixture of alkene products being formed.

Br

DBU

+

1

92:8 ratio
(100% yield)

N N

DBU (1,8-diazabicyclo[5.4.0]undec-7-ene)

DBU is a strong non-nucleophilic base.

7.4

Draw a curly arrow mechanism for the reaction above, in which DBU acts as a base.

Guidance 1

We don't need to actually explain the different outcome at this point. We just need to think about the curly arrows.

Where are you going to start drawing your first curly arrow?

I would say there are three possible correct answers (lone pairs, π bond) but one of them is 'more correct' than the others.

Of the three possibilities, which electrons are not involved in any bonding?

We can exclude the π bond. As soon as we draw a curly arrow from the π bond, we break a bond and that 'costs' us energy.

Here is the structure of DBU again, this time with the lone pairs drawn in clearly.

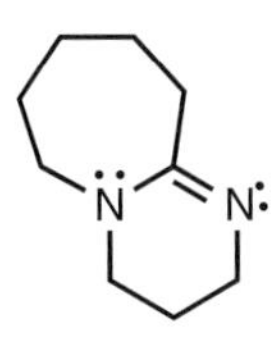

You will need to consider which nitrogen atom acts as the base. They are both drawn with a lone pair, but are both lone pairs equally available?

Think about steric and electronic factors.

You **always** need to think about both factors, even if you are able to dismiss one of them as being less important. In fact, this is probably the best reason why you should think about both factors.

QUESTION 2

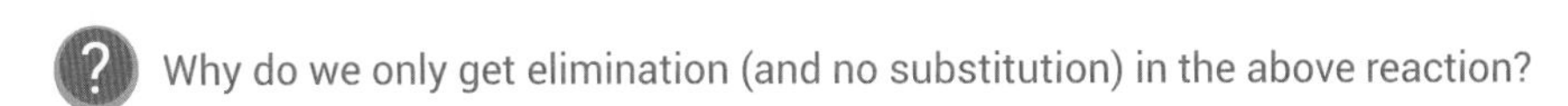

Why do we only get elimination (and no substitution) in the above reaction?

Guidance 2

We have drawn a perfectly good mechanism for elimination. There's no point arguing that elimination happens because it is favourable. What we really need to see is why substitution is **less** favourable.

Draw a mechanism for substitution. What do you get? How favourable is this process?

In the mechanism for elimination shown above, you could ask the same question.

Could the reasons be steric?

QUESTION 3

What does the change from ethanol to DBU do to the pathway?

Guidance 3

You should always answer a question of this type in terms of a mechanism. Either there is a change in the mechanism that is operating, or it is the same mechanism but with 'differences' (vague, I know!). In this case, it's a change in mechanism.

QUESTION 4

Why does the E2 elimination mechanism give higher selectivity for the more stable alkene compared to the E1 mechanism?

Guidance 4

I always wanted this to be an 'honest' book.[2] This question was not intentional. In fact, I cannot remember the last time I really thought about the reasons for this selectivity. But with the Hammond postulate and the shape of the reaction profile firmly in the back of my mind, the question became inevitable!

So, there's your hint!

QUESTION 5

We saw the following two reactions in **Applications 4**. We didn't explain why the elimination reaction of compound **4** is slower than the reaction of compound **6**. It's time to look at this now.

[2] That is not to suggest that other organic chemistry books are not honest!

4 $\xrightarrow{\text{NaOH}}$ 5

6 $\xrightarrow{\text{NaOH}}$ 7

Guidance 5

First of all, consider which of the two alkene products will be most/least stable.

Remember, if it ain't electronic, it's steric. Sometimes, it is both!

Can both phenyl groups in compound **5** be coplanar with the alkene so that good orbital overlap is achieved across the conjugated π system? Make a model!

7.5

> *Perhaps you need to refer back to the Newman projections in* **Applications 4**. *If we want to work out why one reaction is slower than another, we need to work out why that reaction has a higher activation energy (***Basics 15***).*

Sodium hydroxide is a base. We are definitely (still) talking about an E2 mechanism.

WORKED PROBLEM 4

Acid Catalysis in Organic Reactions Part 2

INTRODUCTION

7.6

We started looking at this problem in **Basics 22**. Now we have looked at the thermodynamics of reactions involving charged species (**Basics 35**), we can consider the implications of acid catalysis with a strong acid and with a weak acid.

This is a largely mathematical problem. I must admit that I don't really like these problems. The important thing is to work through this once (or twice) and become familiar with the method. Doing this will help you internalize the conclusions.

Once again, we will look at the formation of a carbocation from an alcohol. Here is the reaction again.

$$R{-}OH \rightleftharpoons R^{\oplus} + {}^{\ominus}OH$$

Here it is with acid catalysis.

$$R{-}\overset{\oplus}{O}H_2 \rightleftharpoons R^{\oplus} + H_2O$$

We compared the two processes before (**Basics 22**), and we established that the activation energy from the protonated alcohol is significantly lower.

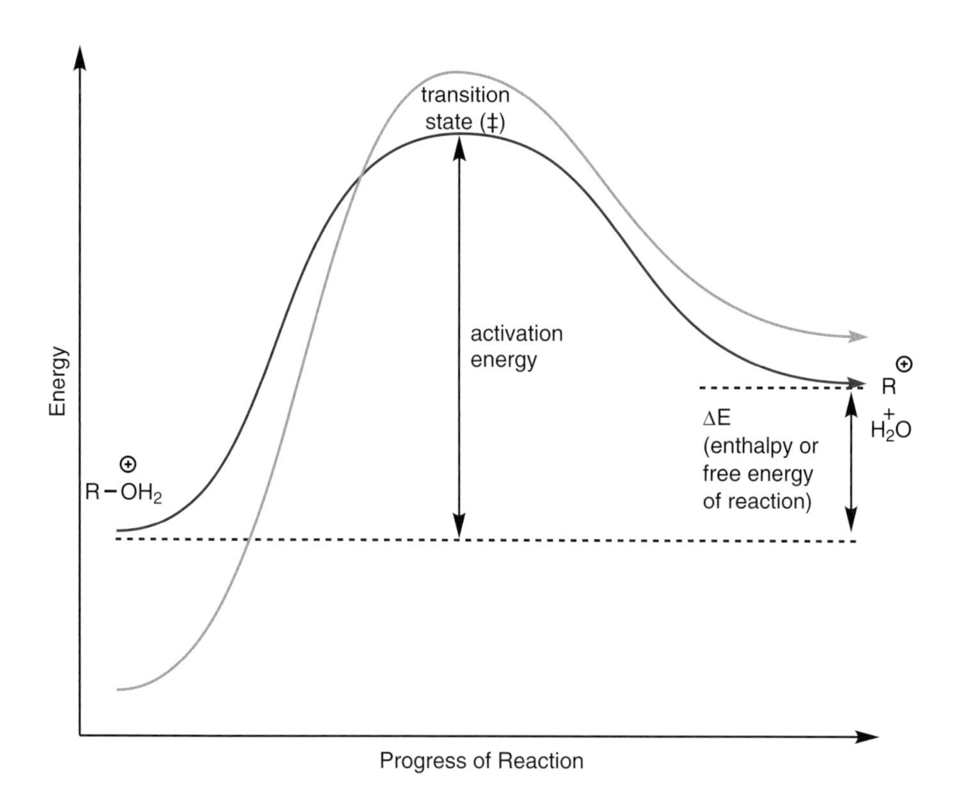

Now it's time to quantify some aspects of this. We won't be able to work out the absolute activation energy in either case.

> *We can estimate the pK_a of the protonated alcohol,[3] and we know the pK_a of water. We can use the methodology in* **Basics 35** *to calculate energy differences from these. We can use the Hammond postulate to relate the difference in energy of the two transition states to those of the 'products' for this endothermic process.*

QUESTION 1

Estimate the energy difference between the following two species.

$R{-}OH$ **1** $\quad\quad$ $R{-}OH_2^{\oplus}$ **2**

[3] We could look it up, or measure it, but it's nice to be able to estimate it from something you already know.

Let's start with an initial assumption. An alcohol isn't a million miles from water.[4] *The protonated alcohol will have a similar pK_a to that of the hydronium ion, H_3O^+.*

$$\underset{\mathbf{2}}{R{-}\overset{\oplus}{O}H_2} \rightleftharpoons \underset{\mathbf{1}}{R{-}OH} + H^{\oplus}$$

Estimated pK_a −1.74

Guidance 1

You need to know some basic thermodynamics. In particular you need to be able to relate an equilibrium constant with a free-energy change.

The equation you need is in **Basics 35**.

QUESTION 2

What is the energy difference between the carbocation plus hydroxide and the carbocation plus water on the right-hand side?

Guidance 2

It's a trick question! Well, not really, but the carbocation stability won't change, so all we need is water versus hydroxide. It's another acid-base equilibrium.

$$H_2O \rightleftharpoons H^{\oplus} + \overset{\ominus}{O}H$$

pK_a 15.7

QUESTION 3

You need to look at the answers to **QUESTION 1** and **QUESTION 2** before you have a go at this.

How can we estimate the energy difference between the transition states in each case? Will it be closer to 10 kJ mol^{-1} or to 90 kJ mol^{-1}?

Guidance 3

If you need guidance on this, go back to the introduction to this chapter. There is a hint there somewhere! The further we get through the problems, the less direct guidance you should need.

4

[4] This also applies to the beer in some bars I have been in!

QUESTION 4

 How does the reaction profile change when we use a **strong** acid catalyst?

We will use H_3O^+. Now you will see why we were happy to assume that the protonated alcohol has the same pK_a as the hydronium ion. It simplifies the maths.

Guidance 4

Reactions such as proton transfer may not have very high barriers (activation energies) but they do have barriers. We need to consider the following equilibrium.

$$\underset{\mathbf{1}}{R{-}OH} + H_3O^{\oplus} \rightleftharpoons \underset{\mathbf{2}}{R{-}\overset{\oplus}{O}H_2} + H_2O$$

If you add enough acid (excess hydronium ion) you can always drive the equilibrium to completion. Assume you have one equivalent of alcohol and one equivalent of hydronium ion.

QUESTION 5

It's taken us a while to get to it, but in **Basics 22** we questioned whether a weak acid catalyst might slow the reaction down.

 How does the reaction profile change when we use a **weak** acid catalyst?

Let's use acetic acid, pK_a 4.76, for this example.

$$H_3C{-}C(=O)OH \underset{pK_a\ 4.76}{\rightleftharpoons} H_3C{-}C(=O)O^{\ominus} + H^{\oplus}$$

Guidance 5

If you are comfortable with the maths, you can try this calculation with a range of acids of varying pK_a. How high would the pK_a need to be before this made a difference?

WORKED PROBLEM 5

Epoxide Opening Reactions

INTRODUCTION

An epoxide is a compound with a three-membered ring containing an oxygen atom. We encountered them in Basics 28 and Applications 2.

An epoxide is an ether in a ring.

All of the ring atoms in an epoxide are sp^3 hybridized. The ideal bond angles are 109.5°. The actual bond angles are 60°. The deviation from ideal bond angles represents a lot of strain.

You don't normally attack ethers with nucleophiles, but epoxides are much more reactive because the ring is strained.

I have often heard the carbonyl group referred to as the backbone of organic chemistry. I'm not sure which part of your anatomy an epoxide would be, but it's definitely something you cannot manage without!

QUESTION 1

Here is the reaction of an alkyne with an epoxide. A base is used.

Cl—CH₂—C≡C—H (**1**) + epoxide with H and n-C_6H_{13} (**2**) $\xrightarrow{\text{base}}$ Cl—CH₂—C≡C—CH₂—CH(OH)—n-C_6H_{13} (**3**)

Draw a curly arrow mechanism for this transformation. Assume that you add water at the end of the reaction, so that you have a source of protons.

Guidance 1

What is the purpose of the base? I know this sounds like an obvious question, but it is worth articulating it.

Look at it this way. If you mix an alkyne with an epoxide, and don't add a suitable base (we will get to that in a minute!), nothing happens. If you drew a curly arrow going from the alkyne π bond to the epoxide, you're wrong.

7.7

QUESTION 2

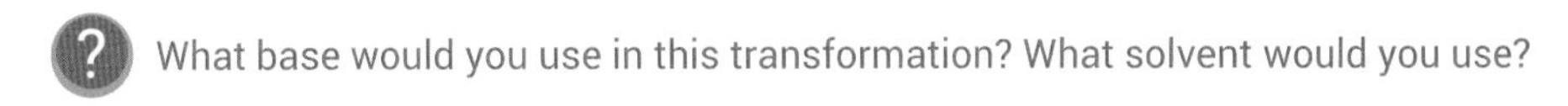

What base would you use in this transformation? What solvent would you use?

Guidance 2

These are not 'fair' questions! In reality, you will learn what base to use, and what solvent to use, by gradually assimilating knowledge as you read a large number of organic chemistry research papers, possibly during a research project or after you graduate.

However, we can rephrase the question.

What are the criteria for a suitable base and solvent for this transformation?

QUESTION 3

What is the absolute configuration of epoxide **2**?

Have another read of Habit 6*, and perhaps repeat the questions in* Practice 10.

Guidance 3

Here is epoxide **2**, drawn in two different ways. In both cases, the oxygen is on 'top'. Assign the stereochemistry in each structure.

H O *n*-C_6H_{13} **2**   H O *n*-C_6H_{13} **2**

Do you get the same answer in each case?

These can be a bit tricky. Clockwise and anticlockwise look different depending on how the structure is drawn. The key point here is to be aware of the problem. As always, your fall-back position is to make a model of the compound.

5

QUESTION 4

This is the 'big' question. There are two possible sites at which the nucleophile could attack the epoxide.

7.8

 Why does it attack at the site shown above?

Guidance 4

The 'simple' answer is that it is less hindered, but you can be sure that since I am asking the question, there might be more to it.

If it ain't steric, it's electronic!

You can be certain this is an S_N2 substitution. You won't form a carbocation under strongly basic conditions. If you did, would you get attack to form the product shown?

QUESTION 5

The initial product of the above reaction is the alkoxide anion **4**. This intermediate has a propargylic chloride, which is a pretty good substrate for S_N2 substitution.

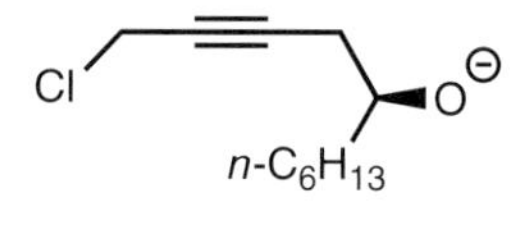

4

7.9

 Why doesn't the alkoxide anion attack the propargylic chloride ('Propargylic' is the equivalent of allylic, but with a triple bond) to form a ring?

Guidance 5

Make a model!

WORKED PROBLEM 6

Is *cis*-Cyclohexane-1,2-diol Really Achiral?

This chapter is somewhere between a worked problem and a more detailed insight.

> *In Habit 7 we discussed the cis isomer of cyclohexane-1,2-diol, and recognized that it has a plane of symmetry.*
>
> *In Practice 12 we reinforced the point, and we considered that sometimes it is easier to see symmetry in one representation or another.*

We will now add a layer of complexity to the discussion in order to reach a deeper understanding.

QUESTION

 Is *cis*-cyclohexane-1,2-diol really achiral?

Here is *cis*-cyclohexane-1,2-diol again, shown in two orientations. There is absolutely no doubt that these are the same molecule. There is a plane of symmetry through the structure in either representation.

OH OH

OH OH

So why are we asking what seems to be a silly question?

 **Guidance**

There isn't a lot of guidance I can give you.

 What shape is cyclohexane?

> *Yes, we are in Basics 31 territory.*

Play around with drawing the structures and consider the relevant chair conformers. Do they meet the definition of non-superimposable mirror images?

> *Now we need* Basics 24*. Hopefully you are understanding why organic chemistry can be hard to learn. You need to pull everything together.*

6

WORKED PROBLEM 7

The Fürst–Plattner Rule

INTRODUCTION

This is a topic that you don't normally encounter at this level. It concerns the ring-opening of cyclohexane epoxides by nucleophiles.

The fact that there is a 'rule' associated with it doesn't make it more complicated.

If you understand cyclohexane conformers, you will be able to follow the discussion. We are going to work through a couple of problems, and then we will formalize the Fürst–Plattner rule. Here is the 'standard' example. A 1:1 mixture of epoxides **1** and **2** is treated with morpholine in water. Only epoxide **1** reacts, giving product **3**. Epoxide **2** is recovered unchanged.

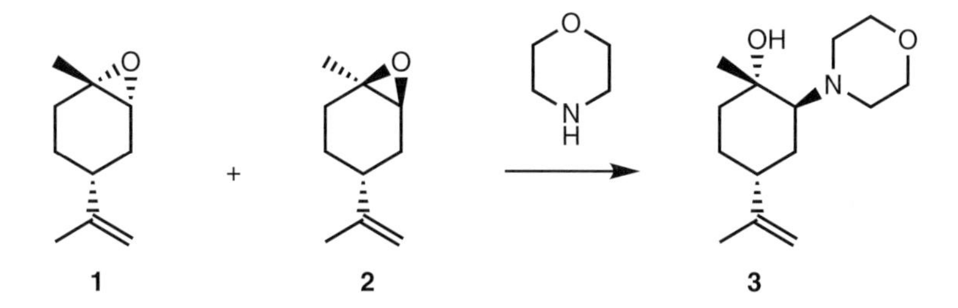

QUESTION 1

? Which substitution mechanism is operating in this reaction?

Guidance 1

It's a substitution reaction. Fundamentally you have two choices. We aren't worrying about why only compound **1** reacts just yet.

? If the reaction was S_N1, where would you expect the nucleophile to attack?

? If the reaction was S_N2, where would you expect the nucleophile to attack?

Perhaps you can exclude one of the mechanisms?

I suggest having another look at **Fundamental Reaction Type 1**, **Basics 16**, *and* **Reaction Detail 1**. *You may also want to recap* **Basics 19**, *just because you can!*

QUESTION 2

Draw a curly arrow mechanism for the formation of compound **3** from compound **1**.

Guidance 2

You need to know the mechanism before you draw the curly arrows, but make sure you see the bigger picture associated with this mechanism. It is an opportunity to practise, and make sure you draw the correct structures with no errors, and you draw stereochemical structures that are fully defined and have no ambiguity.

This is where working with a few friends will help. Collectively, you have more chance of spotting errors. If you did make a mistake, that's okay. Just have another go. By re-drawing the structures and mechanisms, you are quite literally 'rewiring' your brain to become good at this. It takes time and repetition.

QUESTION 3

Assign the configuration of all stereogenic centres in compounds **1**–**3**.

This is all about Habit 6. *Remember why it was presented as a habit.*

Guidance 3

Epoxides can be a bit sneaky. We saw this in **Worked Problem 5**. Trying to visualize what is 'up' and what is 'down' isn't always easy. Make a model of the compound! This will allow you to check your working.

QUESTION 4

Draw both chair conformers of product **3**. Work out which one is more stable.

Guidance 4

This isn't a trivial example.

Have another look at Basics 32. *You won't find 'A' values for all of the substituents here, but you could possibly estimate them.*

Consider how many substituents will be axial and how many will be equatorial, and work from there.

QUESTION 5

Let's take a slight digression for a moment. Here is compound **3**. Alongside this is compound **4**, which would be the product formed from compound **2** if it actually reacted. We are going to look at these reactions in reverse.

As a prelude to this, draw **both** chair conformers of compound **3** and of compound **4**. Determine which is the most stable in each case.[5]

Hopefully by now this sort of thing is getting easier for you.

QUESTION 6

You are going to have to suspend disbelief for a moment. You will **never** have an amide anion as a leaving group in a substitution reaction. Here is a curly arrow mechanism for the process, starting with compound **3**. The 'amide anion' is structure **5**, which of course does not contain an amide functional group. It's another of those horrible nomenclature things you will get used to in time.

Amines have a pK_a > 30, so the anion **5** is unstable. It is a terrible leaving group!

[5] You already did compound **3** in **QUESTION 4**!

7.12

Nevertheless, **if** you were to form an epoxide from compounds **3** or **4** by this type of cyclization, which conformer would need to react for each compound?

> *Have another look at* Applications 2, *and* **Applications 5**. *Forming an epoxide and forming an alkene are actually rather similar. To form an epoxide, you need the oxygen and the leaving group anti periplanar.*

QUESTION 7

Now let's get to the heart of the matter. You should probably check your answers to **QUESTION 1** to **QUESTION 6** before moving on.

Why does compound **1** react faster than compound **2** under these conditions?

Guidance 5

There isn't a lot of additional guidance I can give at this point. Try drawing reaction profiles based on the above conformer energies.

This is not a simple problem by any stretch of the imagination. You should not expect to get to the right answer on your own.

What you should expect is to be able to make a model of compounds **1** and **2** and think about what the best conformer is and where the nucleophile will attack. Then, when you read the explanation, you will be ready to understand it.

> *Have a look at the cyclohexene conformers in* Basics 33. *The epoxide conformers are quite similar to those of cyclohexene. We then need to work out whether the isopropenyl group is pseudo-equatorial or pseudo-axial.*

Once we have looked at the problem at a basic level, we will add some subtlety in the answer.

WORKED PROBLEM 8

S_N2' Stereochemistry and Conformations

INTRODUCTION

The next problem focuses on key skills, and it introduces one more important idea.

We have been considering shape in reactions throughout. Reactions such as E2 elimination are very clear in the sense that they only proceed through one specific conformational isomer. Other reactions may also exhibit conformational preferences, and you need to 'automatically' consider such aspects.

I find that if I draw a structure in a particular way in a question, many students make a model of 'exactly that shape' and only consider the reaction from that given shape. They do not consider that you could rotate a bond to provide a more stable conformer.

It is only a relatively small adjustment to make in your process. Once you do this for one reaction, you will do it all the time. You will no longer have to remember the specific cases where it matters!

SETTING THE STAGE

- *All aspects of this problem relate to the S_N2' reaction that we saw in* **Reaction Detail 5**.

For the following two compounds, **assume** that you get only S_N2' substitution (*i.e.* the nucleophile attacks the double bond).

Ph Cl CH$_3$ **1**

Ph Cl CH$_3$ **2**

We are simplifying the situation by only considering one process.

Furthermore, **assume** that the attack takes place only on the face of the double bond **opposite** to the Cl leaving group.

Again, this is an assumption designed only to give a single outcome.

QUESTION 1

Draw a curly arrow mechanism for the attack of hydroxide ion onto the double bond of compound **1**.

Guidance 1

It is always worth drawing a curly arrow mechanism. Make sure your curly arrows make sense. Make sure you know where the hydroxide is attacking.

QUESTION 2

7.13

What product will be formed by attack of hydroxide anion in each case?

Guidance 2

I have deliberately left this vague, so we can think about what this means. We know what the product will be, but we have not yet fully defined the structure. We need the stereochemistry, and the double bond geometry.

The double bond geometry is also stereochemistry!

We know that the S_N2' reaction requires there to be overlap of the π orbital of the alkene and the σ^* orbital of the C–Cl bond. Therefore, we need only consider reaction from conformers in which the C–Cl bond is directly perpendicular to the alkene π bond.

How you show this is up to you. As always, I would recommend making a model.

If we ignore the phenyl[6] group on the double bond, here is a Newman projection.

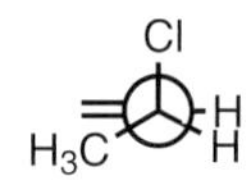

Hopefully this gives you a bit of a clue as to what's coming next! There are two possible conformers in which the C–Cl bond is perpendicular to the alkene π bond.

Did you spot this? Take a moment to draw the 'other' conformer.

QUESTION 3

8

Reaction of compound **1** could give either compound **3** or compound **4**. Which of the possible products would you expect to predominate?

[6] Not phenol! It's been a while since I reminded you!

Ph OH CH_3 Ph CH_3 OH

3 **4**

Guidance 3

Here is the logic that students often use. 'An (*E*) double bond is more stable, so if a reaction can give an (*E*) or (*Z*) alkene, it will give the (*E*) alkene.'

In as far as it goes, the logic is fine. However, we really need to be considering the transition state energy rather than the product stability.

> *Have another look at* **QUESTION 4** *in* **Worked Problem 3** *to see how we approached a similar problem with elimination reactions. Have another look at* **Basics 34** *for the factors that determine which alkenes are more stable.*

QUESTION 4

What products will be formed in a similar reaction of compound **2?**

QUESTION 5

Which of compounds **1** and **2** would you expect to be more selective in this S_N2' reaction?

Guidance 5

We have already established that each reaction can give two possible products. We established that compound **1** could give compounds **3** and **4**, and that compound **4** is expected to be the major product. What we have not yet considered is how selective (what ratio of **3**:**4** we might expect to be formed) the reaction will be.

There is a good reason for this. We don't have a frame of reference. However, we can use the related reaction of compound ***2*** *as a comparison.*

WORKED PROBLEM 9

Complex Substitution Stereochemistry

INTRODUCTION

There is a classic experiment that illustrates many points about stereochemistry in substitution reactions. It is covered at some point in almost every textbook, and with good reason.

In this chapter, we are going to describe the results, but the key point will be to define a series of questions we can ask in order to understand how this reaction takes place.

Learning to ask the right questions is extremely important. Once you've asked the right question, the answers pretty much take care of themselves!

SETTING THE STAGE

Reaction of the following enantiomerically pure tosylate **1** with acetic acid gives a racemic mixture of enantiomers **2** and **3**.

OTs AcOH OAc + OAc

1 **2** **3**

1:1 mixture of **2** and **3**

▌ *We saw what a tosylate group is in* **Reaction Detail 1**.

QUESTION 1

Assign the stereochemistry of all stereogenic centres in compounds **1**–**3**.

Guidance 1

There is a reason for doing this, above and beyond practice. We definitely need to consider the stereochemical outcome, so it is useful to have the labels so that we don't have to work too hard at visualizing the changes.

QUESTION 2

If this reaction gives a racemic product, can it be a 'simple' S_N1 substitution?

Guidance 2

I am guessing you have already realized that there is going to be more to it than this! How can we exclude a simple S_N1 reaction pathway?

Of course, if you cannot exclude it, then perhaps it is!

This is a fundamental point. You can **never** prove a mechanism is correct. The best you can do is design experiments to 'test' a mechanism. The more experiments you design that produce results consistent with your proposed mechanism, the more confident you will be that the mechanism is correct.

In this case, so far, we only have one piece of experimental data.

It would only take one contradictory result to disprove a mechanism.

QUESTION 3

I assume you have either got **QUESTION 2** correct (and know you have!) or you have checked the answer.

How can both stereogenic centres be affected during this reaction?

Guidance 3

I'd say this is the wrong question to ask. Try focusing on the formation of compound **2** first.[7] In the formation of enantiomer **2**, neither stereogenic centre has changed. It is easy to see why stereogenic centre 2 would not change. The stereochemistry at carbon 1 is also unchanged.

[7] Of course, I have the advantage of knowing the answer. Therefore, I know this will be a productive way to proceed. You may feel that this is all very artificial. It's hard to argue against that, but I can assure you that as your experience grows, you will begin to 'just know what to do'.

The last time we saw this was in Applications 2. Perhaps the reasons are the same?

2

Do we have something in compound **1** that could act as an 'internal nucleophile' in the same manner as we saw in Applications 2?

You know what a nucleophile is.

 Which electrons could it use?

You need something that you can draw a curly arrow from.

QUESTION 4

We have to give the game away! There is no other way to do this. You have looked at the answer to **QUESTION 3**. Structure **4** is the intermediate in this reaction.

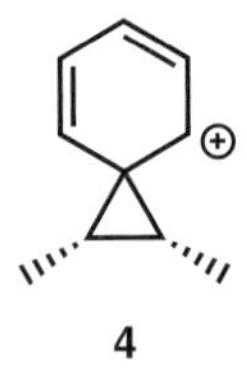

4

7.14

 What do you notice about intermediate **4**?

Guidance 4

What a rubbish question! Let's put it another way. Structure **4** is only one resonance form. Based on your understanding of what resonance forms actually mean (**Basics 10**), can you spot anything 'special' about intermediate **4**?

QUESTION 5

 Which carbon, 1 or 2, in intermediate **4** will be most readily attacked by an acetate nucleophile?

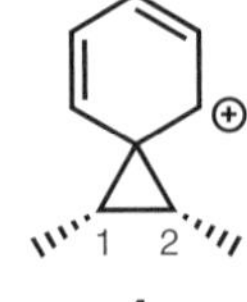

4

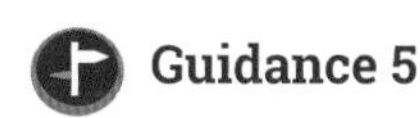

Guidance 5

Is this a trick question?

QUESTION 6

What rate law will this reaction follow? Is this really S_N2, or is it a 'strange' S_N1 reaction?

Guidance 6

I don't actually know the definitive answer to this question.[8] But we know that stable carbocations aren't really that stable (**Basics 16**). We know how much energy aromaticity contributes (**Basics 10**). You are in a position to speculate about the likely activation energy for the formation of intermediate **4**.

QUESTION 7

What will be the stereochemical outcome of the corresponding reaction starting with stereoisomer **5**?

Of course, this is the ultimate test for a mechanism. You can compare the predicted outcome with the experimental outcome.

ADDITIONAL QUESTIONS

There is a lot more we could do with this reaction. Here are some questions I am not going to give the answers to, although there are a couple of hints along the way.

1. Would you expect the same outcome if you used a better nucleophile such as PhS^-, and a solvent such as DMSO?
2. How might you accelerate this process? Where might you put a methoxy group on the benzene ring? **HINT** you want to stabilize the intermediate **4**, because this will stabilize the transition state that precedes it.

[8] I could look it up, but I don't really need to!

WORKED PROBLEM 10

Cyclization Reactions

ADVANCE WARNING!

This is a problem without many clear answers. That doesn't mean we cannot use it to practise the basics.

In this case, the basics are correct curly arrows and consideration of the shape of a compound when it reacts. Let's have a look at the problem and some of the basic stuff, and then we will look at the reaction in more depth.

THE PROBLEM

Here are two reactions which produce two different stereoisomers of a biologically active natural product. One gives a higher yield than the other.

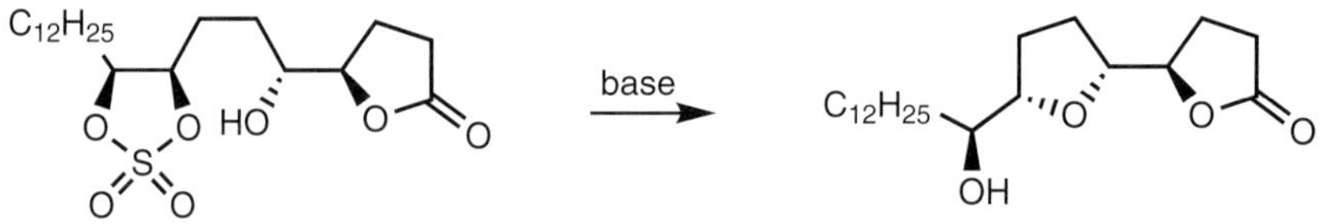

$C_{12}H_{25}$ … **1** —base→ $C_{12}H_{25}$ … OH **3**, 67%

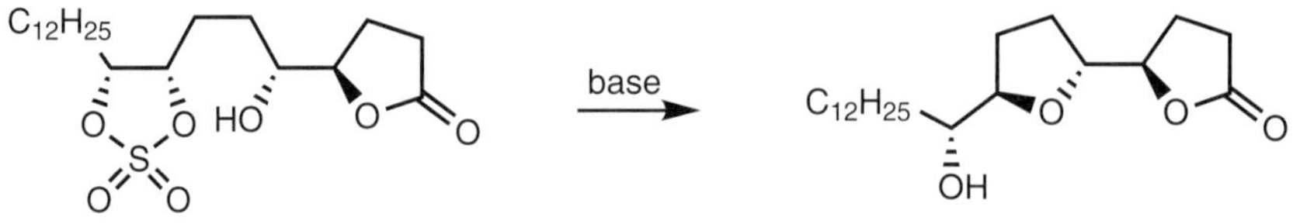

$C_{12}H_{25}$ … **2** —base→ $C_{12}H_{25}$ … OH **4**, 83%

QUESTION 1

 Draw a curly arrow mechanism for this transformation.

 **Guidance 1**

Compounds **1** and **2** are cyclic sulfates.

> *You saw a tosylate ester in* **Reaction Detail 1**. *A cyclic sulfate is a good leaving group for exactly the same reasons.*

It's always a good idea to draw some curly arrows. It helps you focus on what is going on. If nothing else, it helps you get better at drawing curly arrows! Don't worry too much about how you get the OH group. Formally you lose SO_3.

QUESTION 2

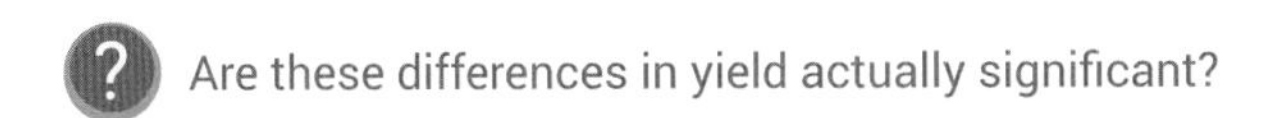

Are these differences in yield actually significant?

Guidance 2

This is another one of those unfair questions! We haven't mentioned this aspect at all. For now, we are just looking at the practical aspect.

When you and your colleagues are making compounds in the lab, do you all get the same yield?

QUESTION 3

Does the fact that compound **4** is formed in higher yield than compound **3** mean that the formation of compound **4** is 'better'?

Guidance 3

Let's suppose the reaction of compound **1** to give compound **3** is slower (has a higher activation barrier) than the reaction of compound **2** to give compound **4**. What does this say about the yields?

QUESTION 4

Compound **3** has the two substituents on the newly formed ring in the *cis* orientation. Compound **4** has them *trans*. Does it automatically follow that compound **4** is more stable than compound **3**?

7.16

Guidance 4

Look at the following two cyclohexanes. Make a model of them. Which do you expect to be the more stable?

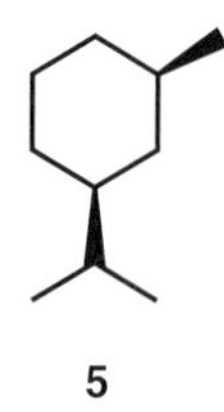

5

6

QUESTION 5

You should look at the answers to **QUESTION 1** to **QUESTION 4** before carrying on.

Suspend disbelief for a moment. We have established that this result probably has no numerical significance. But, it **does** seem reasonable that the formation of compound **4** is 'better' than the formation of **3**. Why is this?

Guidance 5

In any reaction, we have to consider the shape of the compound reacting. The shape may include chirality, or it may just relate to the size of various substituents. This is relatively easy to consider—most of the time you only need to look at one part of a molecule.

> *In* **Worked Problem 8** *we considered how rotation around one bond can affect the outcome of a reaction.*

The next level up from this is to consider the rotation of all bonds in a system, and to identify the important conformers for reaction.

> *We did this for elimination reactions in* **Applications 4**. *Elimination is a great example for this, as the requirements in an E2 elimination are clear.*

Considering all of the conformers of a compound with many bonds is not at all easy. There are just so many of them! Even a relatively simple molecule can have thousands of different possible conformers.

We can simplify the situation!

In some S_N2 substitution reactions, you can form a ring.

> *We have seen this already in Applications 2.*

The transition state for ring formation very often resembles the ring, as all of the atoms need to become 'pre-arranged' to make ring formation possible.

> *We have spent quite a bit of time looking at the conformers of rings, particularly in Basics 31 and Basics 33. We have applied this chemistry in Applications 3 and* **Applications 5**.

10

You need to get your molecular model kit out and make a model of compound **1** *and compound* **2**. *Orient the model so that the nucleophilic oxygen and the leaving group are on opposite sides. Twist the model around to remove any unfavourable interactions. And then try to work out if one is better than the other.*

Remember, this isn't about whether you get the correct answer. It is about the process involved. Even if you don't get this one right, you will still learn from it.

SOLUTION TO PROBLEM 1

Curly Arrows and Reaction Profiles

ANSWER 1

Assuming a given substrate has the option of either a substitution reaction (**purple**) or an elimination reaction (**blue**), the two possible mechanisms are as shown below.

We aren't balancing charges—assume the nucleophile has a proton it can lose!

We will see this with ethanol in **ANSWER 3**.

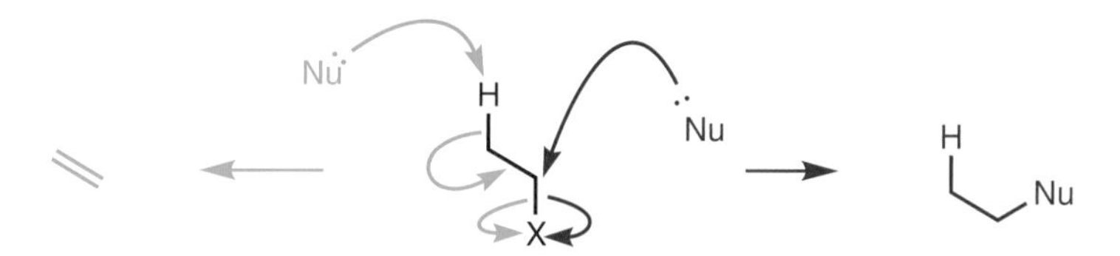

ANSWER 1 REFLECTION

When you draw a curly arrow going from one atom to another, you are forming a bond between those atoms. When you draw a curly arrow going from a bond, you are breaking that bond (or reducing the bond-order if it is a multiple bond). When you draw a curly arrow going to an existing bond, you are increasing the bond-order of that bond.

In the early stages, students often draw curly arrows connecting atoms, without considering whether it makes sense. Take your time, make sure you can explain each curly arrow. After a while, you'll just draw the correct arrows instinctively.

ANSWER 2

Assuming the two products cannot be interconverted, the ratio of substitution:elimination will depend **only** on the relative rates of the two competing processes. This depends on the activation energies for the processes.

Of course, if the products can be interconverted, you will get more of the most stable product.

It is really important to spot a fundamental question and give a focused answer. This really isn't about passing an exam. If you understand what is being asked, you will get good at organic chemistry and the exam will take care of itself.

ANSWER 2 REFLECTION

There isn't really much more to say about this. Have a look at your answer. If you gave more information, or used more complicated reasoning, ask yourself whether the added information allowed you to answer the question better. Be your own worst critic. It will help you out in the long run.

ANSWER 3

Isn't here! Sorry! I have already given you hints.

> *The S_N1 mechanism is in* **Fundamental Reaction Type 1**. *The E1 mechanism is in* **Fundamental Reaction Type 2**. *Just make sure you are drawing the correct mechanism in each case.*

Remember that there is a key difference compared to **QUESTION 1**. Now we have ethanol as the nucleophile, so we will need an extra curly arrow for loss of the proton. Make sure this is a separate step!

ANSWER 3 REFLECTION

Yes, I am asking you to do the same thing three times. Don't take short cuts or get bored with this. The only difference (in the structures) is the number of methyl groups. Check your answers a couple times, and make sure you haven't missed anything.

ANSWER 4

We looked at the individual reaction profiles in Basics 14. Don't over-think this. You just need to put two of them on the same diagram! Substitution is the major pathway. Therefore, it will have the lowest activation barrier. Remember, this is the barrier for reaction of the carbocation. It will be the same up to this point for formation of either product.

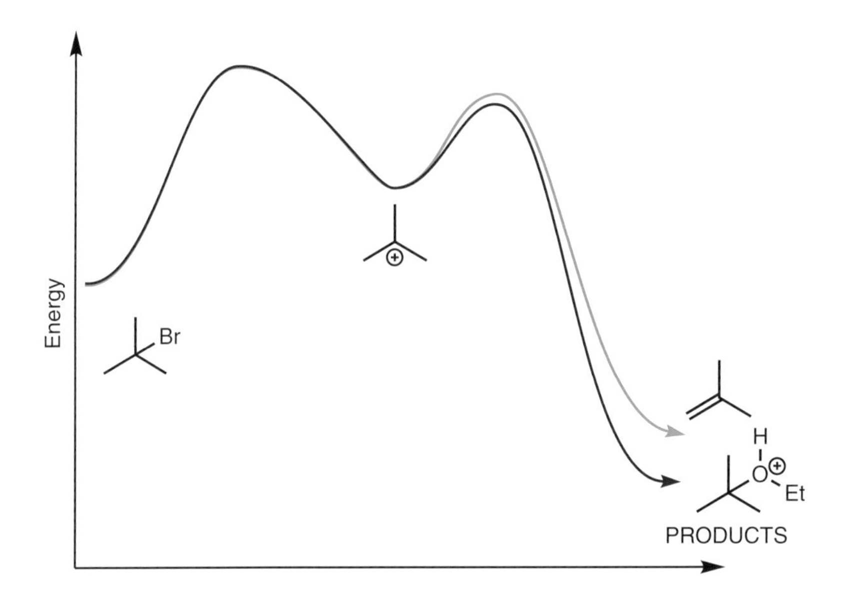

For substitution, this does not **yet** give the final product.

Modify the reaction profile to add the loss of a proton. You will need another transition state!

ANSWER 4 SELF-ASSESSMENT

It is really easy to confuse major/minor product and higher/lower activation barrier. It is also very easy to make the profile different in each case prior to carbocation formation. Make sure you haven't made any mistakes.

ANSWER 5

You don't need it. No, really, you don't need it. You did **QUESTION 4. QUESTION 5** adds a little complexity, but I am confident that you just drew 'the same thing' but with one more profile for the additional product, and you made sure that you had the major product with the lowest barrier and the minor product with the highest barrier.

You're starting to get it! Good, isn't it?

REFLECTION ON THE SERIES OF PROBLEMS

We are still working at a relatively basic level. The curly arrows should be (or need to be) flowing naturally. I use the word 'flowing' because if you have got to grips with this, they will 'flow'.

You **need** to be thinking about the energy of transition states, intermediates, and products. Most of the time you won't need to do much with them, but you need this to be part of your process, so that you can apply it when it matters.

Hopefully you have already started thinking about **why** an increasing proportion of elimination is observed, and less substitution, going from compound **1** to compound **4** to compound **8**. We are going to get to that next!

You can definitely predict how the alkene product stability changes from the first reaction to the second and then to the third.

> *You need to look at* **Basics 34**.

There would be no point trying to use bond dissociation energies to calculate the difference, as all carbon–carbon double bonds are equal in these calculations. Of course, that's a key limitation of the method.

SOLUTION TO PROBLEM 2

Competing S_N1 Substitution and E1 Elimination

ANSWER 1

In **Worked Problem 1**, we established that the three carbocations, **12**, **13**, and **14**, are of broadly similar stability, as long as we are considering electronic effects (hyperconjugation).

They are all tertiary carbocations.

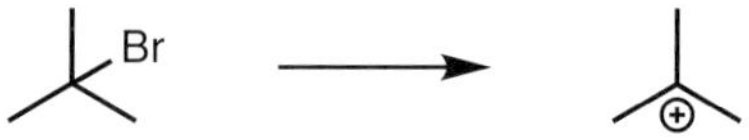

Br

1 **12**

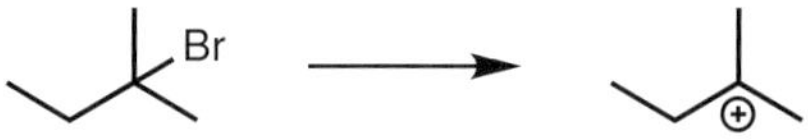

Br

4 **13**

Br

8 **14**

Therefore, we need to be looking for steric effects. We should automatically look for steric effects in the starting material, the product (in this case a carbocation), and along the entire reaction coordinate.

*Compound **8** is more crowded than compound **4**. Compound **4** is more crowded than compound **1**.*

Now let's look at the carbocations.

Let's call the carbocation carbon the α-carbon and the other one the β-carbon. Here is the shape of the carbocation in each case, shown as both a Newman projection and a flying wedge projection. In the Newman projection, the carbocation carbon atom is at the front. The positive charges have been omitted for clarity.

12 **13** **14**

We have seen Newman projections in Habit 4 and Basics 30, but we haven't used them for carbocations. That doesn't mean we cannot, but we do have to refine them a bit as shown above.

When considering the conformer, I drew a conformer in which a C–H bond on the β-carbon was lined up with the empty p orbital of the carbocation.

We saw this in **Applications 4**.

This is so we can get ready for one possible next step—the elimination reaction. Don't worry about it for now.

Carbocation ***14*** *is more crowded than carbocation* ***13****. Carbocation* ***13*** *is more crowded than carbocation* ***12****.*

What will this crowding do to the rate of carbocation formation?

Carbon atoms α- and β- are tetrahedral (bond angle 109.5°) in the alkyl halide precursors. Carbon atom α- is trigonal planar (bond angle 120°) in the carbocations.

The alkyl halides are more significantly affected by steric crowding than are the carbocations.

As we form the carbocation, we are relieving steric strain that is present in the alkyl halide precursor.

The transition state leading to the carbocation will resemble the carbocation (Hammond postulate, **Basics 19***).*

Therefore, you can assume that the destabilization of the starting material means that the barrier to carbocation formation will be lower. It doesn't have as far uphill to go.

 Draw this on a reaction profile.

ANSWER 1 REFLECTION

This doesn't really make any difference to the outcome (product distribution), but it may affect the rate of the reaction, since carbocation formation is rate-determining.

As with everything else, it is all about the process of analysis. We get used to looking at problems in this way, so that we can determine what is important/relevant and what is not.

ANSWER 2

We should consider each process in turn, and try to work out whether it will be faster, slower, or unchanged upon moving from carbocation **12** to **13** to **14**.

Yes, we are talking about reaction of the carbocation, not the alkyl halide. After all, both products in each case are formed from the carbocation.

Substitution is addition of a nucleophile to the carbocation.

 Would addition of a nucleophile to carbocation **14** be slower than for carbocation **12**?

The additional methyl groups provide a steric barrier. This will slow down addition of the nucleophile. Furthermore, we would expect this to be more significant for bigger nucleophiles.[9]

The product will be more crowded, and less stable. The transition state leading to the product will also be more crowded, and less stable. We do indeed expect that substitution will be slower for the second and third reactions.

Up to now, we have probably done 'enough'. If substitution is slower and elimination is either faster or unchanged, we have rationalized the outcome. But we should still aim to do a complete job.

What about elimination? Do we expect elimination to be faster with increasing alkyl substitution?

[9] We are immediately in a position to make a prediction. If the experimental data does not match the prediction, you can then refine your mechanism.

We will only consider the major alkene product from each reaction. Alkene **9** is more stable than alkene **5**, which in turn is more stable than alkene **2**.

*More substituted alkenes are more stable (***Basics 34***).*

So, we expect the product stability of the alkene to increase upon going from compound **2**, to compound **5**, to compound **9**. We expect the product stability of the ether to decrease upon going from compound **3** to compound **7** to compound **11**.

This doesn't quite tell us about the relative rates of the three elimination reactions, but it is a step in the right direction.

We really need to be thinking about the transition states for the elimination reactions. We expect formation of the product from the carbocation to be *exothermic*. Therefore, the transition state leading to the product will resemble the carbocation more than it does the product.

This is the Hammond postulate again, **Basics 19**.[10]

So, we can make our lives a little easier by thinking about the carbocations, and putting them in the correct orientation to lose a proton. We already did that in **ANSWER 1**, but here they are again.

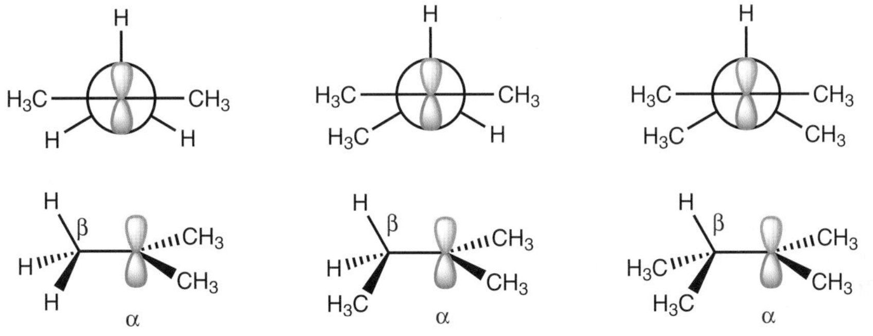

12 **13** **14**

We expect carbocation **14** to be the least stable, due to steric crowding, but this is by a relatively small margin.

As we form an alkene, the hybridization changes. The bond angles on the β-carbon change from 109.5° to 120°. This represents a relief of steric strain at the β-carbon which becomes more significant as this carbon atom becomes more substituted.

The substituents on this carbon atom want to get further apart!

[10] I *really* like the Hammond postulate. Can you tell? I don't think you will appreciate why at first, and then you'll 'get it'. Then you'll understand why I like it!

Think about this in terms of the progress along the reaction coordinate. As you proceed from the carbocation to the transition state, the more substituted carbocation (**14**) will have a lower barrier.

> *The change in geometry is already starting to happen as we move towards the transition state (Basics 23).*

The transition state does not, therefore, have much 'double bond character', but it will have some. There will be **some** stabilization of the transition state as the π bond is starting to form.

In conclusion, the transition state for formation of the more substituted double bond is lowered in energy due to two distinct factors—relief of steric strain as the hybridization of the β-carbon atom changes, and electronic stabilization as the π bond starts to form.

So we do expect elimination to become more favoured for compound **8** compared to compound **1**.

ANSWER 2 REFLECTION

Couldn't you just consider the stability of the products in each case to justify this answer? It certainly gets you where you need to be. It just isn't as rigorous as considering the transition states.

It does take a while to always think about transition states, but then it becomes natural. You need to appreciate why it is necessary to make this shift.

We can even take this a little further.

The carbocation intermediate is planar, sp^2 hybridized. The alkene product from an elimination reaction has two sp^2 hybridized carbon atoms. By definition, the alkene has eclipsed bonds.

What happens if the substituents are too big? Do we reach a point where the amount of energy we gain by moving from sp^3 hybridized (109.5°) to sp^2 hybridized (120°) on the β-carbon is more than offset by the energy we lose by having the bulky substituents on the alkene eclipsed?

Now we are thinking like organic chemists! Don't get me wrong, this is actually quite deep thinking. The answer to the question might not be that straightforward, but the process behind it is not too bad once you are used to it.

Of course, if the substituents are 'too big' to form the alkene, then they will be too big and will also hinder attack of the nucleophile.

Once again, we have a prediction. Adding more substituents will slow down S_N1 substitution and speed up E1 elimination, right up to the point where it also starts to slow down E1 elimination. At that point, it isn't at all easy to predict which pathway will be favoured.

ANSWER 3

This **might** look like an unfair question. There has been nothing in the discussion so far that would seem to relate to temperature. We do have the equation $\Delta G = \Delta H - T\Delta S$. An elimination reaction results in an increase in the number of molecules. A substitution reaction does not. Therefore, the entropy change for a substitution reaction will be close to zero, while that for an elimination reaction will be positive.

Elimination reactions become more exergonic as temperature increases. Therefore, the proportion of elimination will increase.

ANSWER 3 REFLECTION

You don't need to know **much** basic physical chemistry to be good at organic chemistry, but you do need to know the really fundamental equations. It is important, when you are carrying out reactions, to consider whether changing parameters such as temperature or pressure will be beneficial.

A GOOD TIME TO RECAP

We have applied quite a lot of basic knowledge here. This isn't easy. The key to success is to always use this systematic approach, and to consider all aspects. If you get used to doing this, it will become second nature. If you don't get used to it, you will always be memorizing reasons.

Trust me, it's worth making the effort in the early stages in order to make the later stages of your organic chemistry learning a lot easier.

SOLUTION TO PROBLEM 3

Competing S_N2 Substitution and E2 Elimination

ANSWER 1

Here are the resonance forms, showing that the lone pair on the nitrogen in the seven-membered ring is conjugated with the C=N double bond. It isn't available.

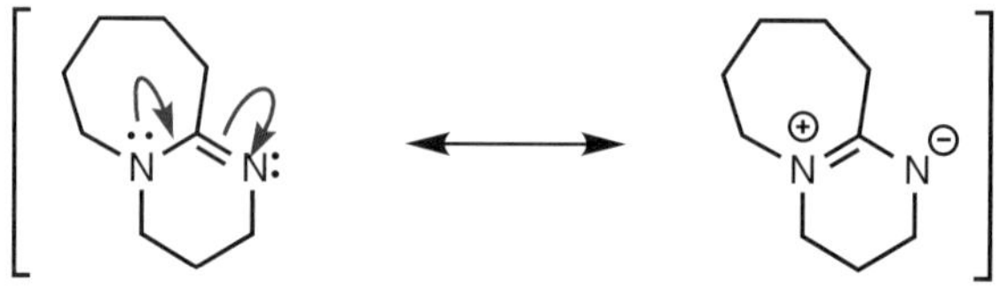

Conjugation and resonance are covered in **Basics 10**.

You cannot draw any meaningful, correct curly arrows in which the lone pair on the nitrogen atom on the right overlaps with anything else! This lone pair is available for bonding!

So, if we want to draw curly arrows for DBU acting as a base, we should start the curly arrow on the lone pair on the right.

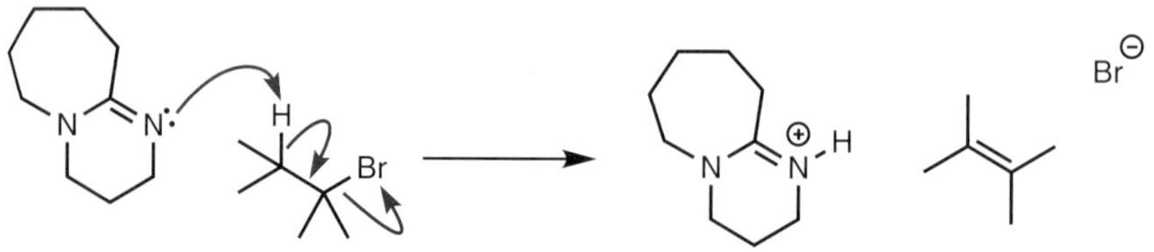

This might all seem excessively pedantic. Does it really matter which lone pair you use as the base?

The point is that you will encounter many more structures, some of which are far more complicated. You need to have the right habits in place.

ANSWER 1 REFLECTION

Did you draw the resonance forms for DBU? If you didn't, is it because you didn't think about it, or because you knew exactly what to do instinctively?

*Inexperienced students might not consider the need to draw the resonance forms. Gaining experience is all about drawing **everything** out carefully.*

But after a while, you might find you can visualize the orbital overlap that the resonance forms represent, so you draw the correct curly arrows without really thinking about it. I'd say this is a good thing.

The important thing at this stage is to know exactly where you are along this journey.

ANSWER 2

Here is the curly arrow mechanism for substitution.

I have explicitly drawn in four hydrogen atoms. We called Section 5 'Bonds Can Rotate'.

Well, these ones can't!

That's the benefit of using a rigid bicyclic base. Structure **2** looks incredibly crowded. This cannot be favourable.

If formation of compound **2** is *endothermic* then the transition state for the formation of compound **2** will resemble compound **2**.

*That sounds like the Hammond postulate (**Basics 19**) to me!*

Compound **2** is crowded. The transition state for the formation of compound **2** will therefore also be crowded, and high in energy.

If the transition state is high in energy, compound **2** will be formed slowly.

Draw a reaction profile for the formation of compound **2**, in order to reinforce this.

ANSWER 2 REFLECTION

I think the key skill here is knowing when to draw the relevant hydrogen atoms in. Underpinning this is automatically knowing that they are there.

One of the most common reasons that students do poorly in organic chemistry is that they forget about the hydrogen atoms.

ANSWER 3

Ethanol favours carbocation formation, so with a *tertiary* halide, carbocation formation will be favoured.

In fact, we saw the reaction with ethanol in **Worked Problem 2**. It was S_N1/E1.

We have seen that the reaction with DBU is E2, but it would be good to formalize this.

DBU does not assist carbocation formation at all. The barrier to carbocation formation will be higher under these reaction conditions. Even if we did get a carbocation, we still don't have a nucleophile.

So, the fundamental point is that by raising the barrier to carbocation formation, and getting rid of the possibility of substitution reactions completely, you are left only with E2 elimination. E2 elimination reactions tend to be a little more selective for the more substituted alkene.

Of course, DBU is a base, and we didn't have a base in the reaction in ethanol. The addition of a base makes a deprotonation reaction more favourable, so we are also lowering the barrier to E2 elimination at the same time.

ANSWER 3 REFLECTION

There isn't a lot to add here. If you raise the barrier to one process, another process may become the more favoured. If you keep the barrier to one process the same, and lower the barrier to another process, it may then be the more favoured pathway.

ANSWER 4

In an E2 elimination, the transition state is halfway between the starting material and the product. Compare this to an E1 elimination, in which the transition state for alkene formation closely resembles the carbocation intermediate.

So, there is more alkene character in the transition state for E2 elimination reactions than in E1 elimination reactions. Alkene stability is more of a driving force in E2 eliminations.

Of course, this is explained in every good organic chemistry textbook, yet it is remarkably easy to forget. Fortunately, with a good understanding of the basics, you can 'reinvent' it without too much trouble.

ANSWER 4 REFLECTION

Take the time to consider a few more scenarios. What if the E2 elimination is particularly *exothermic*? What if it is *endothermic*? Draw reaction profiles and consider how the Hammond postulate can be applied in these situations.

In a particularly *exothermic* E2 elimination, the transition state will be significantly closer in energy to that of the substrate. In this case, the transition state will be much more 'substrate-like'.

ANSWER 5

As we remove H and Br, the molecule is becoming flattened. In the case of compound **4**, this is pushing the two Ph groups closer together. This requires energy, and so is slow (the activation barrier will be higher).

As I hinted, you do not 'recoup' all of this energy by overlap of the π orbitals.

With compound **6**, the interactions are Ph with H and Ph with Me. Since the methyl group is smaller, the elimination is less crowded (and you can therefore get better π overlap), and so is faster.

Here are the Newman projections from **Applications 4**.

particular crowding as Ph groups move closer together

H, Ph, H, Ph, Me, Br

4

H, H, Ph, Ph, Me, Br

6

ANSWER 5 REFLECTION

In this chapter, we have considered stereoisomers, conformers, Newman projections, steric hindrance, kinetics, thermodynamics, and orbital overlap. It might seem

3

complicated. Trust me. It gets easier with practice. The problem is, you won't get very far if you don't consider all relevant aspects. That's why we need to build those habits!

SUMMARIZING THE TRENDS

If we consider the following group of compounds:

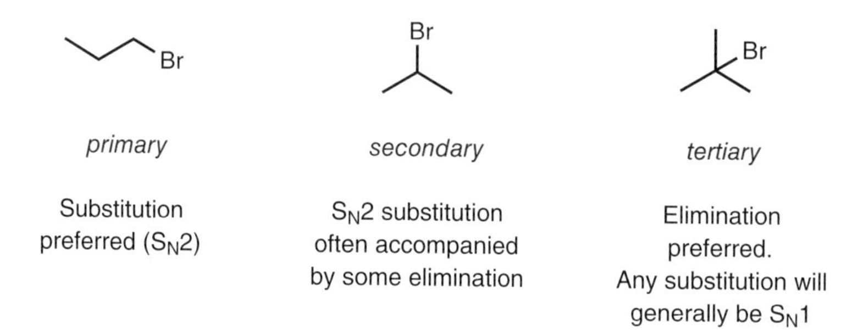

CLOSING REMARK

Getting the highest possible yield from a reaction is important. If you get mixtures of products, you have to separate them, and you get less of what you wanted. On an industrial scale, this costs money! Don't think of these discussions as being esoteric academic ideas.

3

SOLUTION TO PROBLEM 4

Acid Catalysis in Organic Reactions Part 2

ANSWER 1

Here is the equilibrium again.

$$\underset{\mathbf{2}}{R-\overset{\oplus}{O}H_2} \rightleftharpoons \underset{\mathbf{1}}{R-OH} + H^{\oplus}$$

Estimated pK_a −1.74

We know that

$$pK_a = -\log_{10} K_a$$

So that K_a for the above equilibrium will be $10^{1.74}$, which is 55.

We also know that

$$\Delta G = -RT \ln K$$

So, at 298 K, the protonated alcohol is about 10 kJ mol^{-1} (calculate it!) less stable than the unprotonated alcohol.

We now have the energy difference on the left-hand side of the reaction profile in the introduction to this worked problem.

ANSWER 1 REFLECTION

I said this was largely mathematical. It helps if you are comfortable with logarithms.

I often talk to my students about the 'shape of an equation'. What I mean by this is how the output changes with the input. The only way you will get used to this is by plugging some numbers into the equations.

ANSWER 2

Hydroxide plus a proton is approximately **90 kJ mol^{-1}** less stable than water. Make sure you can get this value using the above methodology. I'm not going to work through the calculation with you this time.

We now have the energy difference on the right-hand side of the reaction profile.

ANSWER 3

Before you can answer this question, you need to know where the transition state is along the reaction profile. Is it closer to the starting materials, or to the carbocation?

Remember that formation of the carbocation will be endothermic!

You need to start sketching a reaction profile. We will fully answer this in the next question.

ANSWER 3 REFLECTION

How did you do with this one? It isn't easy to tell, as I haven't given you the answer yet. But if you did okay, you will know already!

Did you start drawing? If so, you probably did okay!

ANSWER 4

Here is the equilibrium again.

$$\underset{\mathbf{1}}{\text{R}-\text{OH}} + \text{H}_3\text{O}^{\oplus} \rightleftharpoons \underset{\mathbf{2}}{\text{R}-\overset{\oplus}{\text{O}}\text{H}_2} + \text{H}_2\text{O}$$

We would calculate the equilibrium constant for this reaction to be approximately 1, assuming the hydronium ion and the protonated alcohol have the same pK_a. Of course, they won't be exactly the same, but they will be close enough for us not to worry.

The other way to look at it is that the two sides of the equation have the same stability/energy. Since we are trying to draw a reaction profile, this is much more productive.

Here is the profile.

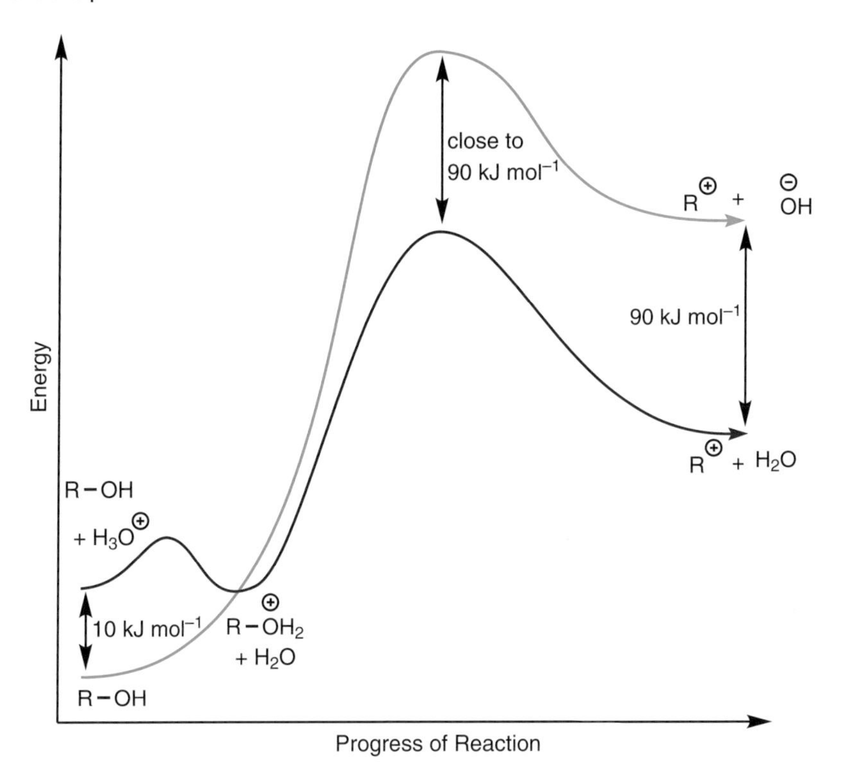

The reagents in the protonated profile (purple) are 10 kJ mol^{-1} less stable. The products are 90 kJ mol^{-1} more stable. Since the transition state more closely resembles the product, we expect the transition state in the acid-catalysed pathway to be almost 90 kJ mol^{-1} more stable (this answers **QUESTION 3**). Therefore, in this case the acid catalysis lowers the barrier by close to 100 kJ mol^{-1}.

This is a significant effect!

ANSWER 4 REFLECTION

None of this is easy. Perhaps there is a lot more maths in this than you were expecting. But you have to consider stability, and the sooner you do this in kJ mol^{-1}, the better.

For me, it's all about having more understanding and less to 'learn'.[11]

4

[11] By this, I mean, 'memorize without understanding it'.

ANSWER 5

Once again, we need to consider the position of the equilibrium for protonation of the alcohol. We are still going to assume that the alcohol has pK_a −1.74, and we have a 1:1 mixture of the alcohol and acetic acid.

$$\underset{\mathbf{1}}{R{-}OH + H_3C{-}C(=O)OH} \rightleftharpoons \underset{\mathbf{2}}{R{-}OH_2^{\oplus} + H_3C{-}C(=O)O^{\ominus}}$$

First of all, this equilibrium will definitely be towards the left. This means that ΔG will be positive.

It is really important to establish facts such as these before you do the calculation. If you don't do this, you will simply be trusting your fingers and your calculator!

The difference in pK_a values is 6.5 units. Therefore, $K_a = 10^{-6.5} = 3.2 \times 10^{-7}$.

There won't be much alcohol protonated at any given time!

The other way to look at this is that the protonated alcohol **2** is a strong acid, and it can readily protonate the acetate anion.

$$\Delta G = -RT \ln K$$

Therefore, we calculate ΔG = 37 kJ mol^{-1} at 298 K. Here is the resulting energy profile.

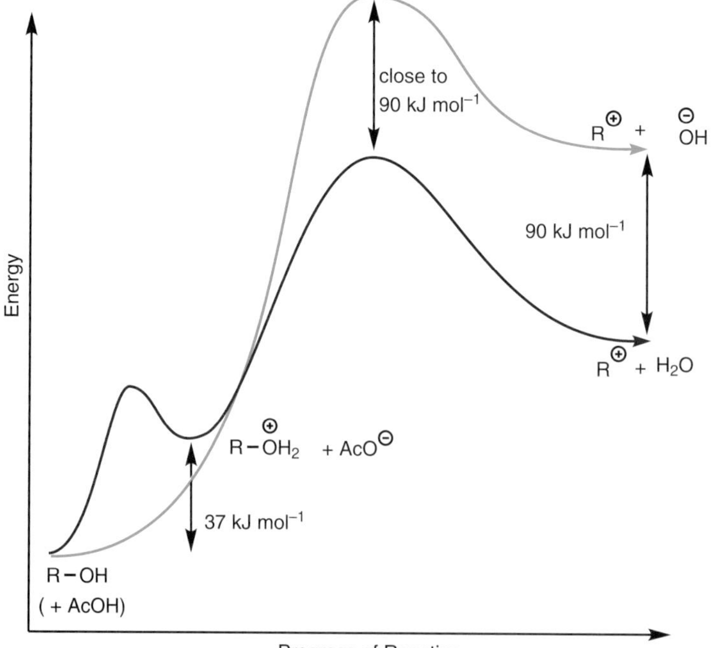

What we can see is that the 90 kJ mol^{-1} trumps everything else. Even with the weaker acid, carbocation formation will still be faster.

Note that we aren't necessarily assuming that the alcohol and acetic acid have the same energy as the alcohol alone. We have two different profiles with two different barriers.

ANSWER 5 REFLECTION

You won't need to do calculations such as these very often.

Personally, I didn't find this easy at all, as I haven't done one of these calculations in many years!

The point is, if you understand the basic principles and the key equations, you can do quite a lot with a limited amount of information. You just need to know exactly what to do.

Of course, once I got started, it seemed to get easier!

You should make sure you understand all of the methodology in this chapter, and you can refer back to it whenever you need it.

SOLUTION TO PROBLEM 5

Epoxide Opening Reactions

ANSWER 1

We haven't really covered this before, but we know that an alkyne hydrogen is moderately acidic (pK_a 25).

We saw this in **Basics 18**.

Yes, it's difficult to keep track of all the things you have to 'remember'. That's why you need to build things up gradually, so that these 'facts' simply become part of your natural process. Here is a curly arrow mechanism.

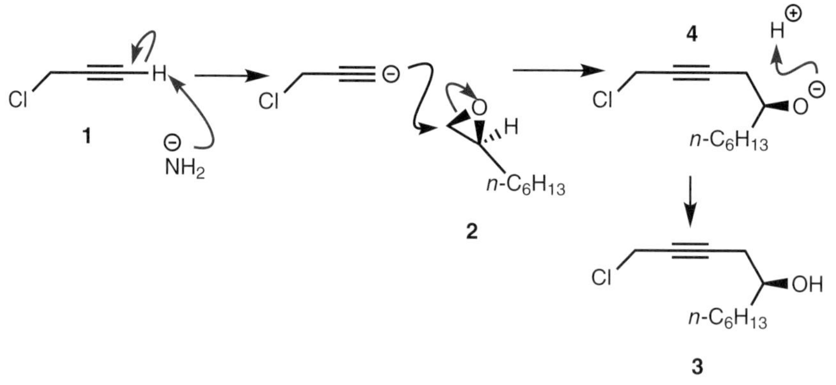

ANSWER 1 REFLECTION

The most common error on problems of this type is not to use the base to deprotonate the alkyne.

You are then wondering 'which electrons do I use?' to attack the epoxide. And then it all goes badly wrong.

At this point, you need to take a little time to analyse the problem and look at your own solution. If you made this mistake, draw out formation of the acetylide anion a few times, just to reinforce it.

ANSWER 2

You need a base with a 'higher pK_a' than 25.[12] Sodium amide ($NaNH_2$) is quite commonly used to deprotonate an alkyne. The 'default' base in many transformations is butyllithium, which you can think of as a butyl anion and a lithium cation.

When the 'butyl anion' deprotonates the alkyne, butane is generated as a by-product. This is a gas and will be lost from the reaction. Therefore, butyllithium is a very 'clean' base.

In terms of solvents to use, you want something that you cannot deprotonate. Protic solvents such as ethanol are completely out of the question, but a solvent such as acetone also has acidic hydrogen atoms. The most commonly used solvent in these transformations is tetrahydrofuran (THF).

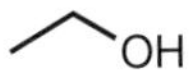

OH O O

ethanol 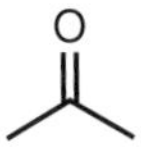 acetone  tetrahydrofuran (THF)

Ethanol has a pK_a of about 16. Acetone has a pK_a of 19. We consider ethanol to be a protic solvent, as it has a hydrogen atom attached to oxygen or nitrogen. Acetone is only slightly less acidic than ethanol, and is considerably more acidic than ammonia. Learn the definition but understand the principles!

ANSWER 3

The absolute configuration is *S*. Make sure you can see this. At this level, I am not going to tell you the priorities of each substituent.

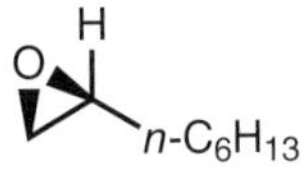

O H *n*-C_6H_{13} H *n*-C_6H_{13} O

2 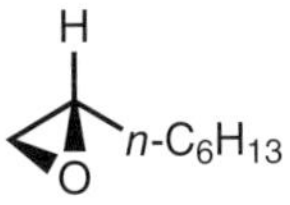 **2**

If you feel you need this level of help, working through the problem will take you a little longer, and you will then benefit far more than me giving you a direct answer.

[12] As we have noted before, when we talk about the pK_a of a base, we are really talking about the pK_a of the conjugate acid. In the present discussion, ammonia (the conjugate acid of the amide anion) has a pK_a of 38. Butane has a pK_a of >50, but butyllithium isn't *really* a butyl anion.

ANSWER 4

Consider how the negative charge that is building in the S_N2 transition state is destabilized by a hexyl group.

> *Everything you need is in* **Reaction Detail 1**.

ANSWER 5

Don't over-think this one. For the alkoxide to attack the carbon and displace the chloride, it has to be able to get close to it. Make a model of the compound and you will hopefully see that there is no way the oxygen atom can get close to the relevant carbon atom.

We will be exploring this aspect in more detail in ***Worked Problem 8***.

The other way to look at it is 'can you have a triple bond in a six-membered ring?'

SOLUTION TO PROBLEM 6

Is *cis*-Cyclohexane-1,2-diol Really Achiral?

ANSWER

As usual, it's all about the method—making systematic adjustments to how you think about organic chemistry so that you see all aspects of a problem. In this case, we are asking about chirality, so we need to think about shape. These are cyclohexanes, so we need to think about chair representations.

Here are the original structures again.

OH OH OH OH

Here are the chair forms for the structure on the left.

They relate to the structure on the left in the sense that both OH groups are 'down' in each structure.

HO OH OH OH

1 **2**

They are also the representations of the structure on the right if we rotate them 180° so that the OH groups are up instead of down.

Each structure is a chair with one equatorial OH and one axial OH. They definitely have the same energy.

The structures can be interconverted by flipping the chair. They are conformers, since we do not need to break any bonds in order to interconvert them.

But look closer. Structure **1** is the mirror image of structure **2**. Furthermore, they are not superimposable unless we rotate bonds to flip the chair form.

*When considered as static structures, these are enantiomers! If we couldn't flip the chair forms, structures **1** and **2** would represent different molecules.*

HOW DO WE INVERT THE CHAIR CONFORMERS OF CYCLOHEXANES?

You have done this with your molecular models. You would have flipped one CH_2 group on one side of the molecule, and then another CH_2 group on the other side of the molecule. You would find it very difficult to flip both of them at the same time.

We saw in Basics 33 that if we rotate one CH_2 in a cyclohexane, we get a **boat** conformer. Very often, this boat is distorted slightly to give us a conformer we would refer to as a **twist boat**.

Don't worry too much about this distinction. Focus on the principles. Cyclohexane chairs are more stable because they minimize torsional interactions between bonds. Cyclohexane boats are higher in energy because there are lots of eclipsed bonds. A small twist removes some of the eclipsing, so that the structure is most like a boat, but is slightly lower in energy.

Thinking about how you would do this with a molecular model is useful, because it gives you an insight into what happens at a molecular level. This helps when we try to calculate the energy barriers.

CALCULATED BARRIERS TO RING INVERSION

Here is the data, calculated rather than measured. The twist boat conformer in the middle is 28 kJ mol^{-1} higher in energy than either chair conformer on the left and right. The highest point on the profile (yes, this is a transition state!) is 44 kJ mol^{-1} higher in energy. This represents the barrier to interconversion of the two 'enantiomers'.

This barrier is significant, but it is accessible at room temperature. The 'enantiomers' will interconvert, so you would not be able to separate them or to prepare a single enantiomer in this case.

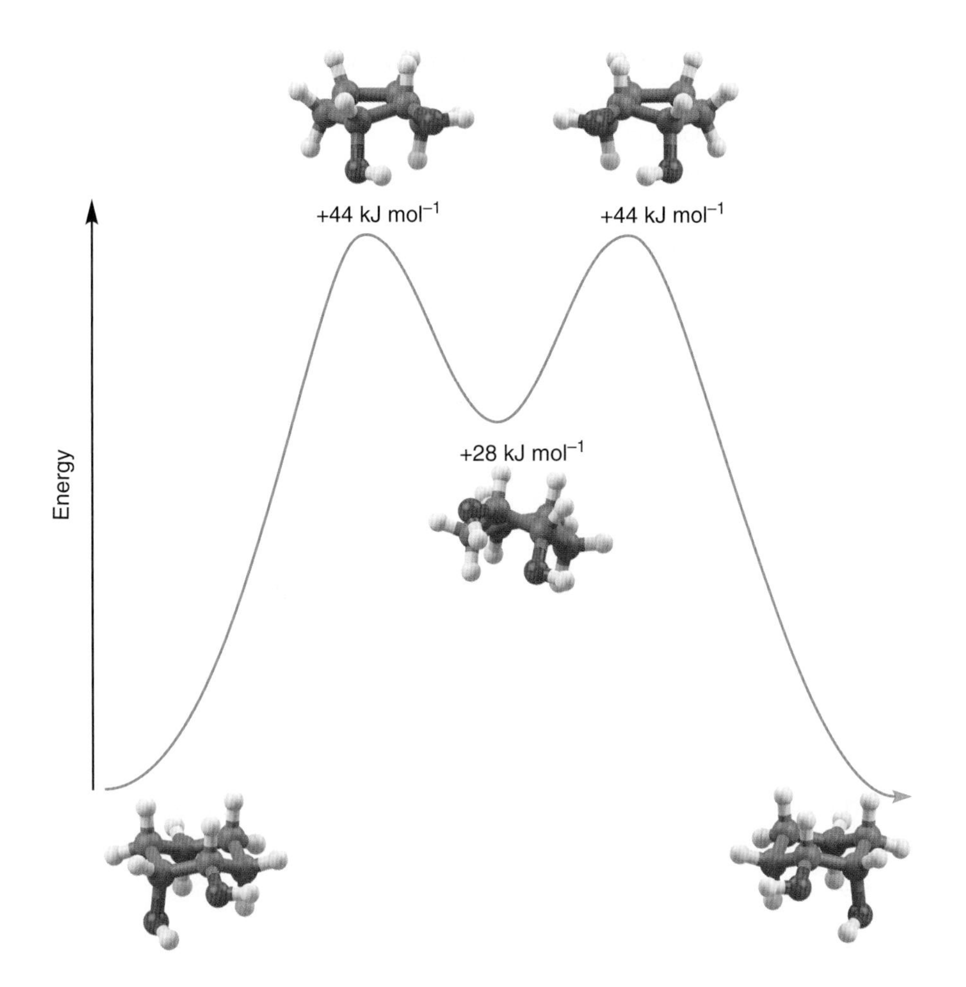

IN CONCLUSION

Yes, *cis*-cyclohexane-1,2-diol is really achiral, in a dynamic sense because the barrier to interconversion of the two chair conformers is low enough for this to be a rapid process at room temperature.

It is therefore safe to look at the 'flat' representation of the molecule and see the symmetry.

If the barrier were higher, the ring flip would be slower, and we might be able to separate the two 'enantiomers'. If we cool down a *meso* cyclohexane of this type, the ring flip will also be slower.

Has this ever been done for a meso-1,2-disubstituted cyclohexane?

I don't think it has, but then again this is pretty esoteric and it isn't the easiest thing in the world to search for.

If it could be done, you would have to consider what to change in the structure in order to raise the barrier to the ring flipping. Then you would have to consider whether it was actually worth doing in the first place!

But it **is** a theoretical possibility, and many of the ideas we take for granted now started out as theoretical possibilities.

Here is a neat example of a compound that is chiral only because a bond rotation is slow (has a high barrier). As a result of hindered rotation around the bond between the two aromatic rings, the enantiomers of BINAP are configurationally stable.

'Configurationally stable' simply means that the two enantiomers are not being rapidly interconverted at a 'useful' temperature.

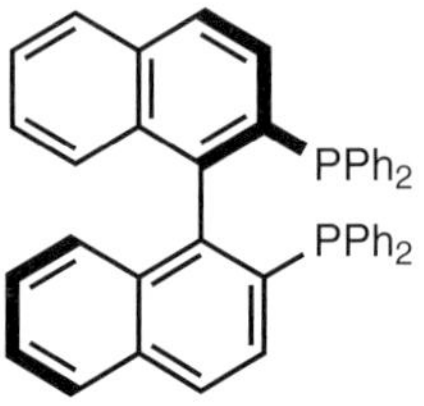

(R)-BINAP 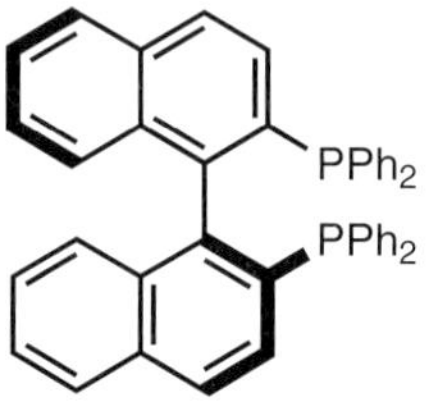 (S)-BINAP

Make a model of this, and see what is blocking the rotation!

If you coordinate the phosphorus atoms to a transition metal and use the resulting metal complex to catalyse organic reactions, the processes can be very stereoselective.

SOLUTION TO PROBLEM 7

The Fürst–Plattner Rule

ANSWER 1

We looked at epoxide opening reactions in **Worked Problem 5**.

In terms of mechanism, if we have an S_N1 mechanism, we would expect attack at carbon 1, since this is where we would form a more stable carbocation. Product **3** is formed by attack at carbon 2 with inversion of configuration.

It must be S_N2!

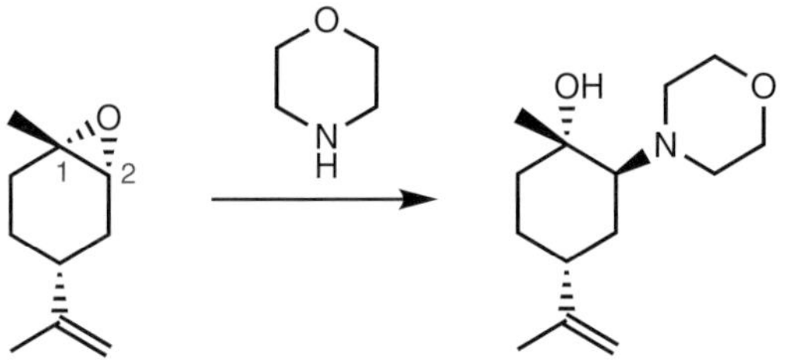

1 → **3**

ANSWER 2

Having established what the mechanism is, the curly arrows are pretty straightforward. A key point for me is how we deal with the proton transfer steps. We tend not to worry too much about what removes the proton from the nitrogen, and what delivers the proton to the oxygen. There are plenty of lone pairs!

Perhaps we can also draw the proton transfer directly from N to O. The transition state for this process has a five-membered ring, which is not unreasonable.

ANSWER 2 REFLECTION

There are a couple of points that we need to consider. It is quite common for students in the early stages to dissociate the amine as follows, so that they have a negatively charged nucleophile.

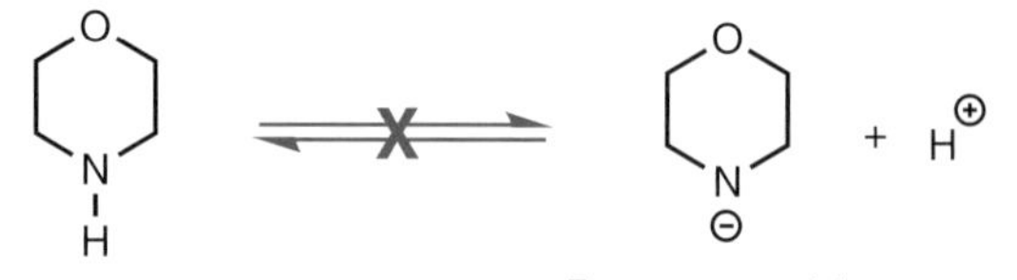

Far too unstable to happen spontaneously

There is absolutely no need to do this, and since the amine is not remotely acidic, it would be wrong to draw this.

You will also see in the final reaction scheme in **ANSWER 2** that I drew the morpholine ring as a chair.

It gets the bonds out of the way, and why use arbitrary, untidy, bond angles when we have a perfectly good representation to use?

ANSWER 3

Here are the structures, with the stereochemical descriptors shown. I am not going to give you a full worked solution.

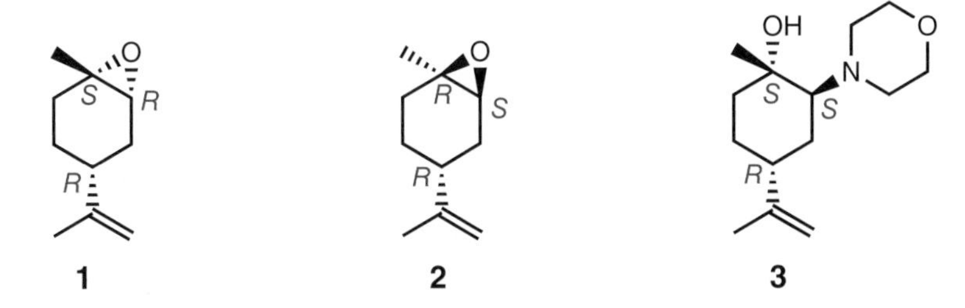

ANSWER 3 REFLECTION

Hopefully by now, you are starting to spot the stereochemical outcome of reactions without needing the descriptors. Some reactions are harder to follow than others, so it is still good to have a reliable backup plan.

ANSWER 4

Both of your conformers should have two axial substituents and two equatorial substituents. The morpholine substituent is largest, so this will be equatorial in the lowest energy conformer, shown on the left. The less stable conformer is on the right. Of course, you could have drawn either of these in several different orientations.

 Is there anything else that could favour the conformer on the left?

By now you must be getting used to these rubbish questions! The problem is, if I ask a 'better' question (such as 'could you draw a hydrogen bond?'!) I would have given you the answer.

In the structure on the left, the N and O look quite far apart. The structure below shows the proximity much better.

You need to look at the structure above left, and 'see' the structure below.

ANSWER 4 REFLECTION

Make sure the chair structures you have drawn are the same enantiomer as the flying wedge projections. You may need to assign stereogenic centres to make sure.

If you don't do this, following stereochemical changes through a reaction will be almost impossible.

ANSWER 5

Here are the conformers in each case.

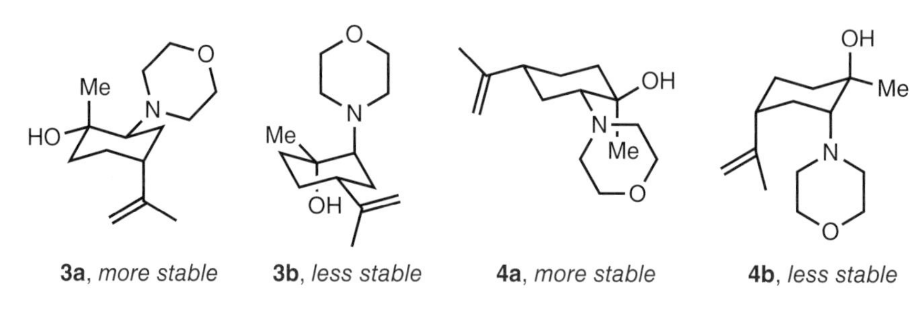

3a, *more stable* **3b**, *less stable* **4a**, *more stable* **4b**, *less stable*

Compound **3** Compound **4**

Conformer **4a** has three equatorial substituents and one axial. Conformer **4b** has three axial substituents and one equatorial. This is a very clear-cut case.

Conformers **3a** and **3b** each have two equatorial and two axial substituents. They will be considerably closer in energy, although **3a** has the largest group equatorial.

Hopefully by now you are used to drawing cyclohexanes. Perhaps this problem emphasizes the importance of developing this skill. If you are not confident, have another go at the problems in Practice 12 *and* Practice 13*.*

ANSWER 6

We have already done the hard work. This is an S_N2 substitution. The nucleophile **must** attack from the back of the leaving group. Therefore, we need a transition state that resembles conformer **3b**. Let's not worry about the energetics (it is the less stable conformer) for now.

3b $\rightarrow$ **1**•H^+ + **5**

Yes, we need the nucleophile and the leaving group to be anti periplanar, which means both of them must be axial.

We didn't worry about the energetics of this situation, because we didn't have anything to compare this reaction to. Now let's look at the same reaction of conformer **4b**.

4b $\rightarrow$ **2**•H^+ + **5**

There is a key difference here. In the reacting conformer of **4b**, the isopropenyl group is axial. If we were to form 'compound' **2**·H^+ from compound **4**, we would require compound **4** to adopt the higher energy conformer.

That is, we need its own higher energy conformer, which is less stable than the higher energy conformer, ***3b****, of compound* ***3****.*

ANSWER 6 REFLECTION

We are pretty much there now. There is going to be a little added subtlety, but we have applied basic knowledge to a problem. We have drawn sensible things at every stage, and we have reached the conclusion that the formation of **2**·H from amino alcohol **4** will have a higher barrier than the formation of **1**·H from amino alcohol **3**.

Of course this isn't the problem we set out to solve, but it is very close. In preparation for **QUESTION 7**, draw a reaction profile for these two reactions. Try to put conformers **3a** and **4a** on as well, including sensible energy differences.

Use the 'A values' in Basics 32.

I am not going to give you the answer to this. You will know if it is right.

ANSWER 7

The formation of **3** from **1**·H^+ is the reverse of the formation of **1**·H^+ from **3**.

This is an S_N2 substitution. It has a transition state and no intermediates.

We do have to make a minor refinement—protons! We saw a curly arrow mechanism for the formation of compound **3** from compound **1** in **ANSWER 1**.

Assuming we protonate epoxide **1** or **2** first, and then add an amide anion nucleophile, we would expect the formation of compound **3** to be more favourable than the formation of compound **4**.

We would reach exactly the same conclusion if the proton was on nitrogen rather than on oxygen!

Basically, that answers the question. The key point here is that we can look at the reaction in both directions. Sometimes looking at the reverse reaction allows insight that looking at the forward reaction does not.

ANSWER 7 REFLECTION

It would be tempting to look at compound **1** and compound **2** and to claim that in the case of compound **2** the nucleophile must attack from the same side as the isopropenyl group, which is more hindered.

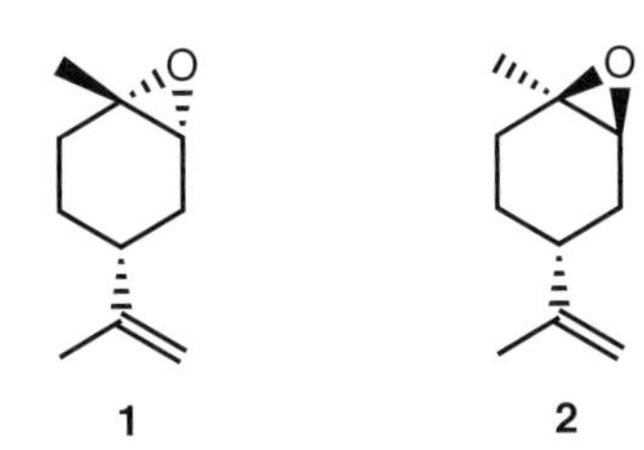

Here is the lowest energy conformer of compound **1**.

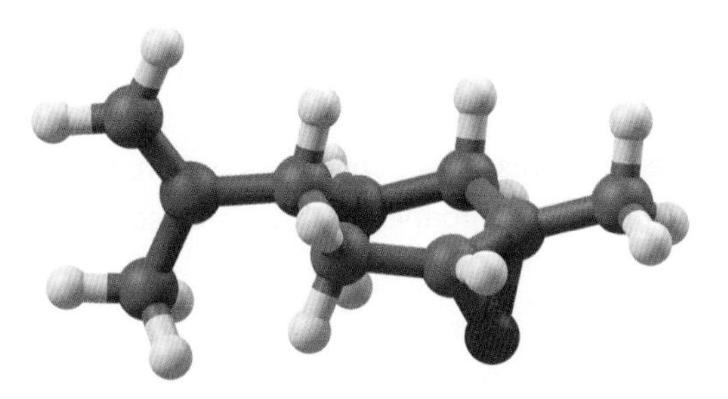

Hopefully you can see that the isopropenyl group is *pseudo* equatorial (the hydrogen atom on the same carbon is more or less axial), and it would not block the approach of a nucleophile.

Here is a similar representation of compound **2**. Again, the isopropenyl group is *pseudo* equatorial, and is nicely out of the way. It would be difficult to make a case for it blocking the attack of a nucleophile.

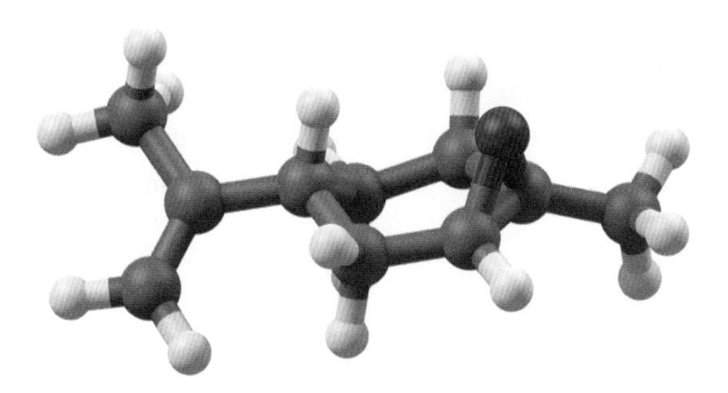

So, it definitely isn't a 'simple' steric reason. We really did need to draw the mechanisms and the conformers.

THE FÜRST–PLATTNER RULE

There is just one more level of subtlety as we formalize this as the Fürst–Plattner rule.

Everything we have done so far assumes that chair conformers are the only conformers we need to consider. This isn't true.

We have seen boat and twist boat conformers in Basics 33 and in **Worked Problem 6**.

Fundamentally, the Fürst–Plattner rule states that cyclohexene derivatives, such as epoxides, react with nucleophiles to give *trans* diaxial products.

This relates to the initial reaction. The product may undergo conformational change to give a product with equatorial substituents after the reaction.

Here is how it relates to the reactions in this chapter. First of all, the reaction of compound **1** is shown below. As the nucleophile attacks, the carbon atom labelled 1 moves up and the one labelled 2 moves down. This gives the product in a chair conformer, albeit not the most stable chair conformer for this stereoisomer.

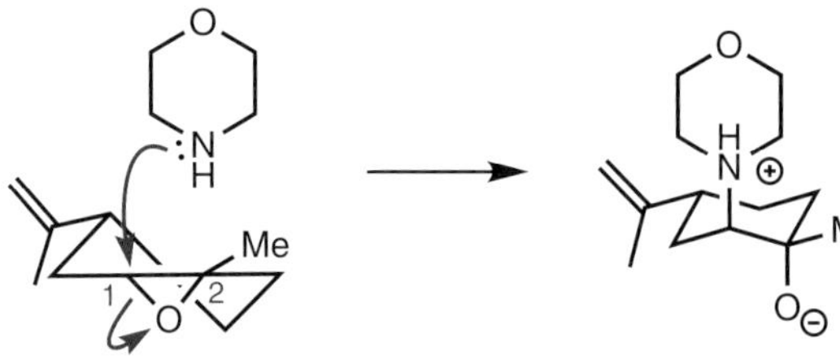

1

Now let's look at the corresponding reaction of compound **2**. As the nucleophile attacks, the carbon atom labelled 1 moves down and the one labelled 2 moves up. This gives the product in a twist boat conformer.

We saw, in **Worked Problem 6**, that a twist boat conformer can be in the region of 30 kJ mol^{-1} higher in energy than a chair conformer. This gives you an indication of the difference between the activation energy for compound **1** and compound **2** to undergo the same reaction.

2

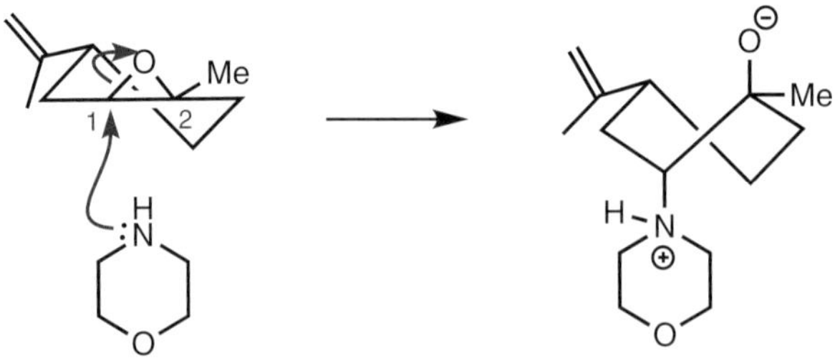

This is slightly more complicated than in **ANSWER 6** and **ANSWER 7**. The added complication is that the first explanation assumes a less stable chair conformer with an axial isopropenyl group, and the 'better' explanation finds that there is an alternative twist boat conformer that isn't good, but is better than this chair conformer.

Spotting the potential for 'alternative' conformers isn't easy, and it takes experience. However, you will only gain experience if someone first points out the potential for this.

IN CLOSING

This is a tough chapter. I can guarantee that you won't have followed everything the first (or even second) time through. That's okay. The main point is still to draw curly arrow mechanisms and cyclohexane chairs well. You have had a little more practice assigning stereogenic centres. We have added some discussion of pK_a values, and linked it with leaving-group ability.

The icing on the cake is consideration of reactions going forwards and backwards as two sides of the same coin. Having a problem like this is a good way to gauge your progress. Each time you work through it, you will 'get' a little more. The important thing is to keep going over the problem until you are comfortable with it, but to accept that it is perfectly okay if this takes some time.

SOLUTION TO PROBLEM 8

S_N2' Stereochemistry and Conformations

ANSWER 1

Here it is, without any stereochemistry. We don't yet know which double bond isomer or OH stereoisomer will be formed, so I have used wiggly bonds to indicate ambiguity.

Ph, Cl, CH_3, $HO^{\ominus}$ → Ph, CH_3, OH

1

ANSWER 2

There are four possible products given the constraints we imposed on the question—two products from each compound.

Here is the conformer of compound **1** in which the Cl is 'up'.

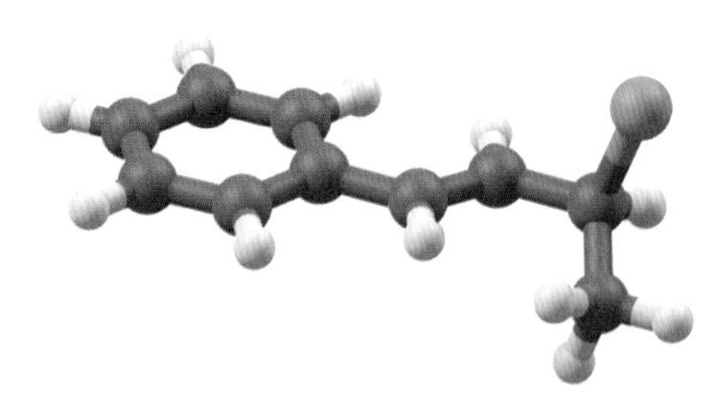

If hydroxide attacks opposite to the Cl leaving group, it must attack the lower face of the alkene double bond in the orientation shown. This has very clear stereochemical consequences.

Make a model and work with it until you can visualize the outcome.

As we lose the Cl, the carbon to which it is attached becomes flattened. We are forming a π bond, so the methyl group cannot move. We will get a (*Z*) double bond, because the methyl group is on the same side as the alkene carbon that is being attacked by hydroxide. Here is the stereochemical outcome for this reaction.

Ph, Cl, CH_3 **1** $\xrightarrow{\ominus OH}$ Ph, OH, CH_3 **3**

It's horrible if you can't see ***why*** *this is the outcome. You need to connect the conformer in the 3D representation with the stereoisomer and double bond isomer shown in the reaction scheme above. There will be a point where you just accept this, but for now we will work through the other conformer.*

Here is the conformer with the C–C bond rotated so that the Cl is 'down'.

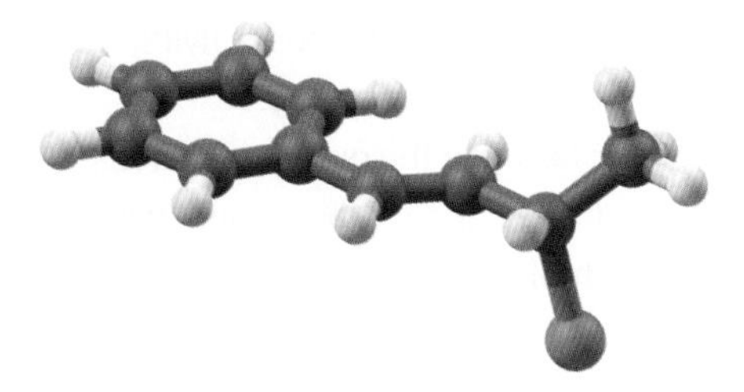

Remember, it is the ***same*** *stereoisomer, but a different conformer.*

If the hydroxide attacks opposite to the Cl group, it must now attack the upper face of the alkene double bond.

Because we are comparing attack of the upper and lower faces of the double bond with the molecule in the same orientation, we can be confident that attack on the upper face will give a different stereochemistry to attack on the lower face!

We should also look at the position of the methyl group as the Cl leaves. Here is the complete reaction scheme.

Ph, Cl, CH_3 **1** $\xrightarrow{\ominus OH}$ Ph, CH_3, OH **4**

We get product **4**, which has a different stereochemistry at the OH group and a different double bond geometry.

ANSWER 2 REFLECTION

Remember, we are talking about a reaction type that works, but an example that does not necessarily give the outcome we are considering. We can expect some direct S_N2 displacement of the Cl.

Perhaps we can also expect some attack on the same side of the double bond as the leaving group. This reaction is not quite as constrained as an E2 elimination reaction, although I hope you can see the similarities.

We are focusing on developing a key skill—understanding the stereochemical outcomes possible in reactions where bonds can rotate.

We did this for elimination reactions in **Applications 4**.

We are emphasizing the methods (molecular models, flying wedge projections, Newman projections) you can use to work out the outcome.

The main point is that I can show you the outcome of a reaction, and you can learn it, but if you understand the principles and methods, you will have less to learn, and you will be able to apply the same tools to many other reactions.

After a while, you will 'just see it' and you won't need to use your molecular models. But don't underestimate the value of using your molecular models to get you to this point.

ANSWER 3

The distinction between product stability and transition state stability may seem subtle or pedantic, but again, it is one of the adjustments you need to make to your process. Once you make this adjustment, considering the factors that stabilize or destabilize transition states is no more difficult than considering the same factors for starting materials or products.

The 'required' conformer with the Cl pointing up has the CH_3 group close to one of the alkene hydrogen atoms. This is an unfavourable steric interaction.

As the Cl leaves and the carbon becomes sp^2 hybridized, this gets even worse.

We have no such interaction for the formation of compound **4**. Therefore, we expect compound **4** to be favoured.

There is a specific name for the interaction of a substituent in an allylic position with a substituent on an alkene. It is $A^{1,3}$ strain. You don't need to know this in order to anticipate it.

Here is structure **3** again. The CH_3 on the double bond cannot rotate out of the way. The best we can do is rotate the bond to the stereogenic centre to place the smaller H close to this CH_3.

Ph
HO
H)(CH_3

3

ANSWER 4

Don't over-think this. Just do the same as you did with compound **1**. Here is the answer, on a single reaction scheme.

Cl, Ph, CH_3 → Ph, OH, CH_3 + Ph, CH_3, OH

2 **5** **6**

*Note that the orientations of compounds **5** and **6** are not the easiest to see. This is deliberate, so you have to work for it!*

ANSWER 5

Compound **1** gives products **3** and **4**. Compound **2** gives products **5** and **6**.

We expect (*Z*) alkene **3** to be formed more slowly than (*E*) alkene **4**.

We expect (*Z*) alkene **5** to be formed more slowly than (*E*) alkene **6**.

You cannot work out which reaction will be more selective until you have worked out which will be the major and minor products in each case.

Now let's look a bit deeper. Here is structure **2** with the Cl rotated below the plane of the double bond. This will give product **6**. This doesn't look great. The CH adjacent to Cl is very close to a hydrogen atom on the benzene ring.

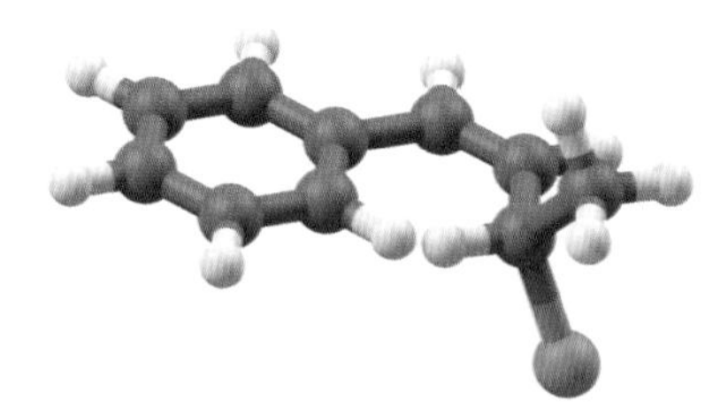

But now look at the alternative conformer, which would lead to product **5**.

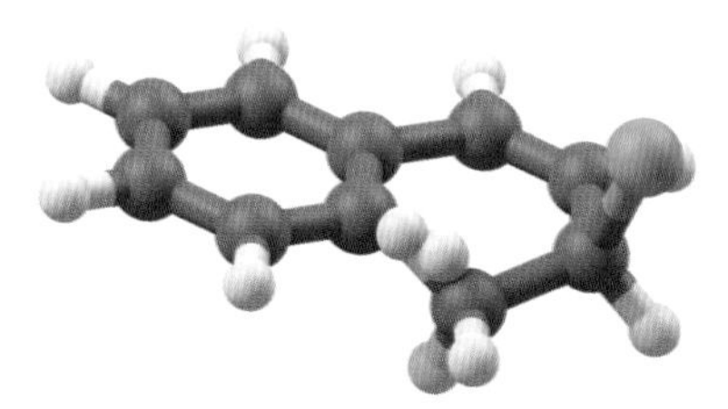

I can't even fit it all in on the model. The CH_3 and the phenyl groups are actually overlapping. This is horrific! You could relieve some of this strain by rotating the phenyl group as follows.

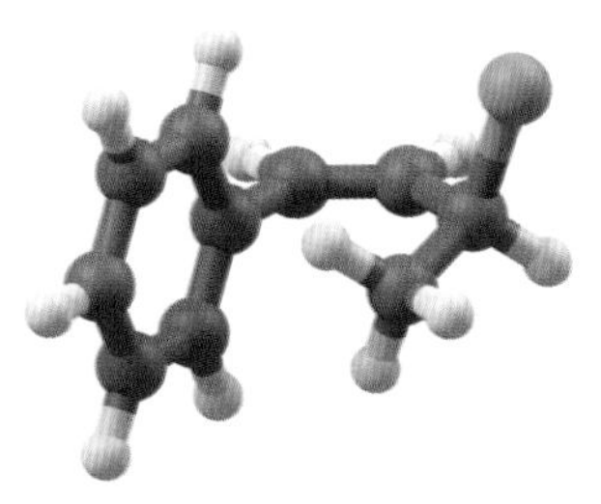

If you used your molecular models, you probably realized this for yourself.

But a phenyl group isn't as 'thin' as your molecular models make it seem. You have a cloud of electron density above and below the plane of the ring. It isn't quite a cube, but it's not far off!

We would be surprised to get any of product **5**. Therefore, this reaction will be very selective for product **6**.

Of the two reactions, reaction of compound **2** will be more selective, in terms of giving a higher yield of one product.

ANSWER 5 REFLECTION

Once again, this is a difficult series of problems. It is also quite artificial. Now you have done the difficult work, let's take a step back and emphasize what it was all about.

> *Some reactions require a particular conformational isomer for reaction to take place. It is important to consider which conformer is required for reaction. The simplest example of this is an elimination reaction (***Applications 4***) but you should be looking to apply the principles in all cases. How do you know it isn't relevant unless you check?*
>
> *If a reaction takes place only from a less stable (higher energy) conformer, the energy difference between the most stable and reacting conformer is effectively added to the energy barrier. We saw this idea for elimination reactions in* **Worked Problem 3***. We applied the same principles in* **Worked Problem 7***.*

It doesn't matter if you don't understand the S_N2' reaction. What matters is that you have worked through the problem, and you are more aware of conformational isomers in a wider range of compounds.

Also, be aware of the limitations of the discussion. Don't forget that we imposed significant constraints on the problem. We cannot predict how much S_N2 substitution and how much S_N2' we will get.

But we could potentially look at a series of similar reactions and decide whether each individual process will be more favoured or less favoured compared to a reference point.

SOLUTION TO PROBLEM 9

Complex Substitution Stereochemistry

ANSWER 1

Here they are.

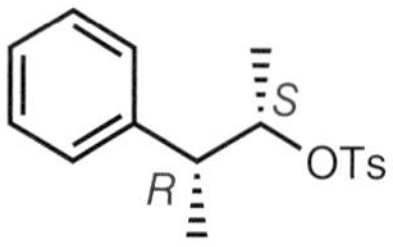

OTs | OAc | OAc

1 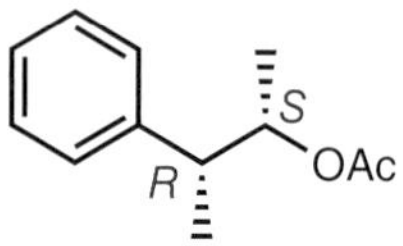 | 2 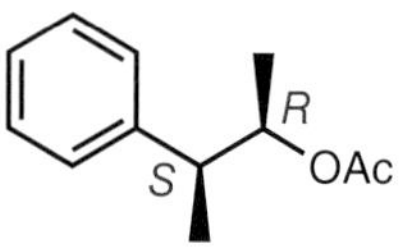 | 3

I only really needed to assign the stereochemistry in one of the compounds. I then did the other two by inspection.

I made a very deliberate choice about which to assign. Can you work out which one, and why?

ANSWER 1 REFLECTION

The easiest stereogenic centres to assign are those with the H pointing away from you. You should recognize these, even when the H are not explicitly drawn.

ANSWER 2

If in doubt, draw it out! You would form carbocation **6**. The nucleophile would attack from either face. There is existing chirality in the carbocation, so that attack on either face gives rise to diastereoisomeric products with potentially different energies. The products would be **2** and **7**, not **2** and **3**.

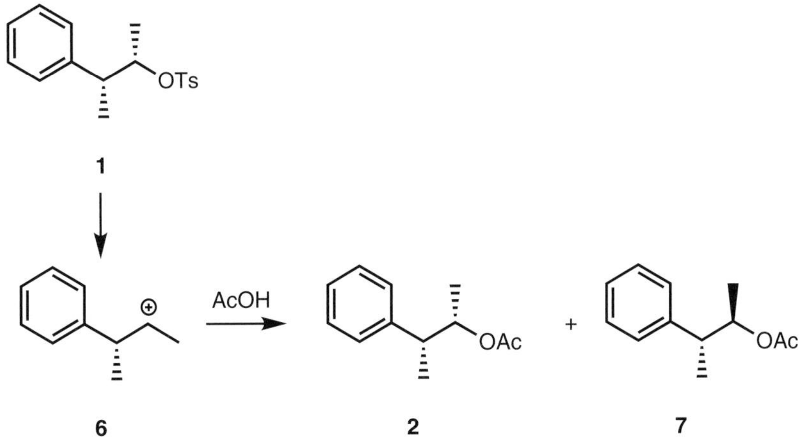

ANTICIPATED 'SIMPLE' S_N1 OUTCOME

We are in the territory of Basics 28.

If this mechanism cannot produce compound **3**, then it cannot be correct. Anyway, carbocation **6** isn't very stable.

In Reaction Detail 2 I warned about linking the S_N1 mechanism automatically with the word 'racemization'. If you do that here, you will leap to the wrong conclusion.

ANSWER 3

We have seen substitution reactions that proceed with retention of configuration before. The most likely scenario is inversion followed by inversion.

Now we should be looking at Applications 2.

We aren't looking at this because it is necessarily the right answer. We are looking at it because the last time we saw a similar stereochemical outcome, it was the right answer.

Again, it's about building patterns.

The substrate **1** doesn't look very much like the compounds we encountered in Applications 2. They all had a nucleophilic atom on the adjacent carbon to the leaving group.

We don't have an obvious nucleophile, but we do have a nucleophile! We can draw a curly arrow from a benzene ring π bond. This gives us intermediate **4**. We need to be careful drawing the stereochemistry of this intermediate.

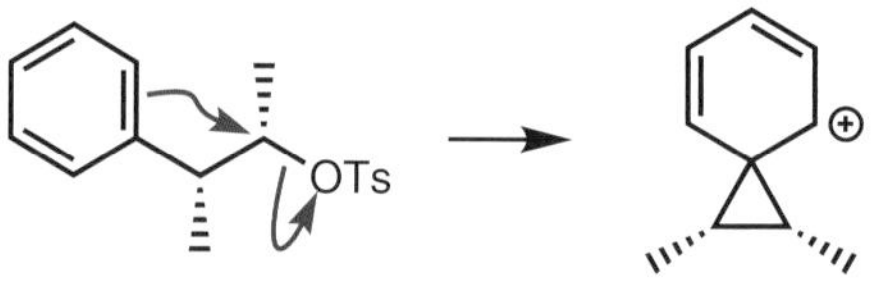

Can we use the stereochemical descriptors?

We can, but we do need to be careful. The starting material (**1**) is (1*S*,2*R*). The intermediate **4** is (1*R*,2*S*), following the numbering below. But stereogenic centre 2 hasn't changed in any meaningful sense.

It's just the Cahn–Ingold–Prelog rules. When we lose the oxygen, the priorities change!

ANSWER 4

Here is another resonance form of carbocation **4**. In this representation, hopefully you can see a plane of symmetry.

We have a molecule with a plane of symmetry, and two identical stereogenic centres, but with opposite stereochemistry.

Now we should be looking at *Habit 7**.*

Yes, this is a *meso* compound.

ANSWER 4 REFLECTION

At this stage, you are probably recognizing the scope of the challenge. You have to be able to pull information from lots of different parts of the book in order to solve a moderately complicated problem.

I can't say anything that will make it easier for you. All I can do is reassure you (yet again!) that it will become easier if you keep reading and drawing, and working through the problems.

ANSWER 5

In **QUESTION 4**, we established that intermediate **4** is symmetrical. There should be absolutely no preference for attack at carbon 1 or at carbon 2. We expect a 1:1 mixture. Now, what product do we form in each case?

This is where the 'key skills' come into play. You follow the stereochemistry. Carefully! Every time!

I would say attack at carbon 1 is easiest to see. We know it was inverted when we formed **4**, and it will be inverted again. This will give stereoisomer **2**.

I know I am using acetate as the nucleophile rather than acetic acid. Acetate is a poor nucleophile. Acetic acid is even worse! We could get into a debate about how much acetate is present in acetic acid, or we could just draw curly arrows and follow the stereochemistry!

What about attack at carbon 2? Here it is!

It gives enantiomer **3**, although it is drawn in a different orientation above.

*So, the natural, **inevitable** outcome of this mechanism is that you get a 1:1 mixture of enantiomers **2** and **3**.*

This is the accepted mechanism/explanation. Of course, that doesn't automatically make it correct. I always advise running through a few checks.

Are all steps in the mechanism plausible? Do the curly arrows make sense? Do you have any particularly unstable intermediates?

If it all seems good, run with it until and unless some data contradicts your proposed mechanism.

ANSWER 5 REFLECTION

It is really important that you understand why I have included this example. Your lecturer probably included this example, or one very much like it, in your course. At first, you will almost certainly learn it by rote. With experience, you will recognize the same patterns in other molecules. What I am trying to do here is to kick-start the process of experience. If you can understand and apply the basic principles correctly, your lecturer won't be able to tell whether you have learned the reaction well, or whether you are making it up as you go along! After all, organic chemistry is very predictable!

ANSWER 6

There is no doubt that we have an intermediate formed, and this reacts further with a nucleophile. Therefore, the reaction profile will look like an S_N1 reaction.

Have another look at **Perspective 4**.

We have a polar protic solvent, which will favour carbocation formation. We start to break the C–O bond in compound **1**. We build the positive charge on carbon 1, but the eventual outcome is that we form the carbocation **4**.

Draw a reaction profile to show this.

Instead of forming the unstable carbocation **6**, the benzene ring can donate electron density that stabilizes the charge as it builds. This stabilization is represented by carbocation **4**.

We have lost the aromaticity in the ring by this point. We expect the activation barrier to be quite high.

The very specific nature of this stabilization has consequences that we observe for the stereochemical outcome of the reaction.

So, although we initially argued against an S_N1 mechanism, I would formally class this as S_N1, but with 'complications'.

A mechanism that inevitably leads to the observed outcome as a result of the symmetry of an intermediate has a certain elegance. It is simpler than any other explanation I can imagine for this outcome. This is very often the case. You draw pages and pages of structures and curly arrows in order to try to explain the outcome of a reaction, and then you suddenly draw something simple that explains everything. At that point, you **just know** that you have the right answer. It just **feels** right.

You need to appreciate and to understand this, as you need to develop your instincts and opinions.

ANSWER 7

You don't need it! The reaction of stereoisomer **5** is exactly the same as that of compound **1**, although the stereochemical outcome will be different. You just need to do all the same things in the same order to reach a logical outcome.

IN CLOSING

We can consider this to be a rearrangement reaction. The phenyl group starts on carbon atom 2, and it finishes up on carbon atom 1 in the formation of stereoisomer **3**.

SOLUTION TO PROBLEM 10

Cyclization Reactions

ANSWER 1

Here is the mechanism. I'm not going to talk you through each curly arrow at this stage of the book. Make sure you can see why each arrow is going in the direction it is, and why it leads to formation of the product shown.

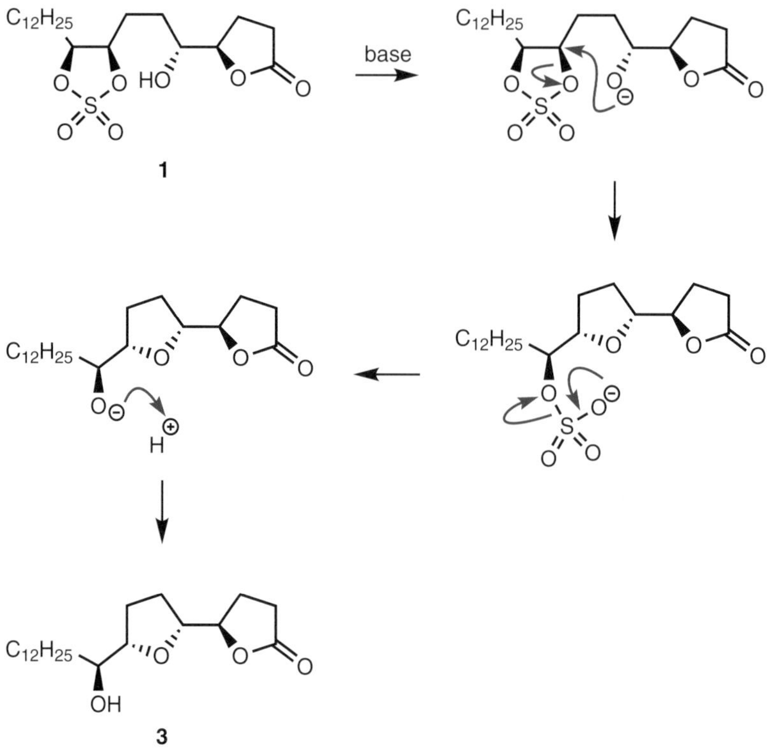

ANSWER 2

This is a really important question. Before you spend too much time thinking about the difference in outcome of two reactions, you need to know whether the data are meaningful.

In this case, almost certainly not! There is a reasonable chance that the chemist who did these reactions only did each one once, and on a small scale. Perhaps if the first reaction was repeated, it would give a higher yield? They could have had a bad day!

It turns out that we can apply basic principles to these two reactions, and we might expect the second reaction to be better.

That's why we are using this example! It fits our agenda! But I also included it so I could make this point!

ANSWER 3

Let's suppose the reaction of compound **1** to give compound **3** is slower (has a higher activation barrier) than the reaction of compound **2** to give compound **4**.

*This says absolutely nothing about the yield you would get from each reaction! Assuming compound **1** only ever gives compound **3**, you could get 100 per cent yield if you leave it long enough.*

ANSWER 4

This is a relatively basic question, but it often helps to think of the simplest possible example in order to gain insight.

> *Do you need more guidance on this question? If you do, read through* Basics 31 *and* Practice 12 *again.*

It is important that you understand that this is a basic question, and you need to get to the point where you no longer need to keep recapping it. But it might take you a little longer to get there and you shouldn't worry about it. Just keep practising and have faith!

Here are the two cyclohexanes, drawn in their most stable chair conformers.

5 **6**

The most stable conformer of compound **5** has both substituents equatorial. Compound **6** will have an axial methyl group.

The compound with the alkyl groups on the same side is more stable!

ANSWER 4 REFLECTION

Relating this to the original problem, it **might** be that compound **4** is more stable than compound **3**. But it doesn't have to be. We are taking what we know (for certain!) about cyclohexanes, and applying it to a five-membered ring with one oxygen atom in it. There will be similarities, but there will also be differences.

ANSWER 5

In order to present it to you, I have carried out a calculation of exactly what the transition state for each cyclization reaction would look like. In order to simplify the problem, I have used modified substrates **7** and **8**.

H_3C Ph O O HO S O O

7

H_3C Ph O O HO S O O

8

The lactone ring on the right in structures **1** *and* **2** *has an additional stereogenic centre. This adds complexity. I replaced this with a phenyl group. The long and flexible twelve-carbon chain on the left added even more complexity. For the purpose of the calculation, I replaced this with a methyl group.*

Here is the transition state for the reaction of compound **7**, the model for compound **1**. This is the reaction that did not work so well.

We have the phenyl ring on the right, and the rest of the cyclic sulfate on the left. The phenyl and methyl groups are both circled. More importantly, they are quite close together.

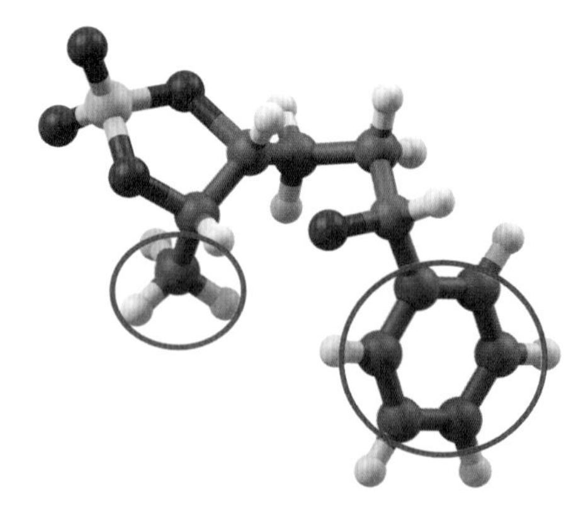

Here is the corresponding transition state for the cyclization of compound **8**. The same two groups are much further apart, so there is no steric clash between them.

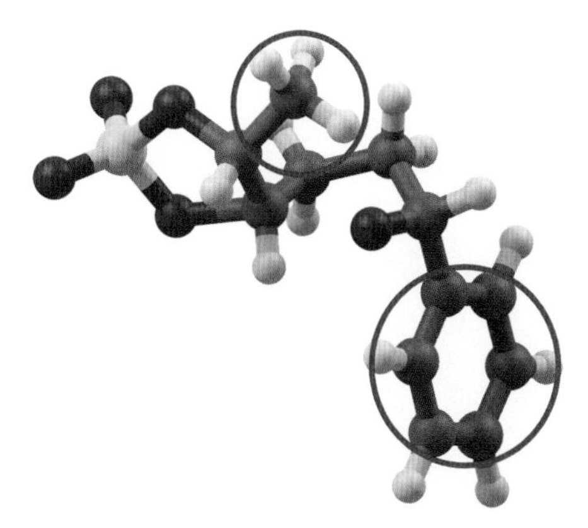

The transition states have a very defined geometry. The nucleophile attacks directly opposite the leaving group.

We cannot simply flex the five-membered ring to alleviate strain in the transition state. Or at least we cannot do this as much as we can in the product.

> *We saw a similar situation in* **Worked Problem 8** *when we considered the transition states for S_N2' reactions rather than the product stability—even though they told us the same thing.*

I'll make the same point I made there (because repetition is good!). It isn't fundamentally harder to consider the factors that stabilize/destabilize transition states. You just need to apply the same ideas to a different point along the reaction coordinate.

This is the beauty of organic chemistry, but it is also the challenge. Change your perspective slightly, and things become easier. We are right back where we started. There aren't all that many ideas in organic chemistry, but you need to know when and how to apply them to a range of situations. That's the challenge!

AFTERWORD

So, you've got to the end. Hopefully you have enjoyed reading this book. More importantly, I hope you have understood all of the explanations and learned quite a lot of organic chemistry.

*Did you 'just' **read** this book?*

If you did, then that's okay. The book covers all of the principles and ideas in year 1 university organic chemistry, and at a level of depth that will allow you to follow your lectures. Sure, not all of the reactions you will see have been covered, but the skills you will need in order to learn those reactions are all here.

If you **did** just read the book, I think you can get more out of it. Go back to the beginning and start again, but this time draw everything out. Reflect on whether you can draw the structures better. Be your own worst critic. It will take a little longer, but you will get quicker!

If it takes a long time to do, you need to do it. If it doesn't take long, there's no harm in doing it!

Sadly, the reality is that organic chemistry **is** difficult to learn, and yet most of the things organic chemists do with structures on a daily basis can be done with little conscious thought, as long as you have really (and I **do** mean really) internalized your understanding of the basics.

Let's return to one of the analogies I used at the start. Most people who learn to drive do so over a moderate period of time, building their skills over a period of months, perhaps even years. You would never have a couple of lessons in order to 'master' the controls, and then book your test, take a break, and have ten lessons in the three days before your test.

You know that this would not be a safe (in every sense of the word!) way to pass your test. And yet many university students do exactly the same thing with organic chemistry.

You often hear of people passing their driving test on the 17th attempt. Clearly, there are people who have little or no natural aptitude for driving! But they persevere, and they pass, eventually. The same is true of organic chemistry. If you persevere, you will pass.

I would prefer that 'your version' of persevering involves recognizing your ability/aptitude very early on, and fixing it quickly by practising key skills.

But you **can** do this!

I hope you have found this book instructive. I hope you have learned to avoid some of the common errors, and in doing so become more successful as an organic chemist.

Keep re-reading it. Your job is not yet finished. Keep building those neural pathways that will make you better at organic chemistry. To use a Japanese word, Gambatte![1]

[1] Roughly translated, it means 'keep on keeping on'!

INDEX

K

L

M

N

O

P

R

S

T

W